Analysis of
Pesticide Residues

CHEMICAL ANALYSIS

A SERIES OF MONOGRAPHS ON
ANALYTICAL CHEMISTRY AND ITS APPLICATIONS

VOLUME 58

A WILEY-INTERSCIENCE PUBLICATION

JOHN WILEY & SONS

New York / Chichester / Brisbane / Toronto

Analysis of
Pesticide Residues

Edited by

H. ANSON MOYE
University of Florida

A WILEY-INTERSCIENCE PUBLICATION

JOHN WILEY & SONS
New York / Chichester / Brisbane / Toronto

Library of Congress Cataloging in Publication Data:

Main entry under title:
Analysis of pesticide residues.

  (Chemical analysis ; v. 58)
  "A Wiley-Interscience publication."
  Includes index.
    1. Pesticides—Analysis. 2. Chromatographic
analysis. I. Moye, H. Anson. II. Series.

TP248.P47A48   628.5'4   80-15216
ISBN 0-471-05461-5

Printed in the United States of America

10 9 8 7 6 5 4 3 2 1

PREFACE

Because of their inherent toxicity and ubiquity pesticides have continued to be of interest to toxicologists, biologists, zoologists, ecologists, analytical chemists, and agriculturalists. Many of these scientists and technologists have devoted their productive years to the research, development, and safe utilization of pesticides, often limiting their activities to a very specific area, such as mode of action, metabolism, efficacy, analytical chemistry, and so on, while others have dealt with the chemicals only casually.

This book deals with the subject of pesticide trace analysis, or what has come to be known as "pesticide residues," an area that has generated its own vocabulary, protocol, literature, and hardware. Many governmental regulatory agencies as well as nonaffiliated organizations have dedicated immeasurable effort and resources to develop, verify, standardize, and publish analytical methods for pesticide residues. In addition, for each pesticide marketed the manufacturer or supplier has been required to provide methods that can be used to monitor their products in foods and environmental samples. Universities, through their agricultural experiment stations, have also contributed to the development of methods for analysis of pesticide residues, usually in an effort to hasten the registration process so that food production can be increased. One would expect that with such widespread effort there would be considerable duplication or redundancy in the pesticide analytical chemistry literature; this is sometimes the case, but more often than not there are gaps rather than multiplicities.

Such gaps result from the fact that methods which have been developed for one pesticide-commodity combination frequently are not satisfactory for another type of commodity. As well, methods that utilize sophisticated and expensive instrumentation of a particular type are of little value to the analyst whose laboratory is not similarly equipped. For these reasons, as well as others, any given set of "standardized" procedures supplied by any of the types of previously mentioned organizations often do not meet the needs of some laboratories. One is frequently forced to do a literature search and subsequently improvise a method as well. It is to meet this sort of need that this book has been written.

Since gas chromatography continues to play a dominant role in the analysis of pesticide residues, special emphasis has been given to that topic in Chapter 1, which deals with the nuts and bolts of column making and selection, and Chapter 2, which treats currently available detectors from both a theoretical and practical viewpoint. Applications of gas

v

chromatographic techniques are discussed in Chapter 5, which treats chlorinated hydrocarbons, Chapter 6 on the acidic herbicides, Chapter 7 on the organophosphates, Chapter 8 on the carbamates, and Chapter 9 on the analysis of insect pheromones and hormones.

Confirmation of pesticide residues has most frequently been done via thin layer chromatography. This very useful technique is discussed comprehensively in Chapter 3.

Growing rapidly in popularity, high-performance liquid chromatography has recently begun to find its place in pesticide residue analysis. Chapter 4 provides a comprehensive review of the literature and discusses the unique characteristics of this technique when applied to trace analysis.

Of special note for those whose residue data will be subject to governmental scrutiny is Chapter 10, which deals with the various requirements imposed by those agencies on the analyst. Too often there have been those, I among them, who have found their labors in the laboratory negated by an oversight regarding the taking of analytical data. Of special interest, since it applies to all previous chapters, is the discussion on sampling and analytical variability between samples.

I would like to express my appreciation to Ms. Sandra Westhart and Ms. Shannon Peacock who typed portions of this manuscript, to Dr. Marjorie Malagodi who assisted in the editing, and to Ms. Susan Scherer whose artwork and editing were of such help.

H. ANSON MOYE

Gainesville, Florida
November 1980

CONTENTS

Analysis of
Pesticide Residues

CHAPTER

1

GAS-CHROMATOGRAPHIC COLUMNS IN PESTICIDE ANALYSIS

JOHN F. THOMPSON* and RANDALL R. WATTS

*Health Effects Research Laboratory, EPA,
Research Triangle Park, North Carolina*

CONTENTS

1.1. INTRODUCTION

The first known form of a chromatographic column was described in 1906 by Tswett (1). It was used for the separation of colored pigments from plant materials. The term "chromatography" was coined at that time and,

* Retired.

literally translated, means color writing. During ensuing years, the basic chromatographic technique was applied to colorless materials so that the name assumed some inconsistency with the practice.

It would be nearly 40 years before techniques were developed for controlling or modifying chromatographic separations. Claesson (2) in 1946 developed the concept of displacement, the process of controlled removal of adsorbed materials from an adsorbent surface by a more highly adsorbed substance. Tiselius (3) had previously conceived the idea of utilizing a continuous stream of the sample itself as a displacing agent, in which each solute achieves its own competitive adsorption equilibrium. This technique is known as frontal analysis.

In 1952 Martin and James (4, 5) described an apparatus fitted with a recording buret for the separation of mixtures of organic acids and bases by the gas–liquid partition process. Two years later Ray (6) published a series of gas–liquid chromatograms obtained with a thermal conductivity detector.

During the era of the late 1950s and early 1960s, after the development of the microcoulometric and electron capture detector (ECD) systems, GC columns were developed that bore some semblance to those in use today (7–14). These employed such single stationary phase materials as SE-30, DC-200, DC high vac grease, DC-11, and QF-1.

The single stationary phase columns presented some distinct disadvantages in attempting to resolve certain pesticidal compounds of closely similar elution characteristics, that is, on the relatively nonpolar DC-200 column, such compound pairs as dieldrin/p,p'-DDE, the common benzene hexachloride (BHC) isomers, and o,p'-DDT/p,p'-DDD would not yield complete peak separations. In 1964 McCully (15) reported a column combining SE-30 with QF-1 in a 2:3 ratio. The elution patterns of pesticide mixtures chromatographed on this column were sufficiently different from the single stationary phase column to provide the pesticide chemist with a means of resolving a number of compounds that could not be separated on the single stationary phase.

In 1966 Burke (16) reported a column of DC-200 combined with QF-1 in a loading ratio of 5:7.5. Expectedly, the elution characteristics of McCully's and Burke's columns were nearly identical, as both combined a methyl silicone with a trifluoropropylmethyl silicone, QF-1, in a basic 2:3 ratio.

In the mid 1960s, a series of stationary phase materials was introduced to the marketplace which was, and still is, designated as the OV series. Some of these were quite similar to materials that had been marketed earlier, but were more highly refined. The older materials such as QF-1 or DC-200 were of a technical or "tank car" quality and contained residual impurities that tended to bleed off the packed column unless great care was taken to thoroughly heat condition or "bake" the column before use. Even then, column operating temperatures had to be rather carefully controlled.

In 1966 Henly (17) combined 7% OV-17, a methyl silicone of 50% phenyl substitution, with 9% QF-1 in a 1:1 ratio. This blend produced an elution pattern for mixtures of pesticides quite different from any other material previously reported. Because of its high loading and the relatively high retentiveness of OV-17, a 1.8-m column operated at a temperature of 200°C was determined to be impractically slow unless the carrier gas velocity was increased to objectionably high levels. Because of the potentially attractive separation characteristics of the basic ratio Thompson (18) in 1969 proportionately reduced the loading of this combination to a ratio of 1.5:1.95, and found the compound elution pattern was identical, efficiency was improved, and total retention time was reduced to a practical range.

Later experience indicated that OV-210, a trifluoropropylmethyl silicone, comparable to but more highly refined than QF-1, would serve as an excellent replacement for the technical grade QF-1. SP-2401, marketed by Supelco, Inc., is also a fluorosilicone, molecularly comparable to QF-1 or OV-210, but reported to be of lower viscosity.

In a series of papers Aue and his coworkers (19–23) demonstrated that Carbowax 20M could be chemically bonded to solid supports. These column packings show very little bleed and therefore can be used with an electron capture detector or a mass spectrometer. In addition, these materials are very inert and may be used as they are, or coated with other liquid phases. Many investigators (24–26) have reported on the use of these column packings for the gas chromatography of pesticides and metabolites. Hall and Harris (27) reported on the gas chromatography of intact carbamate pesticides using commercially available support bonded Carbowax 20M coated with several different liquid phases. Moseman (28) described a rapid and simple method of preparing support-bonded Carbowax 20M and demonstrated its usefulness for the gas chromatography of intact carbamates and polar pesticide metabolites.

Material presented in this chapter is strongly oriented toward the practical day-to-day activities and problems of the working pesticide analytical chemist. A minimum of theory is included, only enough to support certain working concepts. For the reader wishing to pursue the theory and highly involved mathematics of column chromatography, several excellent texts are recommended (19–21).

1.2. COLUMN COMPONENTS

1.2.1. Containing Tube

Borosilicate glass, stainless steel, aluminum, and Teflon are the column container materials most widely used in pesticide analysis, with glass by far the most predominant. Metal columns have not enjoyed widespread use by reason of their tendency to promote decomposition of a number of pesticidal compounds at high temperatures. Teflon columns are quite expensive as compared to glass.

The two main principles of column configuration are U shape and coiled, and at present the preferred dimensions appear to be 1.8 m in length and 6.5 mm o.d., 4.0 mm or 2.0 mm i.d. Some experimentation has been conducted with capillary columns, but these have not been utilized widely for pesticide work. Columns shorter than 1.8 m are often used for specific compound analysis to reduce analysis time, but the shorter column is proportionately less efficient and therefore less capable of the high resolution required in multiresidue analysis.

1.2.2. Support Materials

The "ideal" support should conform to the following criteria:

1. Be available in narrow ranges of particle or mesh size
2. Be inert and have no active adsorption sites
3. Have large surface area per unit volume
4. Have good thermal stability and mechanical strength

There are many supports available in the marketplace. Those in principal use, such as those subsequently listed, currently are prepared from either marine or terrestrial diatoms. The Chromosorb® series is manufactured by Johns-Manville and distributed through many outlets. The Gas-Chrom® series is prepared and distributed by Applied Science Laboratories. The Supelcoport® line is prepared and distributed by Supelco, Inc.; and the Anakrom® series is manufactured by Analabs, Inc. and sold by them directly and through distributors.

The quality of the support is of paramount importance for the performance of the column in terms of efficiency and chromatographic peak configuration. In the ideal situation the support should not interact with the solute, but, in practice, the ideal is not always met. The presence of active adsorption sites results in peak tailing and in the decomposition of certain pesticidal compounds. Endrin can be most troublesome in this respect. A number of supports that are currently available are treated by acid and/or base washing and by silanization. The former treatment is to remove minerals and other extraneous matter, and the latter is to reduce tailing and compound decomposition.

1.2.3. Stationary Phase Material

There are literally hundreds of stationary or "liquid" phase materials available. Relatively few have been determined suitable for pesticide work. Those most widely used are given in Table 1.1 along with their approximate equivalents in terms of their compound elution patterns.

The amount or percentage of stationary phase(s) in a column packing will exert a strong influence over column efficiency. Generally speaking, columns of low loading will be of higher efficiency than those of high loading. In the author's laboratory it has proved possible to prepare 5% OV-210 to yield an efficiency of 830 theoretical plates (TP) per foot, but the yield obtained from typical 10% OV-210 has been around 415 TP/foot.

1.3. PREPARATION OF COLUMN PACKING

Several systems have been devised for making small batches of column packing. It is difficult to say that any one is superior to the others. In addition to the science, there is a certain degree of art involved in the process. Some of the more common techniques are as follows:

1. *Beaker technique.* Liquid phase(s) dissolved in appropriate solvent in a beaker. The solid support is added and the mixture stirred while evaporat-

TABLE 1.1. Stationary Phase Materials Commonly Used in Pesticide Analysis

No. Designation		Chemical Name	Equivalent Designations
1.	DC-200	Methyl silicone	OV-1, OV-101, SE-30, SP-2100, DC-11, SF-96
2.	QF-1	Trifluoropropylmethyl silicone	OV-210, SP-2401
3.	SE-30	(See No. 1)	
4.	SE-52	Methyl silicone, 10% phenyl substituted	OV-3
5.	SF-96	(See No. 1)	
6.	XE-60	Cyanoethylmethyl silicone	OV-225
7.	OV-17	Methyl silicone, 50% phenyl substituted	SP-2250
8.	OV-7	Methyl silicone, 20% phenyl substituted	—
9.	OV-210	(See No. 2)	
10.	DEGS	Diethyleneglycol succinate	—

ing the solvent under a stream of air or nitrogen. It has one strong disadvantage, in that the constant stirring required tends to fracture support particles.

2. *Rotary vacuum technique.* Liquid phase(s) dissolved in appropriate solvent in a small beaker and transferred to a Morton flask with indented sides. The support is added, the flask is placed in variable heat water bath and connected to a rotary evaporator. Mixing and solvent evaporation is carried out by rotating the flask under vacuum and with applied heat.

3. *Filtration technique.* This is an extension of the beaker technique in that the slurry in the beaker is poured on a filter paper held in a Buchner funnel, and the solvent is removed by drawing air through the layer of packing on the filter paper.

4. *Fluidization technique.* This is also a somewhat sophisticated extension of the beaker technique. The slurry in the beaker is transferred to a fluidizer cylinder, so constructed that a high flow of nitrogen can be blown up through the packing from the bottom while heat is applied by an element at the base of the cylinder.

This latter system was developed and the equipment is marketed by Applied Science Laboratories under the brand name of HI-EFF® Fluidizer. The Fluidizer is shown as Figure 1.1.

Whatever method is used, the operator must take meticulous care and not attempt to cut any corners. The author has had excellent success with

Figure 1.1. HI-EFF® Fluidizer. Courtesy Applied Science Laboratories, State College, PA.

Method 2 above in the preparation of 25-g batches, but we have known of other individuals who have used the technique with far less success. One of the more critical points in this technique is extremely slow rotation of the flask during solvent evaporation. This minimizes the probability of fracturing the support particles.

1.4. COLUMN PERFORMANCE CHARACTERISTICS

The most important characteristics of a column packed for multiresidue pesticide analysis are efficiency, sensitivity, retention, compound elution pattern, stability, and resistance to on-column compound decomposition.

1.4.1. Efficiency

This characteristic is very important as high efficiency is generally synonymous with the capacity of the column to yield good resolution between adjacently eluting peaks. Therefore, the high-efficiency column permits the separation of pesticides from each other and from extraneous peaks. A low-efficiency column, on the other hand, results in peaks overlapping, often to the extent that peak identification and quantitation are impossible. Two contrasting chromatograms are shown in Figure 1.2 to illustrate this situation. (A) in the upper half is a chromatogram of an 8-compound mixture obtained on a 1.8-m column of 2% OV-1/3% QF-1, which had only 740 TP. Note that the peaks for p,p'-DDE and dieldrin are partially overlapped, as are those for o,p'-DDT and p,p'-DDD. By contrast, the 1.8-m column on which (B) was chromatographed had an efficiency value of 4530 TP. Here, we note a sharp separation of all peaks for all seven compounds in the mixture.

In spite of the great effect of efficiency, many chromatographers ignore the importance of this characteristic, and conduct their daily chromatography on columns that yield peaks suggesting the gently rolling Great Smokey Mountains—with large mounds rather than sharp, narrow peaks.

During past years it has been the author's good fortune to review the raw data from a sizable number of field laboratories conducting pesticide monitoring studies throughout the United States. In conducting these reviews, we were appalled on numerous occasions to observe chromatograms for columns which were obvious candidates for the trash can. At one point, these casual observations motivated us to make a survey of the 18 laboratories that were participating at that time in order to obtain systematic information on column performance at all the laboratories. We were deeply concerned because all the column packing used in the laboratories was

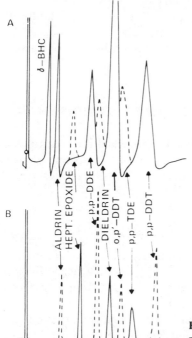

Figure 1.2 Effect of column efficiency on pesticide peak resolution. (See text.)

obtained from a central source, namely, our laboratory. We had purchased bulk lots from a commercial supplier on stringent quality specifications and had checked quality most carefully in our laboratory prior to distribution.

Three mixtures of pure pesticide standards were mailed to each laboratory. The mixtures were:

Mixture #1   Aldrin ------------------------10   Mixture #2   Aldrin ---------10
                   p,p'-DDT--------------------100   Endrin -------100

Mixture #3 α-BHC -----------------------10   Dieldrin ----------50
                   β-BHC -----------------------40   Endrin------------80
                   γ-BHC -----------------------10   o,p'-DDD--------80
                   Heptachlor ------------------10   p,p'-DDD--------80
                   Heptachlorepoxide --------30   o,p'-DDT --------90
                   Aldrin -----------------------20   p,p'-DDT-------100
                   p,p'-DDE --------------------40

(Concentration in picograms per microliter)

A set of instructions was provided to each laboratory. One most important instruction was that at some point *during a workday* these three mixtures would be chromatographed on both of their working columns, a 1.5% OV-17/1.95% QF-1 and a 4% SE-30/6% QF-1. We emphasized that no special preparation was to be made, as we wished the chromatograms to be representative of the output in a normal work situation. Electron capture detectors were used by all, and 1.8-m columns were standard throughout.

We were interested in several other characteristics as well, but for the moment we will address efficiency only. There will be more later concerning the other characteristics. On receiving the chromatograms back in our laboratory, we calculated all the efficiency values from both columns. Table 1.2 shows all values calculated.

The space between laboratory numbers 8 and 9 represent the breaking point in our quality criteria for efficiency. We regard anything less than 3000 TP as unacceptable. Of the 18 laboratories, 8 had satisfactory columns, whereas the remaining 10 laboratories were using inferior columns, based on efficiency.

Carrier flow velocity has a profound effect on efficiency. It is significant that most of the high flow rate reports on the SE-30/QF-1 columns (90 ml/min and above) are centered in the low-efficiency group. We were able to conclude from the calculated retention time for p,p'-DDT that two others in the low-efficiency group of this column, laboratory numbers 9 and 14, must have used flow rates far higher than the values reported.

Another factor influencing apparent efficiency is the amount of polarizing voltage applied to the EC detector. In an issue of *Gas-Chrom Newsletter* published in 1973 by Applied Science Laboratories (22), this subject was given excellent treatment. The publication pointed out that the EC detector gives linear response over a very limited range of applied voltage. When a gross excess is applied, the detector becomes nonlinear and the response-to-concentration slope is very small at low solute concentrations and increases rapidly at high concentrations. This results in an extreme compression of the lower part of the GC peak and an extension of the upper part. The opposite of this occurs when too little voltage is applied, so that if the chromatographer is operating the detector below the optimum voltage range the peaks tend to broaden at the lower part and calculate out to a low efficiency.

The most commonly used equations for calculating column efficiency and peak resolution are given in Figure 1.3.

1.4.2. Sensitivity

More accurately stated, column sensitivity is the influence of the column on the detector response. It is an important characteristic, particularly to

TABLE 1.2. Column Efficiency and Retention Time Characteristics

SE-30/QF-1

Lab. No.	Reported Column (°C)	Computed Column (°C)	Reported Flow Rate (ml/min)	Absolute Retention p,p'-DDT (min)	Efficiency (TP)
1	198	199	70	18	4100
2	195	193–197	75	18–23	3680
3	200	200	85	18	3670
4	190	197	?	14	3540
5	200	207	70	16	3520
6	200	200	60	21	3500
7	200	205	?	17	3480
8	200	197	60	21	3280
9	200	202	90	14	2860
10	191	187	70	12	2860
11	197	199	85	16	2840
12	190	194	87	18	2810
13	200	197	110	16	2680
14	200	197	88	13	2480
15	190	194	120	19	2460
16	195	195	100	15	2100
17	199	194–197	102	22–23	1830
18	200	186	100	20	1220
19	—	—	—	—	—

OV-17/QF-1

Lab. No.	Reported Column (°C)	Computed Column (°C)	Reported Flow Rate (ml/min)	Absolute Retention p,p'-DDT (min)	Efficiency (TP)
15	190	194	50	21	4000
5	200	203	50	19.5	3840
2	195	190	65	20	3820
7	200	208	57	14	3720
6	200	204	70	18	3670
9	200	200	60	16.5	3600
10	191	185	61	23	3310
16	195	194	60	22	3140
3	200	201	55	17	2960
14	200	199	67	16	2850
8	200	200	60	19	2750
1	199	201	60	19	2740
11	197	198	60	20	2600
13	200	195	65	16	2560
4	190	198	60	14	2450
18	200	196–200	60	13–14	2450
12	191	192	100	15	2240
19	200	200	58	18	2100
17	201	202–206	74	14	1940

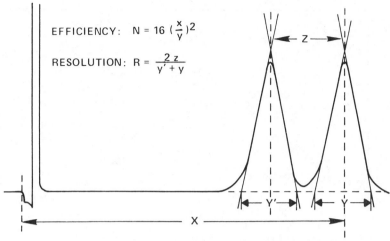

EFFICIENCY: $N = 16 \left(\frac{x}{y}\right)^2$

RESOLUTION: $R = \frac{2z}{y' + y}$

Figure 1.3 Calculation of column efficiency and resolution.

the laboratory analyzing environmental or tissue media where pesticide concentrations are measured in the ppb or even ppt range.

The principal column parameters affecting the electron capture detector response are (1) carrier gas flow rate, (2) amount of stationary phase loading, (3) column temperature, and (4) polarizing voltage applied to detector.

In the selection of a GC column that will provide maximum response, the operator must keep in mind other operational factors that are important in the workday situation. For example, it is possible to achieve a high sensitivity for any given stationary phase loading by simply holding the carrier gas velocity to a low level. This will enhance both efficiency and response, but will not permit the operator a reasonable volume of work output. For example, if a column of 5% DC-200/7.5% QF-1 is operated at 200°C with a carrier flow of 60 ml/min, the column efficiency and sensitivity should be quite acceptable. However, using a 1.8 m column, p,p'-DDT would be expected to elute in something over 30 min, and one could no doubt take an overextended coffee break while awaiting the emergence of a late eluter like methoxychlor. Contrarily, if the carrier flow is increased to 120 ml/min, p,p'-DDT would probably elute in the 15–17 min range, but sensitivity and efficiency would plunge to unacceptable levels. Figure 1.4 shows a comparison of the response of 8 GC columns when each is operated at optimum parameters in terms of a compromise of efficiency, retention, and response. A column of 3% OV-1 was arbitrarily selected as the reference, and all other column sensitivity values are compared to this column.

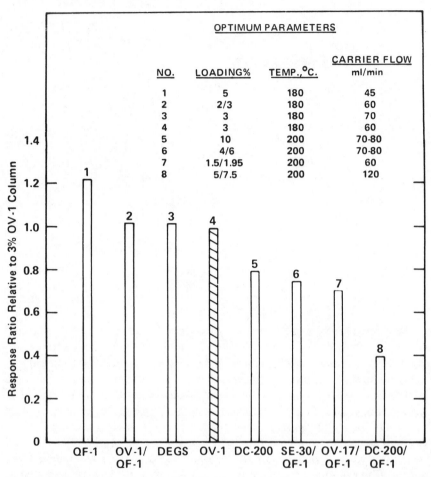

Figure 1.4 Comparative response of 8 GC columns related to the reference column of 3% OV-1, with each column operated at its optimum parameters.

1.4.3.   Retention

Retention is the factor on which the chromatographer relies almost wholly for the tentative identification of compound peaks in a substrate of unknown composition. The main factors influencing retention or elution speed are

1.   Amount of stationary phase loading
2.   Carrier gas flow rate

3. Column temperature
4. Particle size of support material
5. Type of stationary phase
6. Length of GC column

In our applied research on columns, our column construction and operating parameters have targeted an absolute retention time of 16–20 min for *p,p'*-DDT to permit adequate peak separation in a practical time period. Columns of lower stationary phase loading (6% or less) can usually be operated at such parameters that will produce maximum efficiency while still maintaining the 16–20 min time frame. High load columns such as the 10% DC-200, 5% DC-200/7.5% QF-1, and so on, must be operated at elevated temperatures and carrier gas flow rates to obtain the desired absolute retention range. As we stated earlier, the elevated flow rates will result in lowered efficiency and response.

1.4.3.1. Absolute Retention

The term *absolute retention* is used to denote an actual unit of measurement in terms of minutes required from the point of an injection of a solute to the time a given compound elutes from the column. On a chromatogram, the measurement should be made from the injection point to the center of the peak of interest. Figure 1.5 illustrates the method of measurement and provides some pertinent equations.

Many chromatographers, as a matter of convenience, use the upslope of the solvent peak as a point of measurement. This is incorrect and *does not* accurately reflect the entire retention time, as several seconds elapse from the time of injection to the time the solvent impacts the detector. The elution time of different solvents becomes quite important in making relative retention calculations described in Section 1.4.3.2.

We have observed instances where the chromatographer has recorded the number of chart spaces to represent retention. This is not nearly precise enough. Measurements should always be made in millimeters with an accurate ruler. An appropriate equation in Figure 1.5 can then be applied to calculate minutes.

1.4.3.2. Relative Retention

As this value is a ratio rather than an expression of units, the traditional abbreviation RRT (relative retention time) is somewhat of a misnomer, but we will adhere to it by reason of tradition and its familiarity to pesticide chemists.

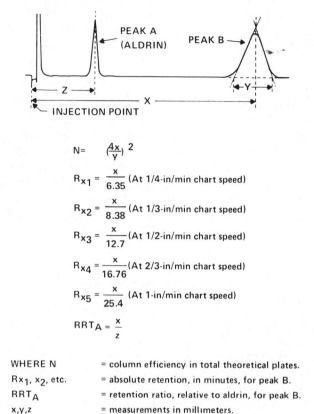

$$N = \left(\frac{4x}{y}\right)^2$$

$$R_{x1} = \frac{x}{6.35} \text{ (At 1/4-in/min chart speed)}$$

$$R_{x2} = \frac{x}{8.38} \text{ (At 1/3-in/min chart speed)}$$

$$R_{x3} = \frac{x}{12.7} \text{ (At 1/2-in/min chart speed)}$$

$$R_{x4} = \frac{x}{16.76} \text{ (At 2/3-in/min chart speed)}$$

$$R_{x5} = \frac{x}{25.4} \text{ (At 1-in/min chart speed)}$$

$$RRT_A = \frac{x}{z}$$

WHERE N   = column efficiency in total theoretical plates.
$R_{x1, x2}$, etc.   = absolute retention, in minutes, for peak B.
RRT_A = retention ratio, relative to aldrin, for peak B.
x, y, z = measurements in millimeters.

Figure 1.5 Retention and efficiency calculations.

The RRT is the ratio of the absolute retention of the compound of interest to that of some reference compound. Aldrin has been most commonly used as a reference compound, particularly when referencing halogenated compounds determined by electron capture detection. Ethyl parathion has been widely used when chromatographing organophosphorus compounds determined by specific detectors such as flame photometric or flame ionization.

The relative retention ratio is far more reproducible than the absolute retention time. Several factors may cause significant variation in absolute retention, whereas the only factor that will influence the RRT is the column temperature. Therefore, when computing retentions on a chromatogram in the process of identifying unknown peaks, it is much better practice to compute the RRT. This is easily done by chromatographing the unknown. Some chromatographers prefer to coinject the reference pesticide with the sample substrate so that the reference compound peak will appear in the same chromatogram. This is acceptable analytical practice, provided that the substrate contains no compound producing a peak at the same elution site as that of the reference compound or that the substrate is unlikely to contain any of the reference compound.

A number of researchers have published RRT data of pesticidal compounds obtained on various GC columns by different modes of detection (15–18, 23–25). In each of these publications, the RRT data have been obtained at one given temperature for each column. While the published data have proved valuable, the chromatographer, in trying to reproduce these RRT data, must adjust the column oven to the exact temperature specified by an author. This is sometimes impractical and time-consuming. With this background in mind, Thompson in 1975 (26) developed tables of RRT_A (aldrin as the reference compound) for 48 compounds on six different GC columns, and also tables of RRT_P (parathion as the reference compound) for 54 organophosphorus compounds on three GC columns. The RRT_A values listed in the tables are applicable over a temperature range from 170 to 204°C, the normal range used by the majority of pesticide chromatographers. See Tables 1.3–1.11 for RRT_A and RRT_P values.

The tables have proved very useful, particularly for tentative peak identifications in the chromatograms of multiresidue samples. First, it is necessary for the operator to establish the prevalent true column temperature. This is determined by chromatographing a standard mixture comprised of the appropriate reference compound (aldrin or parathion) and a relatively late eluting compound such as p,p'-DDT or Imidan. The RRT_A or RRT_P of the late eluting compound is calculated from the chromatogram. Then, by scanning horizontally across the column opposite p,p'-DDT (or Imidan) on the table, the RRT value is located that most closely matches the RRT value obtained by chromatograph. The true column temperature can now be obtained by reference to the top or bottom of the table.

The sample extract is then chromatographed, and the RRT_A or RRT_P values are calculated for all peaks appearing on the chromatogram. By vertically scanning one temperature column on the table the calculated RRT values may be compared with table values to obtain tentative peak identification.

The system saves considerable time in that two chromatograms are generally sufficient to enable the operator to make tentative identifications of a number of the pesticidal compounds present in a sample. Using the more traditional approach, on completion of the exploratory chromatogram on the sample extract, the operator usually selects reference standards on the basis of absolute retention values, or RRT values on the assumption

TABLE 1.3.
1.5% OV-17/1.95% OV-210

Column Temperature, °C.

Compound	170	174	178	182	186	190	194	198	202	204 →
Dimethyl Phthalate	0.25	0.26	0.26	0.27	0.28	0.28	0.29	0.30	0.30	0.31
Mevinphos	0.32	0.32	0.32	0.32	0.32	0.33	0.33	0.33	0.33	0.33
Tecnazene	0.34	0.35	0.36	0.36	0.37	0.38	0.38	0.40	0.40	0.41
Diethyl Phthalate	0.38	0.38	0.39	0.39	0.39	0.40	0.40	0.41	0.41	0.41
2,4-D(ME)	0.44	0.45	0.45	0.45	0.46	0.46	0.46	0.47	0.47	0.47
Hexachlorobenzene	0.42	0.43	0.44	0.44	0.45	0.45	0.46	0.47	0.48	0.49
α-BHC	0.48	0.49	0.50	0.50	0.51	0.52	0.52	0.53	0.54	0.55
CDEC	0.54	0.54	0.55	0.55	0.55	0.55	0.55	0.56	0.56	0.56
2,4-D(IPE)	0.56	0.56	0.56	0.56	0.56	0.56	0.56	0.56	0.56	0.56
Chlordene	0.54	0.55	0.56	0.56	0.57	0.58	0.58	0.59	0.60	0.61
Diazinon	0.67	0.66	0.66	0.66	0.65	0.65	0.65	0.64	0.64	0.64
PCNB	0.65	0.65	0.66	0.66	0.66	0.67	0.67	0.67	0.68	0.68
Lindane	0.66	0.67	0.67	0.67	0.68	0.68	0.68	0.69	0.69	0.69
2,4,5-T(ME)	0.76	0.76	0.75	0.75	0.75	0.74	0.74	0.73	0.73	0.73
β-BHC	0.82	0.82	0.81	0.81	0.81	0.81	0.80	0.80	0.80	0.80
Heptachlor	0.82	0.82	0.82	0.82	0.82	0.82	0.82	0.82	0.82	0.82
2,4,5-T(IPE)	0.94	0.93	0.92	0.91	0.90	0.88	0.87	0.86	0.85	0.83
Aldrin (REFERENCE)	1.00	1.00	1.00	1.00	1.00	1.00	1.00	1.00	1.00	1.00
Dimethoate	1.17	1.15	1.13	1.11	1.09	1.07	1.05	1.02	1.00	0.98
Ronnel	1.17	1.16	1.14	1.13	1.12	1.11	1.09	1.08	1.07	1.05
Dibutyl Phthalate	1.49	1.45	1.41	1.38	1.36	1.32	1.29	1.25	1.22	1.18
1-Hydroxychlordene	1.41	1.39	1.36	1.35	1.33	1.31	1.29	1.27	1.25	1.23
Oxychlordane	1.49	1.47	1.46	1.45	1.44	1.42	1.41	1.39	1.38	1.36
M. Parathion	1.71	1.67	1.64	1.62	1.59	1.55	1.52	1.48	1.45	1.41
Heptachlor Epoxide	1.70	1.68	1.66	1.64	1.63	1.61	1.59	1.57	1.55	1.52
DCPA	1.82	1.78	1.76	1.72	1.68	1.64	1.60	1.56	1.52	1.48

Compound	170	172	174	176	178	180	182	184	186	188	190	192	194	196	198	200	202	204
Malathion	2.07	2.04	2.01	1.98	1.95	1.92	1.89	1.87	1.84	1.81	1.78	1.75	1.72	1.69	1.66	1.63	1.60	1.57
Chlordane, Gamma	1.92	1.91	1.89	1.88	1.86	1.85	1.83	1.81	1.80	1.78	1.77	1.75	1.74	1.72	1.71	1.69	1.68	1.66
Trans-Nonachlor	2.02	2.00	1.99	1.97	1.95	1.93	1.92	1.90	1.88	1.86	1.85	1.83	1.82	1.79	1.78	1.76	1.74	1.73
o,p'-DDE	2.14	2.12	2.09	2.07	2.05	2.03	2.01	1.98	1.96	1.94	1.92	1.90	1.88	1.86	1.84	1.82	1.79	1.77
E. Parathion	2.32	2.28	2.25	2.22	2.19	2.16	2.13	2.09	2.06	2.03	2.00	1.97	1.93	1.90	1.87	1.84	1.81	1.78
Chlordane, Alpha	2.15	2.13	2.11	2.09	2.07	2.05	2.03	2.01	1.99	1.97	1.96	1.94	1.92	1.90	1.88	1.86	1.84	1.82
Endosulfan I	2.20	2.18	2.16	2.15	2.13	2.11	2.10	2.08	2.06	2.05	2.03	2.01	2.00	1.98	1.97	1.95	1.93	1.91
p,p'-DDE	2.75	2.72	2.68	2.64	2.61	2.58	2.54	2.51	2.47	2.43	2.40	2.37	2.33	2.30	2.27	2.23	2.20	2.17
DDA(ME)	2.97	2.93	2.88	2.84	2.79	2.75	2.71	2.66	2.62	2.57	2.53	2.49	2.44	2.40	2.35	2.31	2.27	2.22
Dieldrin	2.80	2.77	2.75	2.72	2.69	2.67	2.64	2.61	2.59	2.56	2.53	2.51	2.48	2.45	2.43	2.40	2.37	2.35
o,p'-DDD	3.34	3.29	3.25	3.20	3.15	3.11	3.06	3.01	2.97	2.92	2.88	2.83	2.77	2.74	2.69	2.65	2.60	2.56
Chlordecone	3.26	3.23	3.19	3.16	3.13	3.09	3.06	3.03	3.00	2.96	2.93	2.90	2.87	2.83	2.80	2.77	2.74	2.70
Endrin	3.47	3.43	3.40	3.36	3.33	3.29	3.26	3.22	3.18	3.15	3.12	3.08	3.04	3.01	2.97	2.93	2.90	2.87
o,p'-DDT	3.98	3.94	3.88	3.83	3.77	3.71	3.66	3.60	3.54	3.48	3.43	3.38	3.32	3.27	3.21	3.16	3.10	3.04
p,p'-DDD	4.65	4.57	4.49	4.41	4.33	4.26	4.18	4.10	4.02	3.94	3.87	3.79	3.71	3.64	3.61	3.48	3.40	3.32
Endosulfan II	4.45	4.39	4.34	4.28	4.23	4.17	4.11	4.05	3.99	3.94	3.88	3.82	3.76	3.71	3.65	3.59	3.54	3.48
p,p'-DDT	5.57	5.48	5.39	5.29	5.20	5.11	5.01	4.92	4.83	4.74	4.64	4.55	4.46	4.36	4.27	4.18	4.09	4.00
Ethion	6.1	5.97	5.85	5.73	5.61	5.49	5.36	5.24	5.12	5.00	4.88	4.76	4.64	4.52	4.40	4.28	4.16	4.04
Carbophenothion	6.4	6.2	6.1	5.99	5.88	5.76	5.64	5.52	5.40	5.28	5.16	5.04	4.92	4.80	4.68	4.56	4.44	4.32
Mirex	7.7	7.6	7.5	7.3	7.3	7.1	7.0	6.9	6.8	6.7	6.6	6.5	6.4	6.3	6.2	6.1	6.0	5.85
Endrin Ketone "153"	10.7	10.5	10.3	10.1	9.9	9.7	9.5	9.3	9.1	8.9	8.7	8.5	8.3	8.1.	7.9	7.7	7.5	7.3
Dioctyl Phthalate	13.1	12.7	12.4	12.0	11.6	11.2	10.8	10.4	10.0	9.7	9.3	8.9	8.5	8.1	7.7	7.3	7.0	6.6
Methoxychlor	12.4	12.1	11.8	11.6	11.3	11.0	10.7	10.4	10.1	9.8	9.5	9.3	9.0	8.7	8.4	8.1	7.8	7.5
Tetradifon	16.9	16.5	16.1	15.7	15.3	14.9	14.5	14.1	13.7	13.3	12.9	12.5	12.1	11.7	11.3	10.9	10.5	10.2
Diphenyl Phthalate	22.1	21.5	20.9	20.3	19.6	19.0	18.4	17.7	17.1	16.5	15.8	15.2	14.6	14.0	13.4	12.7	12.1	11.5

Retention ratios, relative to aldrin, of 49 compounds at temperatures from 170 to 204°C; support of Gas Chrom Q, 100/120 mesh; electron capture detector; tritium source, parallel plate; all absolute retentions measured from injection point. Arrow indicated optimum column operating temperature with carrier flow at 60 ml per minute.

TABLE 1.4.

4%SE-30/6%OV-210

Column Temperature, °C.

Compound	170	174	178	182	186	190	194	198	202	204
Dimethyl Phthalate	0.25	0.26	0.26	0.27	0.28	0.28	0.29	0.29	0.30	0.30
Mevinphos	0.27	0.28	0.29	0.29	0.30	0.30	0.31	0.32	0.32	0.33
Tecnazene	0.34	0.34	0.35	0.36	0.37	0.38	0.38	0.39	0.40	0.40
Diethyl Phthalate	0.39	0.39	0.40	0.40	0.40	0.41	0.41	0.42	0.42	0.42
2,4-D(ME)	0.39	0.39	0.40	0.41	0.42	0.42	0.43	0.44	0.44	0.45
Hexachlorobenzene	0.39	0.40	0.41	0.42	0.43	0.43	0.44	0.45	0.46	0.46
α-BHC	0.42	0.43	0.44	0.45	0.46	0.47	0.47	0.48	0.49	0.50
CDEC	0.44	0.45	0.46	0.46	0.47	0.48	0.48	0.50	0.50	0.51
2,4-D(IPE)	0.54	0.54	0.54	0.55	0.55	0.55	0.55	0.55	0.55	0.55
Lindane	0.54	0.55	0.55	0.56	0.57	0.58	0.59	0.59	0.60	0.61
Chlordene	0.54	0.55	0.56	0.56	0.57	0.58	0.58	0.59	0.60	0.61
β-BHC	0.57	0.57	0.58	0.59	0.59	0.60	0.60	0.61	0.61	0.61
Diazinon	0.60	0.60	0.59	0.59	0.59	0.59	0.58	0.58	0.58	0.58
PCNB	0.59	0.60	0.60	0.61	0.62	0.62	0.63	0.63	0.64	0.65
2,4,5-T(ME)	0.66	0.66	0.65	0.65	0.65	0.65	0.65	0.64	0.64	0.64
Heptachlor	0.80	0.80	0.81	0.81	0.82	0.82	0.83	0.83	0.83	0.83
2,4,5-T(IPE)	0.89	0.89	0.87	0.87	0.86	0.85	0.85	0.84	0.83	0.82
Dimethoate	0.96	0.94	0.93	0.92	0.91	0.89	0.88	0.87	0.86	0.85
Ronnel	1.01	1.00	0.98	0.97	0.96	0.94	0.93	0.92	0.91	0.90
Aldrin (REFERENCE)	1.00	1.00	1.00	1.00	1.00	1.00	1.00	1.00	1.00	1.00
1-Hydroxychlordene	1.04	1.04	1.03	1.03	1.03	1.03	1.02	1.02	1.02	1.02
Dibutyl Phthalate	1.41	1.37	1.34	1.31	1.30	1.27	1.24	1.21	1.18	1.14
Oxychlordane	1.43	1.42	1.40	1.39	1.38	1.37	1.36	1.35	1.34	1.33
M. Parathion	1.49	1.47	1.45	1.43	1.41	1.39	1.37	1.35	1.34	1.32
Heptachlor Epoxide	1.53	1.52	1.50	1.49	1.48	1.46	1.45	1.44	1.43	1.42
D C P A	1.64	1.61	1.59	1.56	1.54	1.51	1.49	1.47	1.44	1.41
Malathion	1.70	1.66	1.63	1.59	1.57	1.51	1.48	1.44	1.40	1.38

Compound	170	172	174	176	178	180	182	184	186	188	190	192	194	196	198	200	202	204
o,p'-DDE	1.67	1.66	1.65	1.63	1.62	1.60	1.59	1.57	1.56	1.55	1.53	1.52	1.50	1.49	1.47	1.46	1.45	1.43
Chlordane, Gamma	1.65	1.64	1.63	1.62	1.61	1.61	1.60	1.59	1.58	1.57	1.56	1.55	1.54	1.54	1.53	1.52	1.52	1.50
Chlordane, Alpha	1.84	1.83	1.82	1.80	1.79	1.78	1.77	1.76	1.75	1.74	1.73	1.71	1.70	1.69	1.68	1.67	1.66	1.65
Trans-Nonachlor	1.86	1.85	1.83	1.82	1.81	1.80	1.78	1.77	1.76	1.74	1.73	1.72	1.71	1.69	1.68	1.67	1.66	1.64
E. Parathion	2.09	2.07	2.05	2.02	2.00	1.98	1.96	1.94	1.91	1.89	1.87	1.85	1.83	1.80	1.78	1.76	1.74	1.72
Endosulfan I	1.99	1.98	1.97	1.95	1.94	1.93	1.91	1.90	1.89	1.87	1.86	1.85	1.83	1.82	1.80	1.79	1.78	1.76
p,p'-DDE	2.16	2.14	2.11	2.09	2.07	2.05	2.02	2.00	1.98	1.95	1.93	1.91	1.89	1.86	1.84	1.82	1.80	1.77
DDA(ME)	2.27	2.25	2.22	2.19	2.16	2.13	2.10	2.07	2.04	2.01	1.98	1.96	1.93	1.90	1.87	1.84	1.81	1.78
o,p'-DDD	2.34	2.31	2.29	2.27	2.24	2.22	2.19	2.17	2.15	2.12	2.10	2.07	2.05	2.03	2.00	1.98	1.96	1.93
Dieldrin	2.43	2.41	2.39	2.37	2.35	2.33	2.31	2.29	2.27	2.25	2.22	2.20	2.18	2.16	2.14	2.12	2.10	2.08
o,p'-DDT	3.02	2.97	2.93	2.88	2.84	2.80	2.76	2.73	2.68	2.64	2.60	2.56	2.52	2.47	2.43	2.39	2.35	2.31
Endrin	2.76	2.73	2.71	2.69	2.67	2.64	2.62	2.60	2.58	2.55	2.53	2.51	2.49	2.46	2.44	2.42	2.40	2.37
p,p'-DDD	3.22	3.17	3.13	3.08	3.04	2.98	2.94	2.90	2.86	2.82	2.77	2.73	2.68	2.64	2.59	2.55	2.51	2.46
Chlordecone	2.97	2.94	2.91	2.89	2.86	2.83	2.80	2.78	2.75	2.72	2.69	2.67	2.64	2.61	2.59	2.56	2.53	2.50
Endosulfan II	3.19	3.16	3.13	3.10	3.07	3.04	3.00	2.97	2.94	2.91	2.88	2.85	2.81	2.78	2.75	2.72	2.69	2.66
Ethion	4.08	4.02	3.96	3.89	3.81	3.76	3.68	3.60	3.53	3.47	3.40	3.32	3.27	3.20	3.13	3.05	2.98	2.90
p,p'-DDT	4.04	3.98	3.92	3.86	3.80	3.73	3.67	3.61	3.54	3.48	3.43	3.36	3.30	3.24	3.18	3.12	3.05	2.98
Carbophenothion	4.08	4.02	3.98	3.90	3.83	3.78	3.72	3.66	3.59	3.52	3.47	3.40	3.34	3.28	3.22	3.16	3.10	3.03
Methoxychlor	6.7	6.5	6.4	6.3	6.1	5.98	5.84	5.70	5.57	5.42	5.29	5.16	5.01	4.88	4.73	4.60	4.46	4.32
Mirex	6.1	6.0	5.96	5.87	5.78	5.68	5.60	5.52	5.43	5.33	5.24	5.15	5.06	4.97	4.88	4.79	4.70	4.62
Endrin Ketone"153"	7.3	7.2	7.1	7.0	6.9	6.8	6.7	6.5	6.4	6.3	6.2	6.1	5.98	5.84	5.75	5.64	5.53	5.42
Dioctyl Phthalate	11.0	10.8	10.5	10.2	9.9	9.6	9.3	9.0	8.7	8.4	8.1	7.8	7.5	7.2	6.9	6.6	6.4	6.1
Diphenyl Phthalate	12.2	11.9	11.6	11.3	11.0	10.7	10.4	10.1	9.8	9.5	9.2	8.9	8.6	8.3	8.0	7.7	7.4	7.1
Tetradifon	11.6	11.3	11.1	10.8	10.6	10.3	10.1	9.8	9.6	9.3	9.0	8.8	8.5	8.3	8.0	7.8	7.5	7.3

Retention ratios, relative to aldrin, of 49 compounds at temperatures from 170 to 204°C; support of Gas Chrom Q, 80/100 mesh; electron capture detector; tritium source, parallel plate; all absolute retentions measured from injection point. Arrow indicated optimum column operating temperature with carrier flow at 70 ml per minute.

19

TABLE 1.5.

5% OV-210

Column Temperature, °C.										
170	174	178	182	186	190	194	198	202	204	Compound
.43	.44	.45	.46	.47	.48	.49	.50	.51	.52	Hexachlorobenzene
.51	.51	.51	.52	.52	.53	.53	.53	.54	.54	Dimethyl Phthalate
.52	.53	.54	.55	.56	.57	.58	.59	.60	.61	Tecnazene
.58	.59	.60	.61	.62	.62	.63	.64	.65	.66	Chlordene
.62	.62	.63	.64	.65	.66	.67	.68	.68	.69	α-BHC
.66	.66	.66	.66	.66	.66	.66	.66	.66	.66	Mevinphos
.69	.69	.69	.69	.69	.69	.69	.69	.69	.69	2,4-D (ME)
.69	.69	.69	.69	.70	.70	.70	.70	.69	.70	CDEC
.74	.73	.73	.73	.73	.72	.72	.72	.72	.71	Diethyl Phthalate
.76	.75	.75	.75	.74	.74	.74	.73	.73	.73	Diazinon
.80	.80	.80	.81	.81	.82	.82	.83	.83	.84	Lindane
.86	.86	.86	.86	.86	.86	.86	.86	.86	.86	PCNB
.87	.87	.87	.87	.87	.88	.88	.88	.88	.88	Heptachlor
.88	.87	.86	.86	.85	.85	.84	.83	.83	.82	2,4-D (IPE)
.97	.96	.96	.96	.95	.95	.95	.94	.94	.94	β-BHC
1.00	1.00	1.00	1.00	1.00	1.00	1.00	1.00	1.00	1.00	Aldrin (REFERENCE) →
1.08	1.07	1.06	1.05	1.03	1.02	1.01	1.00	.98	.98	2,4,5-T (ME)
1.36	1.33	1.30	1.27	1.25	1.22	1.19	1.16	1.14	1.12	2,4,5-T (IPE)
1.41	1.39	1.37	1.35	1.33	1.31	1.29	1.27	1.25	1.24	Ronnel
1.43	1.41	1.39	1.38	1.36	1.34	1.33	1.31	1.29	1.28	1-Hydroxychlordene
1.60	1.58	1.56	1.54	1.53	1.51	1.49	1.47	1.45	1.44	Oxychlordane
1.67	1.63	1.60	1.57	1.53	1.50	1.47	1.43	1.40	1.39	o,p'-DDE
1.90	1.87	1.84	1.80	1.77	1.74	1.71	1.67	1.64	1.62	Chlordane, Gamma
1.88	1.85	1.82	1.78	1.75	1.72	1.69	1.66	1.62	1.61	Trans-Nonachlor
2.02	1.98	1.95	1.91	1.87	1.83	1.79	1.75	1.71	1.69	Heptachlor Epoxide
2.08	2.04	2.00	1.97	1.93	1.89	1.85	1.81	1.77	1.75	Chlordane, Alpha

20

Compound	170	172	174	176	178	180	182	184	186	188	190	192	194	196	198	200	202	204
Dimethoate	2.18	2.15	2.12	2.09	2.06	2.04	2.01	1.98	1.95	1.92	1.89	1.86	1.83	1.80	1.78	1.75	1.72	1.69
p,p'-DDE	2.24	2.21	2.18	2.15	2.12	2.10	2.07	2.04	2.01	1.98	1.95	1.92	1.90	1.87	1.84	1.81	1.78	1.75
Dibutyl Phthalate	2.29	2.25	2.22	2.18	2.14	2.10	2.07	2.03	1.99	1.95	1.92	1.88	1.84	1.80	1.77	1.73	1.69	1.66
Endosulfan I	2.63	2.60	2.57	2.54	2.51	2.48	2.45	2.42	2.39	2.36	2.33	2.30	2.27	2.24	2.21	2.18	2.15	2.12
o,p'-DDD	2.74	2.70	2.67	2.63	2.59	2.55	2.51	2.47	2.43	2.39	2.35	2.31	2.27	2.23	2.19	2.15	2.11	2.07
Chlordecone	2.75	2.73	2.70	2.67	2.64	2.61	2.58	2.55	2.52	2.49	2.46	2.43	2.40	2.37	2.34	2.31	2.28	2.25
DCPA	2.84	2.80	2.75	2.71	2.67	2.63	2.59	2.54	2.50	2.46	2.41	2.37	2.33	2.29	2.24	2.20	2.16	2.12
o,p'-DDT	2.92	2.87	2.83	2.79	2.74	2.69	2.65	2.61	2.57	2.52	2.48	2.43	2.39	2.35	2.30	2.26	2.21	2.17
DDA (ME)	3.10	3.05	3.00	2.94	2.89	2.84	2.79	2.74	2.69	2.64	2.59	2.54	2.48	2.43	2.38	2.33	2.28	2.23
M. Parathion	3.17	3.12	3.07	3.02	2.97	2.92	2.88	2.83	2.78	2.73	2.68	2.63	2.58	2.54	2.49	2.44	2.39	2.34
Malathion	3.20	3.14	3.08	3.02	2.96	2.91	2.85	2.79	2.73	2.67	2.61	2.55	2.49	2.44	2.38	2.32	2.26	2.20
Dieldrin	3.21	3.17	3.13	3.09	3.04	3.00	2.96	2.92	2.88	2.83	2.79	2.75	2.71	2.68	2.62	2.58	2.54	2.49
Endrin	3.82	3.77	3.72	3.67	3.61	3.56	3.50	3.45	3.39	3.34	3.28	3.23	3.18	3.13	3.07	3.02	2.96	2.91
Mirex	4.06	4.01	3.95	3.89	3.84	3.78	3.73	3.68	3.62	3.57	3.52	3.46	3.40	3.35	3.29	3.24	3.19	3.13
p,p'-DDD	4.10	4.03	3.96	3.89	3.82	3.75	3.68	3.61	3.53	3.46	3.38	3.32	3.25	3.18	3.11	3.03	2.96	2.89
E. Parathion	4.24	4.17	4.09	4.02	3.94	3.87	3.78	3.71	3.63	3.56	3.48	3.40	3.33	3.25	3.17	3.10	3.03	2.95
p,p'-DDT	4.47	4.39	4.31	4.23	4.15	4.07	3.98	3.90	3.82	3.74	3.66	3.58	3.49	3.41	3.33	3.25	3.17	3.09
Endosulfan II	4.98	4.91	4.83	4.75	4.67	4.59	4.51	4.43	4.35	4.27	4.19	4.11	4.03	3.95	3.87	3.79	3.71	3.63
Carbophenothion	5.29	5.19	5.09	4.99	4.89	4.78	4.68	4.58	4.48	4.37	4.27	4.17	4.06	3.96	3.86	3.76	3.66	3.56
Ethion	6.00	5.89	5.76	5.64	5.51	5.37	5.25	5.12	4.99	4.86	4.73	4.60	4.48	4.35	4.22	4.09	3.96	3.83
Methoxychlor	7.4	7.2	7.0	6.8	6.7	6.5	6.3	6.1	5.94	5.76	5.59	5.41	5.23	5.06	4.88	4.70	4.53	4.35
Endrin Ketone "153"	13.2	13.0	12.7	12.4	12.1	11.8	11.5	11.2	11.0	10.7	10.4	10.1	9.8	9.5	9.2	9.0	8.7	8.4
Dioctyl Phthalate	13.5	13.1	12.6	12.2	11.8	11.4	11.0	10.6	10.1	9.7	9.3	8.9	8.5	8.0	7.6	7.2	6.8	6.4
Diphenyl Phthalate	21.4	20.6	20.1	19.5	18.8	18.2	17.5	16.9	16.2	15.6	14.9	14.3	13.6	13.0	12.3	11.6	11.0	10.3
Tetradifon	21.1	20.6	20.0	19.5	18.9	18.4	17.9	17.3	16.7	16.2	15.6	15.1	14.5	14.0	13.4	12.9	12.3	11.8

Retention ratios, relative to aldrin, of 47 compounds at temperature from 170 to 204°C; support of gas Chrom Q, 80/100 mesh; electron capture detector; ^{63}Ni source; all absolute retentions measured from injection point. Arrow indicated optimum column operating temperature with carrier flow at 50 ml per min.

TABLE 1.6.

10% DC-200
Column Temperature, °C

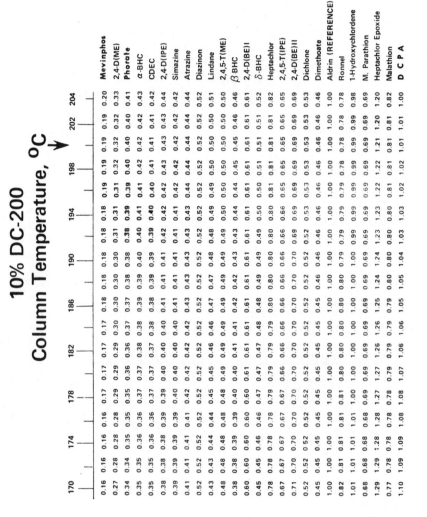

	170	174	178	182	186	190	194	198	202	204	
	0.16	0.16	0.17	0.17	0.18	0.18	0.19	0.19	0.19	0.20	Mevinphos
	0.27	0.28	0.29	0.30	0.30	0.31	0.31	0.32	0.32	0.33	2,4-D(ME)
	0.34	0.35	0.36	0.37	0.38	0.38	0.39	0.40	0.40	0.41	Phorate
	0.35	0.36	0.37	0.38	0.39	0.40	0.41	0.42	0.42	0.43	α-BHC
	0.35	0.36	0.37	0.38	0.38	0.39	0.40	0.41	0.41	0.42	CDEC
	0.38	0.39	0.40	0.40	0.41	0.41	0.42	0.43	0.43	0.44	2,4-D(IPE)
	0.39	0.39	0.40	0.42	0.41	0.41	0.42	0.42	0.42	0.42	Simazine
	0.41	0.41	0.42	0.42	0.43	0.43	0.43	0.44	0.44	0.44	Atrazine
	0.52	0.52	0.52	0.52	0.52	0.52	0.52	0.52	0.52	0.52	Diazinon
	0.43	0.44	0.45	0.46	0.47	0.48	0.49	0.50	0.50	0.51	Lindane
	0.48	0.48	0.48	0.49	0.49	0.49	0.50	0.50	0.50	0.50	2,4,5-T(ME)
	0.38	0.39	0.40	0.41	0.42	0.43	0.44	0.45	0.46	0.46	β-BHC
	0.60	0.60	0.60	0.61	0.61	0.61	0.61	0.61	0.61	0.61	2,4-D(BE)I
	0.45	0.46	0.47	0.48	0.48	0.49	0.50	0.51	0.51	0.52	δ-BHC
	0.78	0.78	0.79	0.79	0.80	0.80	0.80	0.81	0.81	0.82	Heptachlor
	0.67	0.67	0.66	0.66	0.66	0.66	0.66	0.65	0.65	0.65	2,4,5-T(IPE)
	0.71	0.70	0.70	0.70	0.70	0.70	0.69	0.69	0.69	0.69	2,4-D(BE)II
	0.52	0.52	0.52	0.52	0.52	0.52	0.53	0.53	0.53	0.53	Dichlone
	0.45	0.45	0.45	0.45	0.46	0.46	0.46	0.46	0.46	0.46	Dimethoate
	1.00	1.00	1.00	1.00	1.00	1.00	1.00	1.00	1.00	1.00	Aldrin (REFERENCE)
	0.82	0.81	0.80	0.80	0.79	0.79	0.79	0.78	0.78	0.78	Ronnel
	1.01	1.01	1.00	1.00	1.00	1.00	0.99	0.99	0.99	0.98	1-Hydroxychlordene
	0.68	0.68	0.69	0.69	0.69	0.69	0.69	0.69	0.69	0.69	M. Parathion
	1.29	1.28	1.27	1.26	1.25	1.24	1.23	1.22	1.21	1.20	Heptachlor Epoxide
	0.77	0.78	0.78	0.79	0.79	0.80	0.80	0.80	0.81	0.82	Malathion
	1.10	1.09	1.08	1.07	1.06	1.05	1.04	1.03	1.01	1.00	D C P A

Compound	170		174		178		182		186		190		194		198		202	204
Dyrene	1.29	1.28	1.27	1.26	1.25	1.24	1.23	1.22	1.22	1.21	1.20	1.20	1.19	1.18	1.17	1.17		
o,p'-DDE	1.65	1.64	1.62	1.61	1.59	1.58	1.57	1.55	1.54	1.53	1.51	1.50	1.48	1.47	1.46	1.44	1.43	1.41
Chlorbenside	1.43	1.42	1.41	1.40	1.40	1.39	1.38	1.37	1.36	1.36	1.35	1.34	1.33	1.33	1.32	1.31	1.30	1.30
E. Parathion	1.03	1.02	1.01	1.00	1.00	0.99	0.99	0.98	0.98	0.97	0.96	0.96	0.95	0.95	0.94	0.93	0.93	
Endosulfan I	1.64	1.63	1.62	1.61	1.60	1.59	1.58	1.58	1.57	1.56	1.55	1.54	1.54	1.53	1.52	1.51	1.50	
p,p'-DDE	2.09	2.07	2.05	2.02	2.00	1.98	1.96	1.94	1.92	1.90	1.88	1.86	1.83	1.81	1.79	1.77	1.75	1.73
DDA(ME)	1.81	1.79	1.77	1.75	1.73	1.71	1.69	1.67	1.65	1.63	1.61	1.60	1.58	1.56	1.54	1.52	1.50	1.48
Captan	1.21	1.21	1.21	1.20	1.20	1.19	1.19	1.18	1.18	1.17	1.17	1.16	1.16	1.15	1.15	1.14	1.14	1.13
Folpet	1.30	1.29	1.29	1.28	1.27	1.27	1.26	1.26	1.25	1.24	1.24	1.23	1.23	1.22	1.21	1.21	1.20	1.20
Dieldrin	1.99	1.98	1.96	1.95	1.93	1.92	1.90	1.89	1.87	1.86	1.84	1.83	1.81	1.80	1.79	1.77	1.76	1.74
Perthane	2.60	2.56	2.53	2.49	2.45	2.41	2.38	2.34	2.30	2.27	2.23	2.19	2.15	2.11	2.08	2.04	2.00	1.97
o,p'-DDD	2.13	2.11	2.08	2.06	2.04	2.01	1.99	1.97	1.94	1.92	1.90	1.87	1.85	1.83	1.80	1.78	1.76	1.73
o,p'-DDT	2.90	2.86	2.83	2.79	2.76	2.72	2.68	2.65	2.61	2.58	2.54	2.50	2.47	2.43	2.40	2.36	2.32	2.29
Endrin	2.22	2.20	2.19	2.17	2.16	2.14	2.13	2.13	2.11	2.09	2.08	2.06	2.05	2.03	2.01	1.98	1.97	1.95
Chlordecone	3.07	3.03	3.00	2.96	2.93	2.90	2.86	2.83	2.79	2.76	2.72	2.69	2.66	2.63	2.59	2.55	2.52	2.49
p,p'-DDD	2.72	2.69	2.66	2.62	2.59	2.56	2.52	2.49	2.45	2.42	2.39	2.35	2.32	2.29	2.25	2.22	2.19	2.15
Endosulfan II	2.29	2.27	2.25	2.24	2.22	2.20	2.18	2.17	2.15	2.13	2.11	2.10	2.08	2.06	2.05	2.03	2.01	2.00
Ethion	3.04	3.00	2.95	2.91	2.86	2.81	2.77	2.72	2.68	2.63	2.58	2.54	2.49	2.44	2.40	2.35	2.31	2.26
p,p'-DDT	3.72	3.67	3.62	3.57	3.51	3.46	3.40	3.35	3.29	3.24	3.18	3.13	3.08	3.02	2.97	2.92	2.86	2.81
Carbophenothion	3.42	3.37	3.32	3.26	3.21	3.16	3.11	3.06	3.01	2.96	2.90	2.85	2.80	2.75	2.70	2.65	2.60	2.55
Dilan I	3.23	3.19	3.15	3.11	3.06	3.02	2.98	2.93	2.89	2.85	2.80	2.76	2.72	2.67	2.63	2.58	2.54	2.50
Mirex	6.9	6.8	6.7	6.6	6.5	6.4	6.3	6.1	6.0	5.90	5.80	5.69	5.59	5.48	5.38	5.28	5.17	
Methoxychlor	6.1	6.0	5.89	5.77	5.66	5.54	5.42	5.30	5.18	5.06	4.93	4.82	4.69	4.58	4.45	4.33	4.22	4.10
Dilan II	4.07	4.02	3.96	3.89	3.83	3.77	3.70	3.64	3.58	3.52	3.45	3.39	3.33	3.27	3.20	3.14	3.08	3.02
Tetradifon	6.4	6.3	6.2	6.1	5.97	5.85	5.73	5.62	5.50	5.38	5.27	5.15	5.03	4.92	4.80	4.68	4.57	4.45
Azinphosmethyl	6.4	6.3	6.2	6.0	5.92	5.81	5.69	5.58	5.46	5.34	5.23	5.12	5.00	4.88	4.77	4.66	4.54	4.42

Retention ratios, relative to aldrin, of 48 pesticides on a column of 10% DC-200 at temperatures from 170 to 204°C; support of Chromosorb W.H.P., 80/100 mesh; electron capture detector, tritium source, parallel plate; all absolute retentions measured from injection point. Arrow indicates optimum column operating temperature with carrier flow at 120 ml per minute.

23

TABLE 1.7.

5%DC-200/7.5%QF-1
Column Temperature, °C.

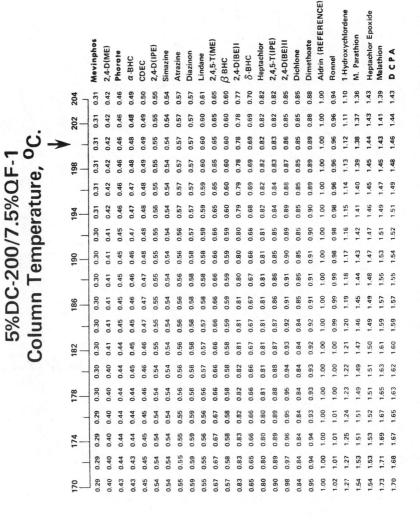

	170	174	178	182	186	190	194	198	202	204
Mevinphos	0.29	0.29	0.30	0.30	0.30	0.30	0.30	0.31	0.31	0.31
2,4-D(ME)	0.40	0.40	0.40	0.41	0.41	0.41	0.41	0.42	0.42	0.42
Phorate	0.44	0.44	0.44	0.44	0.45	0.45	0.45	0.46	0.46	0.46
α-BHC	0.43	0.44	0.44	0.45	0.45	0.46	0.47	0.48	0.48	0.49
CDEC	0.45	0.45	0.46	0.46	0.47	0.48	0.48	0.49	0.49	0.50
2,4-D(IPE)	0.54	0.54	0.54	0.55	0.55	0.55	0.55	0.55	0.55	0.55
Simazine	0.54	0.54	0.54	0.54	0.54	0.54	0.54	0.54	0.54	0.54
Atrazine	0.55	0.55	0.56	0.56	0.56	0.56	0.57	0.57	0.57	0.57
Diazinon	0.59	0.59	0.58	0.58	0.58	0.58	0.57	0.57	0.57	0.57
Lindane	0.55	0.56	0.56	0.57	0.58	0.58	0.59	0.60	0.60	0.61
2,4,5-T(ME)	0.67	0.67	0.66	0.66	0.66	0.66	0.65	0.65	0.65	0.65
β BHC	0.57	0.58	0.58	0.58	0.59	0.59	0.60	0.60	0.60	0.60
2,4-D(BE)I	0.83	0.83	0.82	0.81	0.81	0.80	0.79	0.78	0.78	0.77
δ-BHC	0.65	0.66	0.66	0.67	0.67	0.68	0.68	0.69	0.69	0.70
Heptachlor	0.80	0.80	0.81	0.81	0.81	0.81	0.82	0.82	0.82	0.82
2,4,5-T(IPE)	0.89	0.89	0.88	0.88	0.86	0.85	0.84	0.83	0.83	0.82
2,4-D(BE)II	0.98	0.96	0.95	0.94	0.92	0.90	0.89	0.88	0.86	0.85
Dichlone	0.84	0.84	0.84	0.84	0.85	0.85	0.85	0.85	0.85	0.85
Dimethoate	0.95	0.94	0.93	0.93	0.91	0.91	0.90	0.89	0.88	0.88
Aldrin (REFERENCE)	1.00	1.00	1.00	1.00	1.00	1.00	1.00	1.00	1.00	1.00
Ronnel	1.02	1.01	1.01	1.00	0.99	0.99	0.98	0.96	0.96	0.94
1-Hydroxychlordene	1.27	1.25	1.23	1.21	1.19	1.17	1.15	1.13	1.12	1.10
M. Parathion	1.53	1.51	1.51	1.47	1.45	1.43	1.41	1.39	1.38	1.36
Heptachlor Epoxide	1.53	1.52	1.51	1.50	1.49	1.47	1.46	1.45	1.44	1.43
Malathion	1.73	1.69	1.65	1.63	1.59	1.53	1.47	1.45	1.43	1.39
D C P A	1.70	1.67	1.63	1.60	1.57	1.54	1.51	1.48	1.46	1.43

Compound	170		174		178		182		186		190		194		198		202	204
Dyrene	1.54	1.53	1.51	1.50	1.49	1.48	1.47	1.46	1.45	1.44	1.43	1.41	1.40	1.39	1.38	1.37	1.36	1.35
o,p'-DDE	1.66	1.65	1.63	1.62	1.61	1.59	1.58	1.57	1.55	1.54	1.53	1.51	1.50	1.49	1.47	1.46	1.45	1.43
Chlorbenside	1.62	1.61	1.60	1.59	1.58	1.57	1.56	1.55	1.54	1.53	1.52	1.51	1.50	1.49	1.48	1.47	1.46	1.45
E. Parathion	2.14	2.11	2.09	2.07	2.05	2.03	2.01	1.99	1.97	1.95	1.93	1.91	1.88	1.86	1.85	1.84	1.82	1.80
Endosulfan I	2.00	1.97	1.96	1.95	1.95	1.94	1.93	1.91	1.90	1.89	1.88	1.87	1.86	1.85	1.84	1.84	1.81	1.80
p,p'-DDE	2.15	2.13	2.11	2.09	2.07	2.05	2.03	2.01	1.99	1.97	1.95	1.93	1.90	1.88	1.86	1.84	1.82	1.80
DDA(ME)	2.28	2.25	2.22	2.19	2.16	2.13	2.10	2.07	2.04	2.01	1.99	1.96	1.93	1.90	1.87	1.84	1.81	1.78
Captan	2.35	2.33	2.31	2.28	2.26	2.23	2.21	2.19	2.17	2.14	2.12	2.09	2.07	2.05	2.02	2.00	1.98	1.95
Folpet	2.24	2.23	2.21	2.19	2.17	2.15	2.13	2.12	2.10	2.08	2.06	2.04	2.03	2.01	1.99	1.97	1.95	1.93
Dieldrin	2.40	2.38	2.37	2.35	2.33	2.32	2.30	2.29	2.27	2.25	2.24	2.22	2.21	2.19	2.18	2.16	2.15	2.13
Perthane	2.42	2.39	2.36	2.33	2.30	2.27	2.24	2.21	2.18	2.16	2.13	2.11	2.08	2.06	2.03	2.01	1.99	1.90
o,p'-DDD	2.37	2.35	2.33	2.30	2.28	2.25	2.23	2.21	2.18	2.16	2.13	2.11	2.08	2.06	2.03	2.01	1.99	1.96
o,p'-DDT	2.91	2.87	2.84	2.81	2.78	2.75	2.72	2.69	2.65	2.62	2.58	2.56	2.53	2.49	2.46	2.43	2.40	2.37
Endrin	2.81	2.79	2.76	2.74	2.72	2.69	2.67	2.65	2.62	2.60	2.58	2.55	2.53	2.51	2.48	2.46	2.44	2.41
Chlordecone	2.96	2.93	2.91	2.89	2.86	2.84	2.81	2.79	2.76	2.74	2.71	2.69	2.66	2.64	2.61	2.59	2.57	2.54
p,p'-DDD	3.22	3.18	3.14	3.10	3.06	3.02	2.98	2.94	2.90	2.86	2.82	2.78	2.74	2.70	2.66	2.62	2.58	2.54
Endosulfan II	3.22	3.19	3.16	3.13	3.11	3.07	3.05	3.02	2.99	2.96	2.93	2.90	2.87	2.84	2.81	2.78	2.75	2.72
Ethion	4.10	4.04	3.98	3.91	3.85	3.79	3.73	3.66	3.60	3.54	3.47	3.41	3.35	3.29	3.22	3.16	3.10	3.03
p,p'-DDT	4.07	4.00	3.93	3.86	3.79	3.72	3.65	3.58	3.51	3.44	3.37	3.30	3.23	3.16	3.09	3.02	2.95	2.88
Carbophenothion	4.12	4.06	4.00	3.94	3.88	3.82	3.76	3.70	3.63	3.57	3.50	3.44	3.38	3.32	3.26	3.20	3.14	3.08
Dilan I	6.3	6.2	6.1	5.98	5.87	5.76	5.65	5.53	5.42	5.31	5.20	5.10	5.00	4.88	4.78	4.68	4.56	4.46
Mirex	5.98	5.91	5.83	5.76	5.68	5.61	5.53	5.46	5.38	5.30	5.22	5.15	5.08	5.00	4.92	4.85	4.78	4.70
Methoxychlor	6.6	6.5	6.4	6.2	6.1	5.97	5.83	5.71	5.58	5.43	5.30	5.18	5.03	4.88	4.78	4.64	4.50	4.38
Dilan II	7.5	7.3	7.2	7.0	6.9	6.8	6.6	6.5	6.4	6.2	6.1	5.94	5.80	5.67	5.52	5.40	5.26	5.11
Tetradifon	11.6	11.4	11.2	10.9	10.7	10.4	10.2	10.0	9.7	9.5	9.2	9.0	8.8	8.5	8.3	8.0	7.8	7.6
Azinphosmethyl	12.5	12.2	12.0	11.7	11.4	11.2	10.9	10.6	10.4	10.1	9.9	9.6	9.3	9.1	8.8	8.5	8.3	8.0

Retention ratios, relative to aldrin, of 48 pesticides on a column of 5% DC-200/7.5%QF-1 at temperatures from 170 to 204°C; support of Chromosorb W,H.P., 80/100 mesh; electron capture detector, tritium source, parallel plate; all absolute retentions measured from injection point. Arrow indicates optimum column operating temperature with carrier flow at 120 ml per minute.

TABLE 1.8.

1.6%OV-17/6.4%OV-210
Column Temperature, °C.

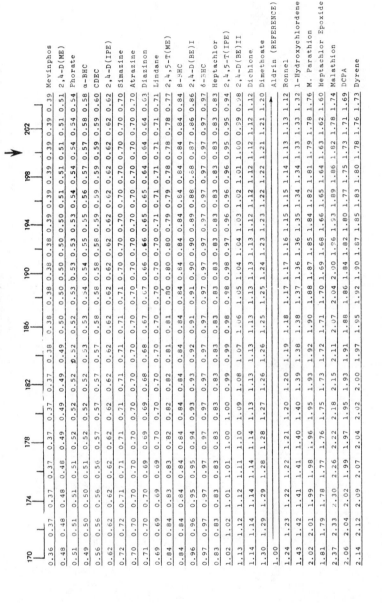

Compound	170	172	174	176	178	180	182	184	186	188	190	192	194	196	198	200	202
Mevinphos	0.36	0.37	0.37	0.37	0.37	0.37	0.38	0.38	0.38	0.38	0.38	0.39	0.39	0.39	0.39	0.39	0.39
2,4-D(ME)	0.48	0.48	0.48	0.49	0.49	0.49	0.49	0.50	0.50	0.50	0.50	0.50	0.51	0.51	0.51	0.51	0.51
Phorate	0.51	0.51	0.51	0.52	0.52	0.52	0.52	0.53	0.53	0.53	0.53	0.53	0.54	0.54	0.54	0.54	0.54
α-BHC	0.49	0.50	0.50	0.51	0.52	0.52	0.52	0.53	0.53	0.54	0.54	0.55	0.55	0.56	0.57	0.57	0.58
CDEC	0.56	0.56	0.56	0.56	0.57	0.57	0.57	0.57	0.58	0.58	0.58	0.58	0.59	0.59	0.59	0.59	0.60
2,4-D(IPE)	0.62	0.62	0.62	0.62	0.62	0.62	0.62	0.62	0.62	0.62	0.62	0.62	0.62	0.62	0.62	0.62	0.62
Simazine	0.72	0.72	0.71	0.71	0.71	0.71	0.71	0.71	0.71	0.71	0.71	0.71	0.70	0.70	0.70	0.70	0.70
Atrazine	0.72	0.72	0.71	0.70	0.70	0.70	0.70	0.70	0.70	0.70	0.70	0.70	0.70	0.70	0.70	0.70	0.70
Diazinon	0.70	0.70	0.70	0.70	0.69	0.69	0.68	0.68	0.67	0.67	0.66	0.66	0.65	0.65	0.64	0.64	0.63
Lindane	0.69	0.69	0.69	0.69	0.69	0.70	0.70	0.70	0.70	0.70	0.70	0.70	0.71	0.71	0.71	0.71	0.71
2,4,5-T(ME)	0.84	0.83	0.83	0.83	0.82	0.82	0.82	0.81	0.81	0.80	0.80	0.79	0.79	0.79	0.78	0.78	0.77
B-BHC	0.84	0.84	0.84	0.84	0.84	0.84	0.84	0.84	0.84	0.84	0.84	0.84	0.84	0.84	0.84	0.84	0.84
2,4-D(BE)I	0.96	0.95	0.95	0.95	0.94	0.93	0.93	0.92	0.91	0.91	0.90	0.90	0.89	0.88	0.88	0.87	0.86
6-BHC	0.97	0.97	0.97	0.97	0.97	0.97	0.97	0.97	0.97	0.97	0.97	0.97	0.97	0.97	0.97	0.97	0.97
Heptachlor	0.83	0.83	0.83	0.83	0.83	0.83	0.83	0.83	0.83	0.83	0.83	0.83	0.83	0.83	0.83	0.83	0.83
2,4,5-T(IPE)	1.02	1.01	1.01	1.00	1.00	1.00	0.99	0.99	0.98	0.98	0.98	0.97	0.97	0.96	0.96	0.95	0.94
2,4-D(BE)II	1.13	1.12	1.12	1.11	1.10	1.09	1.08	1.07	1.06	1.05	1.04	1.04	1.03	1.02	1.01	1.00	0.98
Dichlone	1.14	1.14	1.14	1.14	1.14	1.13	1.13	1.13	1.13	1.13	1.13	1.13	1.12	1.12	1.12	1.12	1.12
Dimethoate	1.30	1.29	1.29	1.28	1.28	1.27	1.26	1.25	1.25	1.24	1.24	1.23	1.23	1.22	1.22	1.21	1.20
Aldrin (REFERENCE)	1.00	→															
Ronnel	1.24	1.23	1.22	1.22	1.21	1.20	1.20	1.19	1.18	1.17	1.17	1.16	1.15	1.15	1.14	1.13	1.12
1-Hydroxychlordene	1.43	1.42	1.41	1.41	1.40	1.40	1.39	1.38	1.38	1.37	1.36	1.36	1.35	1.34	1.34	1.33	1.32
M. Parathion	2.02	2.01	1.99	1.98	1.96	1.95	1.93	1.92	1.90	1.88	1.87	1.85	1.84	1.82	1.81	1.79	1.76
Heptachlor Epoxide	1.81	1.79	1.78	1.77	1.76	1.75	1.73	1.72	1.71	1.70	1.69	1.68	1.66	1.65	1.64	1.63	1.60
Malathion	2.37	2.33	2.30	2.26	2.22	2.18	2.15	2.11	2.07	2.04	2.00	1.96	1.93	1.89	1.86	1.82	1.74
DCPA	2.06	2.04	2.02	1.99	1.97	1.95	1.93	1.91	1.88	1.86	1.84	1.82	1.80	1.77	1.75	1.73	1.69
Dyrene	2.14	2.12	2.09	2.07	2.04	2.02	2.00	1.97	1.95	1.92	1.90	1.85	1.83	1.80	1.78	1.76	1.73

26

Retention ratios, relative to aldrin, of 49 pesticides on a column of 1.6%OV-17/6.4%OV-210 at temperatures from 170 to 204°C; support of Chromosorb W,H.P., 80/100 mesh; electron capture detector, tritium source, parallel plate; all absolute retentions measured from injection point. Arrow indicates optimum column operating temperature with carrier flow at 70 ml per minute.

Compound	170	172	174	176	178	180	182	184	186	188	190	192	194	196	198	200	202	204
o,p'-DDE	2.04	2.02	2.00	1.98	1.95	1.93	1.91	1.89	1.87	1.85	1.83	1.80	1.78	1.76	1.74	1.72	1.70	1.68
Chlorbenside	2.18	2.16	2.14	2.11	2.09	2.07	2.04	2.02	2.00	1.97	1.95	1.93	1.90	1.88	1.86	1.84	1.81	1.79
E. Parathion	2.80	2.76	2.73	2.69	2.66	2.62	2.59	2.55	2.52	2.48	2.44	2.41	2.37	2.34	2.30	2.27	2.23	2.20
Endosulfan I	2.33	2.31	2.29	2.27	2.25	2.23	2.21	2.19	2.18	2.16	2.14	2.12	2.10	2.08	2.07	2.05	2.03	2.01
p,p'-DDE	2.67	2.64	2.60	2.57	2.53	2.49	2.46	2.42	2.39	2.35	2.32	2.28	2.24	2.21	2.17	2.14	2.10	2.06
DDA(ME)	3.07	3.02	2.97	2.92	2.87	2.82	2.78	2.73	2.68	2.63	2.58	2.54	2.48	2.44	2.39	2.34	2.29	2.24
Captan	3.58	3.54	3.49	3.44	3.40	3.35	3.30	3.26	3.21	3.17	3.12	3.07	3.03	2.98	2.94	2.89	2.84	2.80
Folpet	3.51	3.46	3.42	3.37	3.33	3.28	3.24	3.20	3.15	3.10	3.06	3.02	2.97	2.93	2.88	2.84	2.79	2.75
Dieldrin	2.94	2.92	2.89	2.86	2.83	2.80	2.77	2.74	2.71	2.68	2.65	2.62	2.60	2.57	2.54	2.51	2.48	2.45
Perthane	3.24	3.18	3.12	3.07	3.01	2.95	2.90	2.84	2.78	2.72	2.66	2.61	2.55	2.49	2.44	2.38	2.32	2.26
o,p'-DDD	3.26	3.20	3.16	3.10	3.05	3.00	2.96	2.90	2.86	2.80	2.76	2.71	2.66	2.61	2.56	2.51	2.46	2.40
o,p'-DDT	3.81	3.75	3.70	3.64	3.58	3.52	3.46	3.40	3.35	3.28	3.23	3.17	3.11	3.05	3.00	2.94	2.88	2.82
Endrin	3.68	3.64	3.59	3.54	3.50	3.45	3.40	3.36	3.32	3.27	3.22	3.18	3.13	3.09	3.04	2.99	2.95	2.90
Chlordecone	3.18	3.15	3.12	3.09	3.06	3.03	3.00	2.97	2.94	2.91	2.88	2.85	2.82	2.80	2.77	2.74	2.71	2.68
p,p'-DDD	4.63	4.55	4.46	4.38	4.30	4.21	4.13	4.05	3.96	3.88	3.80	3.72	3.63	3.55	3.46	3.38	3.30	3.22
Endosulfan II	4.60	4.54	4.48	4.42	4.35	4.30	4.24	4.17	4.12	4.05	4.00	3.94	3.88	3.82	3.75	3.69	3.63	3.57
Ethion	6.2	6.0	5.92	5.79	5.67	5.54	5.42	5.29	5.16	5.04	4.92	4.79	4.67	4.54	4.42	4.29	4.17	4.04
p,p'-DDT	5.46	5.36	5.26	5.16	5.06	4.96	4.86	4.76	4.66	4.56	4.46	4.36	4.26	4.16	4.06	3.96	3.86	3.76
Carbophenothion	6.35	6.2	6.1	5.96	5.83	5.69	5.56	5.43	5.30	5.16	5.03	4.90	4.77	4.63	4.50	4.37	4.24	4.10
Dilan I	9.1	8.9	8.7	8.5	8.3	8.2	8.0	7.8	7.6	7.4	7.2	7.0	6.8	6.7	6.5	6.3	6.1	5.90
Mirex	7.0	6.9	6.8	6.7	6.6	6.5	6.4	6.3	6.2	6.1	5.97	5.87	5.77	5.67	5.57	5.48	5.38	5.28
Methoxychlor	10.9	10.6	10.4	10.2	10.0	9.7	9.4	9.2	8.9	8.8	8.5	8.3	8.0	7.8	7.6	7.3	7.1	6.8
Dilan II	10.5	10.3	10.0	9.8	9.6	9.4	9.1	8.9	8.7	8.4	8.2	8.0	7.8	7.5	7.3	7.1	6.8	6.6
Tetradifon	17.5	17.2	16.8	16.4	16.0	15.7	15.3	14.9	14.5	14.1	13.8	13.4	13.0	12.6	12.3	11.9	11.5	11.1
Azinphosmethyl	24.5	23.8	23.2	22.5	21.9	21.3	20.7	20.0	19.4	18.8	18.2	17.6	17.0	16.3	15.7	15.1	14.5	13.8

27

TABLE 1.9.

4% SE-30/6% OV-210
Column Temperature, °C.

170	174	178	182	186	190	194	198	202	204	
0.06	0.06	0.06	0.07	0.07	0.08	0.08	0.09	0.09	0.09	Dichlorvos
0.04	0.04	0.05	0.05	0.05	0.06	0.06	0.07	0.07	0.07	TEPP
0.14	0.14	0.15	0.15	0.15	0.16	0.17	0.17	0.18	0.18	Mevinphos
0.15	0.15	0.16	0.17	0.17	0.18	0.19	0.20	0.20	0.21	Demeton Thiono
0.16	0.16	0.17	0.18	0.18	0.19	0.20	0.21	0.22	0.23	Thionazin
0.19	0.20	0.21	0.21	0.21	0.22	0.23	0.23	0.24	0.24	Ethoprop
0.20	0.21	0.21	0.23	0.23	0.24	0.24	0.25	0.26	0.26	Phorate
0.21	0.22	0.22	0.23	0.24	0.24	0.25	0.26	0.26	0.27	Sulfotepp
0.23	0.23	0.24	0.25	0.25	0.26	0.26	0.27	0.27	0.28	Naled
0.23	0.24	0.25	0.28	0.26	0.27	0.27	0.28	0.28	0.29	Oxydemeton Methyl
0.27	0.27	0.28	0.28	0.29	0.29	0.30	0.30	0.31	0.31	Diazinon
0.30	0.30	0.31	0.31	0.32	0.33	0.33	0.34	0.35	0.35	Dioxathion
0.32	0.32	0.33	0.33	0.34	0.35	0.35	0.36	0.36	0.37	Demeton Thiolo
0.30	0.31	0.32	0.33	0.34	0.35	0.35	0.36	0.37	0.37	Disulfoton
0.38	0.38	0.38	0.38	0.38	0.39	0.39	0.39	0.40	0.40	Diazoxon
0.40	0.40	0.41	0.41	0.42	0.43	0.43	0.44	0.45	0.45	Dichlofenthion
0.46	0.46	0.47	0.47	0.47	0.48	0.48	0.49	0.49	0.49	Dimethoate
0.48	0.48	0.49	0.49	0.50	0.50	0.51	0.51	0.52	0.52	Ronnel
0.51	0.51	0.52	0.52	0.53	0.53	0.54	0.54	0.55	0.55	Cyanox
0.55	0.55	0.56	0.56	0.57	0.57	0.58	0.58	0.59	0.59	Ronnoxon
0.60	0.60	0.60	0.60	0.60	0.60	0.61	0.61	0.61	0.61	Monocrotophos
0.58	0.59	0.59	0.59	0.59	0.60	0.61	0.61	0.61	0.62	Chlorpyrifos
0.57	0.58	0.58	0.59	0.59	0.60	0.61	0.61	0.62	0.62	Zytron
0.62	0.62	0.62	0.63	0.63	0.64	0.64	0.64	0.65	0.65	Fenthion
0.67	0.67	0.68	0.68	0.68	0.69	0.69	0.70	0.70	0.70	Malaoxon
0.72	0.72	0.73	0.73	0.74	0.74	0.75	0.75	0.76	0.76	Methyl Parathion
0.81	0.80	0.80	0.80	0.79	0.79	0.79	0.78	0.78	0.78	Malathion
0.86	0.85	0.85	0.84	0.84	0.85	0.85	0.86	0.86	0.86	Fenitrothion

Retention ratios, relative to parathion, of 54 organophosphorous pesticides on a column of 4% SE-30/6% OV-210 at temperatures from 170 to 204°C; support of Gas Chrom-Q, 80/100 mesh; flame photometric detector, 5260 Å filter; all absolute retentions measured from injection point. Arrow indicates optimum operating temperature with carrier flow set at 75 ml per minute.

Compound	170	172	174	176	178	180	182	184	186	188	190	192	194	196	198	200	202	204
Bromophos	0.94	0.94	0.93	0.93	0.93	0.92	0.92	0.91	0.91	0.91	0.90	0.90	0.89	0.90	0.90	0.90	0.88	0.87
Methyl Paraoxon	0.90	0.90	0.90	0.90	0.90	0.90	0.90	0.90	0.90	0.90	0.90	0.90	0.90	0.90	0.90	0.90	0.90	0.90
Phenthoate	0.93	0.93	0.93	0.92	0.92	0.92	0.92	0.92	0.91	0.91	0.91	0.91	0.91	0.90	0.91	0.91	0.90	0.90
Bromophos Ethyl	0.91	0.91	0.91	0.91	0.91	0.90	0.90	0.90	0.90	0.90	0.90	0.91	0.91	0.91	0.91	0.91	0.91	0.91
Schradan	0.85	0.84	0.84	0.83	0.83	0.83	0.82	0.82	0.83	0.84	0.84	0.86	0.87	0.88	0.89	0.90	0.92	0.93
Dicapthon	0.96	0.96	0.96	0.96	0.97	0.97	0.97	0.97	0.97	0.97	0.97	0.97	0.98	0.98	0.98	0.98	0.98	0.98
E. Parathion(Reference)	1.00	1.00	1.00	1.00	1.00	1.00	1.00	1.00	1.00	1.00	1.00	1.00	1.00	1.00	1.00	1.00	1.00	1.00
Amidithion	1.02	1.02	1.02	1.02	1.01	1.01	1.01	1.00	1.00	1.00	1.00	1.00	1.01	1.01	1.01	1.01	1.01	1.01
Iodofenphos	1.02	1.02	1.02	1.02	1.02	1.03	1.03	1.03	1.03	1.03	1.03	1.03	1.03	1.03	1.03	1.03	1.03	1.03
Crufomate	1.11	1.11	1.10	1.10	1.10	1.10	1.10	1.10	1.10	1.09	1.09	1.09	1.09	1.09	1.09	1.09	1.09	1.08
DEF	1.18	1.18	1.17	1.17	1.17	1.16	1.16	1.15	1.15	1.15	1.14	1.14	1.13	1.13	1.13	1.12	1.12	1.11
Phosphamidon	1.19	1.18	1.18	1.17	1.17	1.17	1.16	1.16	1.15	1.15	1.14	1.14	1.14	1.13	1.13	1.12	1.12	1.12
Folex	1.18	1.17	1.17	1.17	1.17	1.16	1.16	1.16	1.15	1.15	1.15	1.14	1.14	1.14	1.13	1.13	1.13	1.12
Ethyl Paraoxon	1.25	1.24	1.24	1.23	1.23	1.22	1.22	1.21	1.21	1.20	1.20	1.19	1.19	1.18	1.18	1.17	1.17	1.16
Methidathion	1.23	1.22	1.22	1.22	1.22	1.21	1.21	1.21	1.20	1.20	1.20	1.20	1.20	1.19	1.19	1.19	1.19	1.19
Tetrachlorvinphos	1.37	1.36	1.36	1.35	1.34	1.33	1.33	1.32	1.32	1.31	1.31	1.31	1.30	1.30	1.29	1.29	1.28	1.28
Ethion	1.87	1.85	1.84	1.83	1.82	1.81	1.80	1.78	1.77	1.76	1.75	1.74	1.73	1.72	1.70	1.69	1.68	1.67
Carbophenoxon	1.89	1.88	1.88	1.86	1.85	1.84	1.83	1.82	1.80	1.79	1.78	1.77	1.76	1.75	1.74	1.73	1.72	1.71
Carbophenothion	1.89	1.88	1.87	1.86	1.85	1.84	1.84	1.82	1.80	1.79	1.78	1.77	1.76	1.75	1.74	1.73	1.72	1.71
Phenkapton	3.18	3.15	3.12	3.09	3.06	3.03	3.00	2.98	2.95	2.92	2.89	2.86	2.83	2.80	2.77	2.74	2.72	2.69
Fensulfothion	3.96	3.92	3.87	3.83	3.79	3.75	3.70	3.66	3.62	3.58	3.53	3.49	3.44	3.41	3.36	3.32	3.28	3.23
Imidan	4.65	4.60	4.55	4.50	4.45	4.40	4.35	4.31	4.26	4.21	4.16	4.11	4.06	4.01	3.96	3.91	3.87	3.82
EPN	4.66	4.61	4.56	4.51	4.46	4.41	4.36	4.31	4.26	4.21	4.16	4.11	4.06	4.00	3.95	3.90	3.85	3.80
Famphur	5.67	5.63	5.59	5.55	5.51	5.47	5.42	5.38	5.34	5.30	5.26	5.21	5.17	5.13	5.09	5.05	5.01	4.96
Azinphos Ethyl	7.48	7.38	7.27	7.17	7.06	6.96	6.85	6.75	6.64	6.54	6.43	6.33	6.22	6.12	6.01	5.91	5.80	5.70
Azinphos Methyl	7.53	7.42	7.32	7.21	7.10	6.99	6.89	6.78	6.67	6.57	6.46	6.35	6.24	6.14	6.03	5.92	5.81	5.71
Coumaphos	17.4	17.0	16.7	16.3	16.0	15.6	15.3	15.0	14.6	14.3	13.9	13.6	13.2	12.9	12.6	12.2	11.9	11.5

TABLE 1.10.

10 % OV-210

Column Temperature , °C.

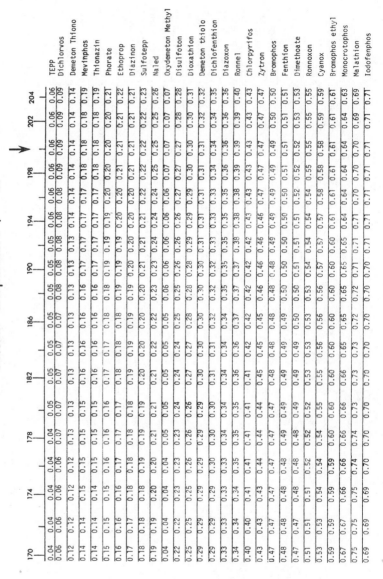

Compound	170	174	178	182	186	190	194	198	202	204
TEPP	0.04	0.04	0.04	0.05	0.05	0.05	0.05	0.06	0.06	0.06
Dichlorvos	0.06	0.06	0.07	0.07	0.07	0.08	0.08	0.08	0.09	0.09
Demeton Thiono	0.12	0.12	0.13	0.13	0.13	0.13	0.13	0.14	0.14	0.14
Mevinphos	0.14	0.15	0.15	0.16	0.16	0.16	0.17	0.17	0.18	0.19
Thionazin	0.14	0.14	0.15	0.16	0.16	0.17	0.17	0.18	0.18	0.19
Phorate	0.15	0.15	0.16	0.17	0.17	0.18	0.19	0.20	0.20	0.21
Ethoprop	0.16	0.16	0.17	0.18	0.18	0.19	0.20	0.21	0.21	0.22
Diazinon	0.17	0.17	0.18	0.19	0.19	0.19	0.20	0.21	0.21	0.21
Sulfotepp	0.18	0.18	0.19	0.20	0.20	0.20	0.21	0.22	0.22	0.23
Naled	0.19	0.19	0.21	0.21	0.22	0.23	0.24	0.25	0.25	0.26
Oxydemeton Methyl	0.04	0.04	0.05	0.05	0.05	0.06	0.06	0.06	0.07	0.07
Disulfoton	0.22	0.23	0.23	0.24	0.25	0.26	0.26	0.27	0.27	0.28
Dioxathion	0.25	0.25	0.26	0.27	0.28	0.28	0.29	0.29	0.30	0.31
Demeton thiolo	0.29	0.29	0.29	0.30	0.30	0.30	0.31	0.31	0.32	0.32
Dichlofenthion	0.29	0.29	0.30	0.31	0.31	0.32	0.33	0.33	0.34	0.35
Diazoxon	0.33	0.33	0.34	0.34	0.34	0.35	0.35	0.35	0.36	0.36
Ronnel	0.34	0.34	0.35	0.36	0.37	0.37	0.38	0.38	0.39	0.40
Chlorpyrifos	0.40	0.41	0.41	0.41	0.42	0.42	0.43	0.43	0.43	0.43
Zytron	0.43	0.44	0.44	0.45	0.45	0.46	0.46	0.47	0.47	0.47
Bromophos	0.47	0.47	0.47	0.48	0.48	0.48	0.49	0.49	0.50	0.50
Fenthion	0.48	0.48	0.49	0.49	0.49	0.50	0.50	0.51	0.51	0.51
Dimethoate	0.47	0.48	0.48	0.49	0.49	0.50	0.51	0.52	0.53	0.53
Ronnoxon	0.51	0.51	0.52	0.53	0.53	0.53	0.54	0.54	0.55	0.55
Cyanox	0.53	0.54	0.54	0.55	0.56	0.56	0.57	0.58	0.58	0.59
Bromophos ethyl	0.59	0.59	0.60	0.60	0.60	0.60	0.61	0.61	0.61	0.61
Monocrotophos	0.67	0.66	0.66	0.66	0.65	0.65	0.64	0.64	0.64	0.63
Malathion	0.75	0.75	0.74	0.73	0.73	0.72	0.71	0.70	0.70	0.69
Iodofenphos	0.69	0.69	0.70	0.70	0.70	0.70	0.71	0.71	0.71	0.71

	170	172	174	176	178	180	182	184	186	188	190	192	194	196	198	200	202	204
DEF	0.77	0.77	0.77	0.77	0.77	0.76	0.76	0.76	0.76	0.76	0.76	0.76	0.76	0.75	0.75	0.75	0.75	0.74
Phenthoate	0.75	0.75	0.75	0.75	0.75	0.75	0.74	0.74	0.74	0.74	0.74	0.74	0.74	0.74	0.75	0.75	0.75	0.75
Folex	0.77	0.77	0.77	0.77	0.77	0.77	0.77	0.77	0.77	0.77	0.77	0.77	0.77	0.77	0.77	0.77	0.77	0.77
Methyl parathion	0.74	0.74	0.75	0.75	0.75	0.75	0.76	0.76	0.76	0.76	0.77	0.77	0.77	0.77	0.78	0.78	0.78	0.79
Schradan	0.81	0.81	0.81	0.80	0.80	0.80	0.80	0.80	0.80	0.80	0.80	0.80	0.80	0.80	0.81	0.81	0.81	0.81
Fenitrothion	0.84	0.84	0.84	0.85	0.85	0.85	0.85	0.85	0.85	0.85	0.85	0.86	0.86	0.86	0.86	0.86	0.86	0.86
Malaoxon	1.02	1.01	1.01	1.00	0.99	0.99	0.98	0.97	0.97	0.96	0.95	0.95	0.94	0.93	0.92	0.92	0.91	0.90
Dicapthon	0.95	0.95	0.95	0.95	0.95	0.95	0.95	0.95	0.95	0.96	0.96	0.96	0.96	0.96	0.96	0.96	0.96	0.96
Crufomate	1.02	1.01	1.01	1.00	1.00	1.00	1.00	0.99	0.99	0.99	0.99	0.98	0.98	0.97	0.97	0.97	0.97	0.96
Parathion (Reference)	1.00	1.00	1.00	1.00	1.00	1.00	1.00	1.00	1.00	1.00	1.00	1.00	1.00	1.00	1.00	1.00	1.00	1.00
Methyl paraoxon	1.04	1.03	1.03	1.03	1.03	1.02	1.02	1.02	1.02	1.01	1.01	1.01	1.01	1.00	1.00	1.00	1.00	0.99
Amidithion	1.03	1.03	1.03	1.03	1.03	1.02	1.02	1.02	1.02	1.02	1.01	1.01	1.01	1.01	1.01	1.01	1.01	1.00
Methidathion	1.13	1.13	1.13	1.13	1.13	1.13	1.13	1.13	1.13	1.13	1.12	1.12	1.12	1.12	1.12	1.12	1.12	1.12
Tetrachlorvinphos	1.22	1.21	1.20	1.19	1.19	1.18	1.18	1.17	1.17	1.16	1.16	1.15	1.15	1.14	1.14	1.13	1.13	1.12
Carbophenothion	1.27	1.26	1.25	1.25	1.24	1.24	1.24	1.23	1.23	1.22	1.22	1.21	1.21	1.20	1.20	1.20	1.19	1.19
Phosphamidon	1.38	1.36	1.35	1.34	1.33	1.32	1.32	1.31	1.31	1.30	1.30	1.29	1.28	1.27	1.27	1.26	1.26	1.24
Ethyl paraoxon	1.42	1.41	1.40	1.39	1.38	1.38	1.37	1.37	1.36	1.35	1.34	1.33	1.32	1.31	1.31	1.30	1.30	1.27
Ethion	1.43	1.42	1.41	1.40	1.39	1.38	1.38	1.37	1.37	1.36	1.36	1.35	1.34	1.33	1.32	1.31	1.31	1.29
Carbophenoxon	1.64	1.63	1.62	1.61	1.60	1.59	1.59	1.58	1.57	1.56	1.56	1.55	1.54	1.53	1.53	1.52	1.51	1.49
Phenkapton	2.09	2.08	2.06	2.04	2.02	2.00	1.98	1.96	1.96	1.94	1.92	1.90	1.88	1.86	1.84	1.82	1.80	1.77
Fensulfothion	4.26	4.21	4.16	4.11	4.06	4.01	3.97	3.92	3.87	3.82	3.77	3.72	3.68	3.63	3.58	3.53	3.48	3.43
EPN	4.37	4.32	4.27	4.22	4.18	4.13	4.08	4.03	3.98	3.93	3.88	3.83	3.78	3.73	3.68	3.63	3.58	3.53
Imidan	4.65	4.59	4.54	4.48	4.43	4.37	4.32	4.26	4.21	4.15	4.10	4.04	3.99	3.94	3.88	3.86	3.77	3.72
Azinphos methyl	6.98	6.88	6.77	6.66	6.55	6.44	6.33	6.22	6.11	6.00	5.89	5.78	5.67	5.56	5.45	5.34	5.24	5.13
Famphur	6.86	6.76	6.66	6.56	6.47	6.37	6.27	6.17	6.08	5.98	5.88	5.79	5.69	5.59	5.49	5.40	5.30	5.20
Azinphos ethyl	7.01	6.90	6.79	6.69	6.58	6.47	6.37	6.25	6.14	6.03	5.92	5.81	5.70	5.59	5.48	5.37	5.26	5.15
Coumaphos	19.1	18.7	18.3	17.9	17.5	17.1	16.7	16.3	15.9	15.5	15.1	14.7	14.3	13.9	13.5	13.1	12.7	12.3

Retention ratios, relative to ethyl parathion, of 54 organophosphorous pesticides on a column of 10% OV-210 at temperatures from 170 to 204°C; support of Gas Chrom Q, 100/120 mesh; flame photometric detector, 5260 A° filter; all absolute retentions measured from injection point. Arrow indicates optimum column operating temperature with carrier flow at 70 ml per minute.

TABLE 1.11.

1.5% OV-17/1.95% OV-210 Column Temperature , °C.

170	174	178	182	186	190	194	198	202	204	
0.04	0.05	0.05	0.06	0.06	0.07	0.07	0.08	0.08	0.09	TEPP
0.06	0.07	0.07	0.08	0.08	0.09	0.10	0.11	0.11	0.12	Dichlorvos
0.12	0.13	0.14	0.15	0.16	0.17	0.18	0.19	0.20	0.21	Mevinphos
0.16	0.17	0.17	0.18	0.19	0.19	0.20	0.21	0.21	0.22	Demeton thiono
0.19	0.19	0.20	0.21	0.22	0.23	0.23	0.24	0.25	0.25	Thionazin
0.20	0.20	0.21	0.22	0.23	0.24	0.24	0.25	0.26	0.26	Ethoprop
0.23	0.24	0.25	0.25	0.26	0.27	0.27	0.28	0.29	0.30	Phorate
0.24	0.25	0.25	0.26	0.27	0.28	0.28	0.29	0.29	0.30	Sulfotepp
0.26	0.27	0.28	0.29	0.30	0.30	0.31	0.32	0.33	0.33	Oxydemeton methyl
0.32	0.33	0.33	0.34	0.34	0.35	0.35	0.36	0.36	0.37	Diazinon
0.31	0.32	0.33	0.34	0.35	0.36	0.36	0.37	0.38	0.39	Naled
0.34	0.34	0.35	0.36	0.36	0.37	0.37	0.38	0.39	0.39	Demeton thiolo
0.36	0.37	0.38	0.38	0.39	0.40	0.40	0.41	0.42	0.43	Disulfoton
0.37	0.38	0.39	0.40	0.40	0.41	0.42	0.43	0.43	0.44	Dioxathion
0.38	0.39	0.40	0.40	0.41	0.41	0.42	0.43	0.44	0.44	Diazoxon
0.44	0.45	0.45	0.46	0.46	0.47	0.48	0.49	0.49	0.50	Dichlofenthion
0.51	0.51	0.52	0.53	0.53	0.54	0.55	0.56	0.57	0.57	Cyanox
0.53	0.53	0.54	0.55	0.55	0.56	0.56	0.57	0.58	0.58	Dimethoate
0.55	0.56	0.57	0.57	0.58	0.59	0.60	0.60	0.61	0.62	Ronnel
0.57	0.57	0.58	0.59	0.59	0.60	0.61	0.62	0.61	0.63	Ronnoxon
0.78	0.76	0.76	0.73	0.72	0.72	0.69	0.68	0.66	0.65	Monocrotophos
0.66	0.67	0.67	0.67	0.68	0.68	0.69	0.70	0.70	0.71	Zytron
0.74	0.74	0.75	0.75	0.75	0.75	0.76	0.76	0.76	0.76	Chlorpyrifos
0.74	0.74	0.75	0.76	0.77	0.77	0.78	0.79	0.79	0.80	Methyl Parathion
0.79	0.79	0.79	0.79	0.80	0.80	0.80	0.81	0.81	0.81	Methyl Paraoxon
0.93	0.92	0.92	0.91	0.90	0.89	0.88	0.88	0.87	0.86	Malaoxon
0.92	0.92	0.91	0.91	0.91	0.90	0.90	0.89	0.89	0.89	Malathion

Pesticide	170	172	174	176	178	180	182	184	186	188	190	192	194	196	198	200	202	204
Bromophos	0.85	0.85	0.86	0.86	0.86	0.86	0.87	0.87	0.87	0.87	0.88	0.88	0.88	0.89	0.89	0.89	0.89	0.90
Fenthion	0.90	0.90	0.90	0.91	0.91	0.91	0.91	0.91	0.91	0.91	0.91	0.91	0.91	0.90	0.90	0.90	0.90	0.90
Fenitrothion	0.90	0.90	0.90	0.91	0.91	0.91	0.91	0.91	0.91	0.91	0.91	0.91	0.91	0.92	0.92	0.92	0.92	0.92
Phosphamidon	1.02	1.01	1.01	1.00	1.00	1.00	0.99	0.99	0.98	0.98	0.98	0.97	0.97	0.96	0.96	0.95	0.95	0.95
Schradon	1.25	1.23	1.21	1.19	1.18	1.16	1.14	1.13	1.11	1.09	1.07	1.06	1.04	1.02	1.00	0.99	0.97	0.95
Parathion (Reference)	1.00	1.00	1.00	1.00	1.00	1.00	1.00	1.00	1.00	1.00	1.00	1.00	1.00	1.00	1.00	1.00	1.00	1.00
Ethyl Paraoxon	1.06	1.06	1.05	1.06	1.06	1.05	1.05	1.05	1.05	1.05	1.05	1.04	1.04	1.04	1.04	1.04	1.04	1.04
Dicapthon	1.05	1.05	1.05	1.06	1.06	1.06	1.06	1.06	1.06	1.06	1.05	1.05	1.05	1.04	1.04	1.04	1.04	1.03
Bromophos Ethyl	1.12	1.12	1.12	1.11	1.11	1.11	1.11	1.11	1.11	1.10	1.10	1.10	1.10	1.10	1.09	1.09	1.09	1.09
Amidithion	1.22	1.22	1.22	1.21	1.21	1.20	1.20	1.20	1.19	1.19	1.19	1.20	1.20	1.20	1.20	1.20	1.20	1.20
Crufomate	1.39	1.38	1.38	1.37	1.37	1.36	1.35	1.35	1.33	1.32	1.32	1.30	1.30	1.29	1.28	1.27	1.26	1.25
Phenthoate	1.35	1.35	1.35	1.34	1.34	1.33	1.33	1.32	1.32	1.31	1.31	1.30	1.30	1.29	1.29	1.28	1.27	1.27
Folex	1.50	1.49	1.49	1.48	1.48	1.47	1.46	1.46	1.44	1.43	1.43	1.42	1.41	1.40	1.39	1.38	1.38	1.37
DEF	1.51	1.50	1.49	1.49	1.48	1.47	1.46	1.45	1.44	1.43	1.43	1.41	1.41	1.40	1.40	1.39	1.38	1.36
Iodofenphos	1.57	1.57	1.56	1.56	1.56	1.55	1.54	1.54	1.53	1.53	1.53	1.52	1.51	1.51	1.51	1.50	1.50	1.49
Tetrachlorvinphos	1.72	1.71	1.70	1.69	1.68	1.67	1.66	1.65	1.65	1.64	1.63	1.62	1.61	1.60	1.59	1.58	1.57	1.56
Methidathion	1.74	1.73	1.73	1.72	1.72	1.71	1.71	1.70	1.70	1.69	1.69	1.68	1.68	1.67	1.67	1.66	1.65	1.65
Carbophenoxon	2.69	2.67	2.64	2.62	2.59	2.57	2.54	2.52	2.49	2.46	2.45	2.42	2.40	2.37	2.35	2.32	2.30	2.27
Ethion	2.88	2.85	2.82	2.79	2.76	2.73	2.70	2.67	2.64	2.61	2.58	2.55	2.52	2.49	2.46	2.43	2.40	2.37
Carbophenothion	2.99	2.97	2.94	2.92	2.89	2.87	2.84	2.82	2.79	2.77	2.75	2.72	2.70	2.67	2.66	2.62	2.60	2.57
Fensulfothion	4.65	4.60	4.56	4.51	4.46	4.42	4.37	4.32	4.27	4.23	4.18	4.13	4.08	4.04	3.99	3.94	3.89	3.85
Phenkapton	5.57	5.50	5.42	5.35	5.27	5.20	5.12	5.05	4.97	4.90	4.82	4.75	4.67	4.60	4.52	4.45	4.38	4.30
Famphur	6.07	5.99	5.90	5.82	5.74	5.65	5.57	5.49	5.40	5.32	5.24	5.15	5.07	4.99	4.90	4.82	4.74	4.65
EPN	6.63	6.53	6.44	6.34	6.25	6.16	6.06	5.97	5.88	5.78	5.69	5.59	5.50	5.41	5.31	5.22	5.12	5.03
Imidan	7.95	7.84	7.72	7.61	7.50	7.38	7.27	7.15	7.04	6.93	6.81	6.70	6.58	6.50	6.36	6.24	6.13	6.01
Azinphos Methyl	10.9	10.7	10.6	10.4	10.3	10.1	10.0	9.8	9.7	9.5	9.3	9.1	8.9	8.7	8.5	8.3	8.1	7.9
Azinphos ethyl	14.4	14.1	13.9	13.6	13.3	13.0	12.8	12.5	12.2	11.9	11.7	11.4	11.1	10.9	10.6	10.3	10.1	9.8
Coumaphos	22.2	21.8	21.3	20.9	20.4	20.0	19.5	19.1	18.6	18.2	17.2	17.3	16.8	16.4	15.9	15.5	15.0	14.6

Retention ratios, relative to ethyl parathion, of 54 organophosphorous pesticides on a column of 1.5% OV-17/1.95% OV-210 at temperatures from 170 to 204°C; column support of Gas Chrom-Q, 100/120 mesh; flame photometric detector, 5260 A° filter; all absolute retentions measured from injection point. Arrow indicates optimum column operating temperature with carrier flow at 70 ml per minute.

33

that the column oven pyrometer readout is correct. If the actual oven temperature is significantly different than that indicated by the pyrometer it may prove necessary to chromatograph a number of candidate reference standards before finding the compounds that produce comparable retention or RRT values.

Observations have been made occasionally that RRT_A values may shift slightly for certain organophosphorus compounds when chromatographed with electron capture detection on a given GLC column. Switching to an alternate column may result in an RRT_A value agreeing very closely with the table value. With this in mind, the chromatographer would be well advised to exercise caution in the identification of organophosphorus compounds determined by electron capture. Such compounds can be reliably identified by flame photometric detection as no such relative retention shifts have been observed in this detection mode when referenced against parathion.

1.4.4. Compound Elution Pattern

Each stationary phase or combination thereof has its own elution pattern or "fingerprint" for the compounds of a given mixture. During the past 20 years or more, the pesticide chemist analyzing environmental media and human or animal tissues has been concerned primarily with a series of

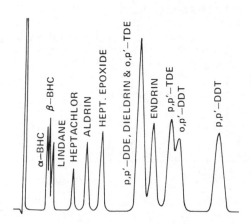

Figure 1.6 The methyl silicones, single stationary phase, i.e., OV-1, OV-101, SE-30, DC-11, SF-96, SP-2100. (*a*) Fair separation of the BHC isomers. (*b*) No separation of *p,p′*-DDE/Dieldrin. (*c*) Poor separation *p,p′*-TDE(DDD)/*o,p′*-DDT.

5% QF-1

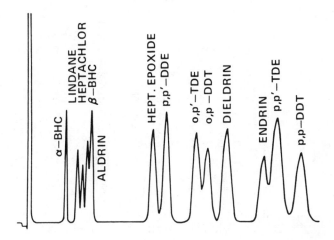

a. Full separation of BHC isomers from each other.

b. Fairly good separation hept. epoxide/p,p'—DDE.

c. Excellent separation p,p'—DDE/Dieldrin.

Figure 1.7 Trifluoropropylmethyl silicones, single stationary phase, i.e., OV-210, SP-2401. (*a*) Full separation of BHC isomers from each other. (*b*) Fairly good separation heptachlorepoxide/*p,p'*-DDE. (*c*) Excellent separation *p,p'*-DDE/Dieldrin.

3% DEGS

a. Excellent separations of the BHC isomers (note elution site of β–BHC).

b. Excellent separations of all common pesticides and metabolites.

c. Note elution sequence of p,p'—DDT and p,p'—TDE (DDD).

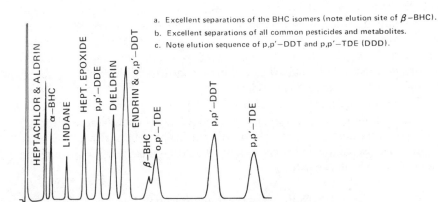

Figure 1.8 A polyester, single phase, no exactly comparable equivalent. (*a*) Excellent separations of the BHC isomers (note elution site of *β*-BHC). (*b*) Excellent separations of all common pesticides and metabolites. (*c*) Note elution sequence of *p,p'*-DDT and *p,p'*-TDE(DDD).

35

4% SE-30/6% QF-1

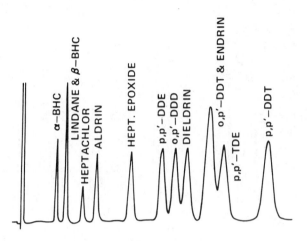

a. No separation β–BHC/γ–BHC.

b. Excellent separation p,p'–DDE/Dieldrin.

c. Rather poor separation o,p'–DDT/p,p'–TDE (DDD).

Figure 1.9 Mixed stationary phases of the methyl silicones with the trifluoropropylmethylsilicones in a basic 2:3 ratio, i.e., DC-200/QF-1, OV-1/QF-1, SE-30/OV-210, SE-30/SP-2401, etc. (a) No separation β-BHC/γ-BHC. (b) Excellent separation p,p'-DDE/Dieldrin. (c) Rather poor separation o,p'-DDT/p,p'-TDE(DDD).

organochlorine compounds such as DDT and its metabolites, heptachlor and its epoxide, BHC and component isomers, PCBs and dieldrin. Several of these compounds are no longer used in the United States, but due to their persistence in the environment at the time of writing this treatise, several of them are still being detected at measurable levels in human media. We conjecture that it would be indeed a rare individual who is carrying less than 5 ppb of p,p'-DDE in his blood plasma.

The chromatograms in Figures 1.6–1.10 represent elution patterns typical of each column type for a mixture of 13 chlorinated compounds. Although the chromatograms were obtained on the particular columns specified in the figures, nearly identical chromatograms would be observed when using the other stationary phases listed.

1.4.5. Stability—Thermal and Injection Loading

Considering the selection of thermally stable stationary phase materials currently available in the marketplace, there should be little need for the chromatographer to be overly concerned about the thermal stability of his or her columns. As more chromatographers update their thinking and move away from the older technical grade materials, this problem should disappear into history.

Instability resulting from injection overloading, however, is another matter which is unlikely to disappear from the chromatographic scene. The situation we have in mind might typically involve a series of substrate injections wherein the extract is relatively " dirty." This may arise when no partitioning cleanup is used, for example, on a high lipid sample; or, in fact, when cleanup is used, but a significant amount of impurities elutes from the cleanup column.

The symptoms of injection overloading are quite often multiple. There may be a gradual depression of response, evidence of peak tailing, lowered column efficiency and resolution, and on-column decomposition of certain compounds such as *p,p'*-DDT. Figure 1.11 provides an illustration of a severe case of injection overloading.

Chromatogram A is from a standard mixture of seven pesticides on a freshly prepared column. The column was then disconnected from the detec-

1.5% OV-17/1.95% QF-1

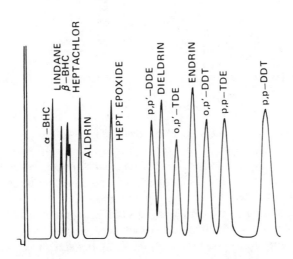

Excepting p,p'—DDE/Dieldrin, good separation

of peaks of all common chlorinated pesticides.

Figure 1.10   Mixed stationary phases of 50% phenyl substituted methyl silicone with trifluoro-propymethyl silicone in a 1.5:1.95 ratio, i.e., OV-17/OV-210, SP-2250/SP-2401, etc. Excepted is *p,p'*-DDE, good separation of peaks of all common chlorinated pesticides.

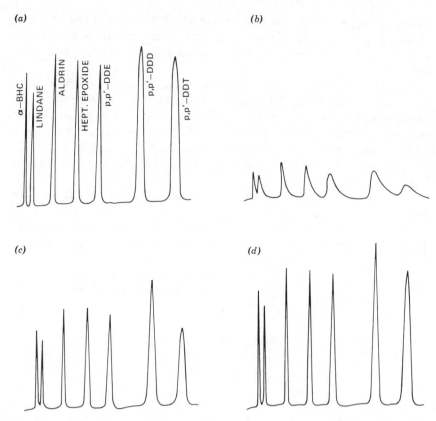

Figure 1.11 Chromatograms illustrating column overloading and subsequent rejuvenation. (see text.)

tor so the exit end vented inside the oven. Then 18 consecutive injections were made of fatty tissue extract after elution with 15% diethyl ether–petroleum ether through a Florisil column, each injection containing the equivalent of 25 mg of fat. The column was reconnected to the detector 30 min after the last injection, the system equilibrated, and an identical volume of the same standard mixture was injected. Chromatogram B shows the results of column overloading: depressed peak heights, peak tailing, peak broadening, and conversion of p,p'-DDT to p,p'-DDD (in actuality, the ratio of these changed from 8:10 to 4:10). A clean Vykor glass insert was then installed in the injection port, the system reequilibrated for 30 min, and another equal volume of standard mixture injected. Chromatogram C shows the dramatic recovery of the system after this single step. Finally, Chromatogram D indicates a complete rejuvenation of the system when the same mixture was injected after overnight purging at normal operating temperature and carrier flow parameters.

This series of chromatograms is striking evidence that damaged columns can often be salvaged by changing the injection insert, forward glass wool plug, and perhaps the first one half or one in. of column packing. More importantly, properly maintained and monitored columns should provide top performance without problems for many thousands of injections.

1.4.6. Resistance to On-Column Decomposition

This subject was touched upon in the foregoing subsection where evidence was presented of the conversion of p,p'-DDT to p,p'-DDD. This particular conversion is very common. We think it a fair estimate that at least half of a random group of laboratories are experiencing this problem to some degree. Admittedly, this view is not pure speculation. In the column performance study mentioned earlier in this chapter, one phase of this study explored this problem. One of the three mixtures given in Section 1.4.1 was comprised of aldrin and p,p'-DDT, both of analytical quality. In our evaluation of the feedback chromatograms, the decomposition was calculated. These data are summarized in Table 1.12.

It will be noted in Table 1.12 that exactly half of the participating laboratories experienced DDT decomposition in excess of 3% on the OV-17/QF-1 column. Several laboratories experienced a sufficient amount of decomposition to significantly alter their quantitative reporting values if they were reporting on the basis of individual DDT metabolites. One laboratory (No. 8) had a whopping conversion totaling 25%, 17% to p,p'-DDE and 8% to p,p'-DDD. This laboratory's chromatogram was copied and is shown in Figure 1.12 as (B). A normal chromatogram is shown in (A).

This phenomenon has often been observed to be related to some problem in or adjacent to the front end of the column. In Figure 1.11, the only change that was made after Chromatogram B was to replace the Vykor glass injection insert (demister tube) with a clean one. This one change resulted in a significant improvement of all characteristics, including the decomposition ratio.

From the data included with Laboratory No. 8's chromatograms, we learned that this chromatographer had not changed the injection insert tube for 3 weeks prior to running these chromatograms. If the chromatographer is injecting directly on-column, this extraneous material responsible for fouling the off-column injection insert will inevitably be deposited in the glass wool plug at the column inlet or onto the first segment of column packing. It therefore behooves the chromatographer to monitor constantly for this breakdown, and when it is observed, change the glass wool plug. In

TABLE 1.12. On-Column Decomposition of p,p'-DDT

	OV-17/QF-1					SE-30/QF-1			
Lab. No.	Column Age (days)	Est. Total No. of Injections	Glass Insert Changed (days)	Total Decomposition (%)	Lab. No.	Column Age (days)	Est. Total No. of Injections	Glass Insert Changed (days)	Total Decomposition (%)
14	21	0	21	None	5	126	500	1	None
17	12	0	1	None	17	13	75	1	None
12	49	?	1	None	15	142	1300	1	None
1	47	20	6	None	12	49	?	1	None
4	257	800	10	Neglig.	6	70	125	2	None
15	144	1300	1	Neglig.	14	46	150	1	Neglig.
19	38	146	1	Neglig.	13	186	1500	1	Neglig.
5	70	250	1	1.4	7	73	60	5	Neglig.
13	186	1500	1	2.2	16	?	3000	5	3.2[a]
9	100	300	1	3.1	4	254	800	10	6.0
3	34	—	1	3.1	2	71	400	5	6.0
6	70	125	2	3.4	10	254	460	22	7.0
16	545	3000	5	3.7[a]	9	110	150	1	8.6[a]
10	253	320	21	8.5[a]	18	4	None	—[b]	11.0[a]
2	71	400	5	14.0	11	5	30	5	13.[a]
7	108	400	5	14.0	1	57	20	0	22.
18	4	45	—[b]	15.0[a]	8	141	200	23	25.[a]
8	141	200	23	—[c]	3	35	?	1	—[c]

[a] Breakdown to p,p'-DDD and p,p'-DDE.
[b] Using on-column injection.
[c] Could not assess due to noise in baseline.

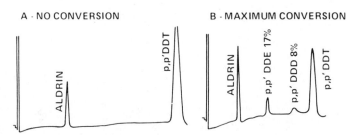

Figure 1.12   On-column decomposition of *p,p'*-DDT on column of 4% SE-30/6% QF-1.

extreme cases, it may prove advisable to discard the first half-inch or so of packing in the front end of the column, replacing it with fresh packing.

Endrin is another compound which will almost always create a decomposition problem on a freshly prepared column. After the column is used for several weeks and becomes "seasoned," the situation generally improves. This decomposition problem is illustrated in Chromatogram *A* in Figure 1.13.

The first decomposition peak shown on the chromatogram as Endrin I elutes shortly after *p,p'*-DDT and is, in fact, endrin aldehyde. Generally, it is far smaller than the second decomposition peak, delta Keto "153," which elutes far out in the chromatogram near methoxychlor.

Chromatogram *B* illustrates the improvement after taking one step with a freshly prepared column to minimize this problem, namely the injection of a silylating agent as a part of the column conditioning. This will be treated in some detail in Section 1.5.2.

This phenomenon was also studied during the column performance survey. We could not seriously fault the laboratories for the high prevalence of this problem as none of this group of laboratories had occasion to chromatograph endrin in their normal work routine. As may be observed in Table 1.13, only 5 of the 18 participating laboratories demonstrated acceptably low endrin breakdown. The decomposition creates no great problem in quantitation. The sum of all three peak areas represents a reasonable assay of the total endrin present. The difficulty is that a long wait is necessary for the elution of delta Keto "153," generally about 30 min on the basis of a *p,p'*-DDT retention of around 18 min.

1.5.   COLUMN CONDITIONING AND EVALUATION

1.5.1.   Heat Conditioning

After the column is prepared, it is by no means ready to use, as several conditioning and evaluation steps are necessary to ensure that the column is truly fit to use.

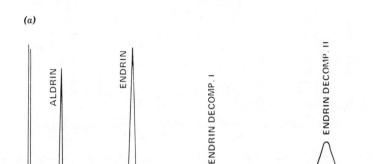

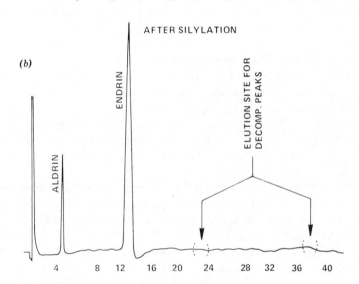

Figure 1.13 Reduction in decomposition of endrin resulting from column silylation. (See text).

It is necessary to heat condition or "cure" the column at a temperature of 40–50° higher than the expected operating temperature. A lengthy conditioning period is mandatory with the older stationary phase materials of technical grade such as DC-200, QF-1, and so on, but is not nearly as critical with the newer reagent grade materials such as OV-1, OV-17, SP-2401, and so on. However, a lengthy heat conditioning period does no harm, and therefore, if convenient, a weekend serves nicely as a conditioning period thus creating no downtime for a GC oven. Some large laboratories use ovens specially designed for conditioning only. If such an oven is available, time is no factor.

During the heat conditioning period the front end of the column should be attached to the inlet port, but the exit end of the column should be vented inside the oven, and most emphatically not attached to the detector. A carrier gas flow of 50–60 ml/min is advised during the conditioning period to dispel volatile impurities.

1.5.2. Silylating Treatment

If endrin is a compound of interest in the reader's laboratory, this treatment is strongly advised. At the end of the heat-conditioning period, the

TABLE 1.13.   On-Column Decomposition of Endrin

	OV-17/QF-1				SE-30/QF-1		
Lab. No.	Est. Total No. of Injections	Column Silylation	Total Decomposition (%)	Lab. No.	Est. Total No. of Injections	Column Silylation	Total Decomposition (%)
19	146	N	Neglig.	6	125	N	None
12	?	N	Neglig.	7	60	N	Neglig.
7	400	N	Neglig.	15	1300	Y	3.6
13	1500	Y	6	9	150	Y	4.4
15	1300	Y	6	13	1500	Y	6.0
14	0	N	7.5	10	460	N	7.0
9	300	Y	7.7	2	400	N	7.5
16	3000	N	8	4	800	N	10.
2	400	N	9	5	500	N	11.
8	200	?	9	12	?	N	11.
4	800	N	15	16	3000	N	11.
10	320	N	15	14	150	N	12.
5	250	Y	17	8	200	?	14.
3	—	N	18	11	30	N	15.
1	20	N	51	1	20	Y	26.
17	0	N	55	18	0	N	—[a]
11	30	N	—[a]	17	75	N	—[a]
18	45	N	—[a]	3	?	N	—[a]
6	125	N	—[b]				

[a] Did not allow chromatogram to run for full retention of last breakdown peak.
[b] Breakdown peaks obscured by noisy baseline.

oven temperature and carrier gas flow rate are adjusted to the appropriate operating levels for the column. Four consecutive 25-μl injections of Silyl 8 (Pierce Chemical Company, Rockford, IL) are made, the injections spaced ca. 30 min apart. Following the final injection, at least 2 hr should be allowed for all traces of the silylation material to elute from the column before attaching the exit end of the column to the detector transfer line.

In the author's laboratory no benefits have been observed from the silylation other than the control of endrin decomposition. So if a laboratory has no concern with this compound, the silylation procedure can be bypassed.

1.5.3.   Treatment by Carbowax Vapor Deposition

In 1970 Ives and Giuffrida (37) reported a technique for Carbowax treating a column to minimize adsorptive difficulties often experienced with chromatographing organophosphorus compounds. It was confirmed in the author's laboratory that a Carbowax-treated column is much more responsive and capable of superior peak resolution than one which is untreated. Figure 1.14 is presented to demonstrate this. In many instances we have observed response improvement of 75% or better, although the improvement varies somewhat, depending on the specific organophosphorus compound and the performance characteristics of the column in use.

Giuffrida placed the Carbowax directly in the front end of the column, and then poured it off after the deposition period. In the author's laboratory it was handled a little differently. A 76 mm length of column tube (borosilicate glass), ¼-in. o.d., was cut and a glass wool plug of ca. 12 mm length was placed in one end, 50 mm of 10% Carbowax 20 M was poured in, then another glass wool plug was placed in the other end. This tube was attached to the inlet end of the 6-foot column to be treated, the column having already been heat conditioned in the usual manner. The connection was made with a standard Swagelok connector. A conventional ¼-inch column nut was placed on the other end of the 76-mm Carbowax column, and this was in turn connected to the instrument injection port, allowing the exit end of the column to vent into the oven. Oven heat was adjusted to 230–235°C and a carrier gas flow of 20 ml/min was applied. The deposition period was 17 hr, and this is somewhat critical. At the conclusion of the deposition period, the Carbowax tube is discarded and the treated column is ready for evaluation. Figure 1.15 is a diagram of the tube design.

The effects of the vapor phase deposition appear to persist for at least 3 months, with a slow decrease in response becoming evident. The operator should monitor frequently, comparing response characteristics with those observed on the column immediately after treatment. A repeat treatment of the same column appears to rejuvenate the response, but has been observed to cause a shift in some of the RRT_A values.

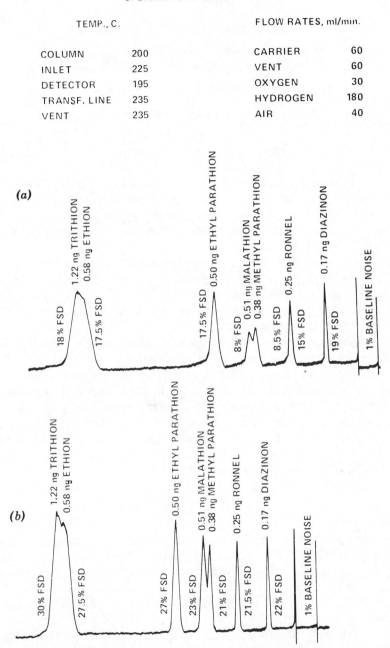

Figure 1.14 Chromatograms of a mixture of 7 organophosphorus pesticides on (*A*) an untreated column of 4% SE-30/6%QF-1, and on (*B*) the same column treated by Carbowax deposition. Amperage full scale 0.8×10^{-8}; 850 V.

45

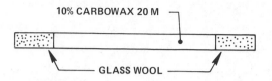

Figure 1.15   Design of Carbowax tube assembly.

1.5.4.   Evaluating the Column

Based on the author's observations of the raw data generated by almost 50 laboratories over a period of years, it has been all too obvious that many chromatographers buy some column packing "off-the-shelf" from a commercial firm, pack it into a column, heat condition it (sometimes) and then start right in using it for routine work.

This modus operandi is tantamount to building a boat and embarking on a transoceanic voyage without a shakedown cruise to determine whether the boat might sink. The innumerable abysmal column performances we have observed have suggested that the columns had indeed "sunk" before the chromatographic cruise ever got under way.

In the foregoing pages, we have discussed various column characteristics and have highlighted various performance criteria such as efficiency, response, compound elution patterns, decomposition, and so on. All these factors should be most carefully evaluated before a column is placed on-line for routine use. Otherwise, the chromatographer may be faced with using an inferior column for its entire use period, which may extend up to a year if the column is cared for properly. The reader is referred to Section 1.7 which gives a list of tentative purchase specifications. These provide some rough guidelines for minimum performance criteria.

1.6.   COLUMN SELECTION

During recent years, the author has scanned many chromatograms submitted with results of interlaboratory check sample analyses of various substrates from over 50 pesticide laboratories throughout the United States. In one such exercise, the 46 reporting laboratories used a total of 25 separate GC column types. For the greater part, these columns were single stationary phases of DC-200, OV-1, OV-17, OV-101, OV-210, QF-1, SP-2401, SP-2250, SE-30, and the mixture columns comprised of an astonishing variety of blend ratios of phases in the foregoing list.

A few laboratories attempted the task of identifying six compounds spiked in a water sample with only one GC column. Those who were lucky used a column which would resolve the six compound peaks. The less fortunate single-column users missed two or three compounds simply because of normal peak overlapping on the particular column they selected. Several other participants used two columns, but of such similarity in compound elution characteristics that they were unable to satisfactorily resolve all the peaks.

When two columns are selected as the principal working columns of a laboratory, several factors should be carefully considered:

1. Do the two columns truly complement one another? Are their compound elution patterns markedly different?

2. Are the stationary phase compositions highly stable or are they likely to bleed?

3. Are there sufficient RRT_A or RRT_P data available to facilitate the task of peak identification?

4. Can the columns be operated at such parameters as to yield maximum efficiency, response, and resolution at a minimal retention time?

5. Do the columns stand up well under sustained periods of injection loading?

Elgar in 1970 (28) designed some ingenious comparisons of RRT data on various columns, drawing from previous data reported by Thompson (18). He plotted RRT_A values of a number of compounds from two similar columns on respective axes and found that the points fell on a relatively straight line with little scatter in evidence. Conversely, when he plotted the RRT_A data of the same group of compounds from two dissimilar columns, the plotted points showed a wide scatter, thus enhancing the probability of reliable peak identification.

Utilizing Elgar's concept, the author prepared plots of a number of different column pairs. Three such plots are shown in Figure 1.16. We used a total of 17 compound RRT_A values obtained on five different columns. It will be observed that when the pair of DC-200 and DC-200/QF-1 was plotted, the scatter is rather restricted. On the other hand, when the DC-200 values were plotted against the DEGS values, the scatter is very great. Similarly, when the OV-210 values were plotted against the OV-17/QF-1, the scatter, particularly of the later eluting compounds, is very good.

1.7. COLUMN PACKING—MAKE IT OR BUY IT?

Up to 1970 this was not a very difficult question to answer. By all means, it was advisable to learn the formulation technique and make your own. The

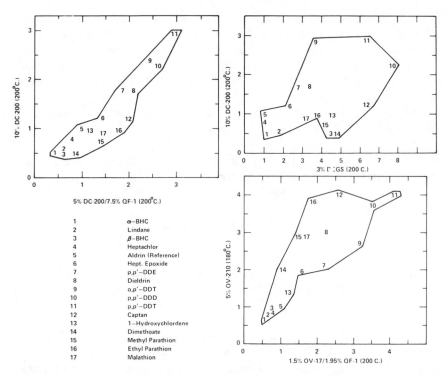

1	α−BHC
2	Lindane
3	β−BHC
4	Heptachlor
5	Aldrin (Reference)
6	Hept. Epoxide
7	p,p'−DDE
8	Dieldrin
9	o,p'−DDT
10	p,p'−DDD
11	p,p'−DDT
12	Captan
13	1−Hydroxychlordene
14	Dimethoate
15	Methyl Parathion
16	Ethyl Parathion
17	Malathion

Figure 1.16 Plots of retention ratios, relative to aldrin, of 17 pesticidal compounds on three column pairs.

quality could not have been much worse than that of the precoated materials on the market at that time.

Today, however, the situation is quite different. At least one government agency has been purchasing for several years in bulk lots of 5 lb, under very stringent purchase specifications. Several commercial suppliers have demonstrated the capability to produce high-quality column packing.

If a laboratory elects to purchase their column packing, it would be advisable to set forth a group of quality specifications that must be met as a basis for purchase. These could include the following:

1. A statement specifying the expected compound elution pattern of a given mixture of pesticidal compounds with attached exhibits showing (1) typical chromatogram and (2) the formulation of the mixture from which the chromatogram was made.

2. A statement of minimum column efficiency acceptable in terms of the theoretical plates in a 6-foot column (3000 TP suggested), the equation for calculating it, and the compound peak used for calculation.

3. A statement of the acceptable absolute retention range, in minutes, for a given compound using specified operating parameters of column temperature and carrier gas flow rate in ml/min.

4. If a mixed stationary phase packing, a statement specifying expected RRT values for all compounds in the test mixture at a prescribed column oven temperature. (This specification would prevent an incorrect mixture ratio.)

5. A statement of compound decomposition limits, that is, endrin to be less than 6% and p,p'-DDT less than 3%.

6. If purchasing large lots, a statement specifying that an advance sample of the lot shall be provided for test before shipment of the entire lot.

If some individual on a laboratory staff has developed the science (and the art) to home brew column packing that will meet minimum quality criteria, it may be advisable for that laboratory to prepare their own. Be well advised, however, that it is not always easy to produce consistently high-quality packing. There are batch-to-batch variations in the quality of support materials, and, although the batch you have on hand at one time may yield extremely good packing, another batch may not turn out as well.

1.8. EPILOGUE

Some chromatographic philosopher uttered the profound observation that *the column is the heart of the gas chromatograph*. Trite as this statement may seem today, no truer statement could possibly have been made. The instrument electronics may be working to perfection, but if the column is bad, it is highly probable that your $12,000 instrument will put out chromatographic data worth less than the chart paper on which the tracings are made.

REFERENCES

1. M. Tswett, *Ber. Beut. Botan. Ges.*, **24,** 316 (1906).
2. S. Claesson, *Arkiv Kemi, Mineral. Geol.*, **23A,** No. 1 (1946).
3. A. Tiselius, *Arkiv Kemi, Mineral. Geol.*, **14B,** No. 22 (1940).
4. A. T. James and A. J. P. Martin, *Biochem. J.*, **50,** 679 (1952).
5. A. T. James and A. J. P. Martin, *Analyst*, **77,** 915 (1952).
6. N. H. Ray, *J. Appl. Chem.*, **4,** 21, 82 (1952).
7. G. Yip, *J. Assoc. Off. Anal. Chem.*, **45,** 2, 367 (1962).
8. L. R. Mattick, *J. Agric. Food Chem.*, **11,** 1, 54 (1963).
9. H. J. Wesselman, *J. Agric. Food Chem.*, **11,** 2, 173 (1963).

10. J. Burke, *A. D. A. Chem. J.*, **46**, 2, 177 (1963).

11. J. Burke, *A. D. A. Chem. J.*, **46**, 2, 198 (1963).

12. A. K. Klein, *J. Assoc. Off. Anal. Chem.*, **46**, 2, 165 (1963).

13. J. E. Barney, C. W. Stanley, and C. E. Cook, *Anal. Chem.*, **35**, 13, 2206 (1963).

14. H. Shuman and J. R. Collie, *J. Assoc. Off. Anal. Chem.*, **46**, 6, 992 (1963).

15. K. A. McCully and W. P. McKinley, *J. Assoc. Off. Anal. Chem.*, **47**, 652 (1964).

16. J. A. Burke and W. Holswade, *J. Assoc. Off. Anal. Chem.*, **49**, 374 (1966).

17. R. S. Henly, R. F. Kruppa, and W. F. Supina, *J. Agric. Food Chem.*, **14**, 667 (1966).

18. J. F. Thompson, A. C. Walker, and R. F. Moseman, *J. Assoc. Off. Anal. Chem.*, **52**, 6, 1263 (1969).

19. W. A. Aue, C. R. Hastings, and S. Kapila, *J. Chromatogr.*, **77**, 299 (1973).

20. C. R. Hastings and W. A. Aue, *J. Chromatogr.*, **89**, 369 (1974).

21. W. A. Aue, M. M. Daniewski, E. E. Pickett, and P. R. McCullough, *J. Chromatogr.*, **111**, 37 (1975).

22. M. M. Daniewski and W. A. Aue, *J. Chromatogr.*, **147**, 119, (1978).

23. W. A. Aue and M. M. Daniewski, *J. Chromatogr.*, **151**, 22 (1978).

24. E. J. Lorah and D. D. Hemphill, *J. Assoc. Off. Anal. Chem.*, **57**, 570 (1974).

25. W. L. Winterlin and R. F. Moseman, *J. Chromatogr.*, **153**, 409 (1978).

26. H. L. Crist and R. F. Moseman, *J. Chromatogr.*, **160**, 49 (1978).

27. R. C. Hall and D. E. Harris, accepted for publication, *J. Chromatogr.* (1980).

28. R. Moseman, *J. Chromatogr.*, **166**, 397 (1978).

29. A. B. Littlewood, *Gas Chromatography, Principles, Techniques, and Applications*, 2nd ed., Academic Press, New York, 1970.

30. J. C. Giddings, *Dynamics of Chromatography*, Part I, *Principles and Theory*, Marcel Dekker, 1965.

31. S. Dal Nogare and R. S. Juvet, Jr., *Gas–Liquid Chromatography, Theory and Practice*, Wiley-Interscience, 1965.

32. *Gas-Chrom® Newsletter*, Applied Science Laboratories, State College, PA, March/April, 1973.

33. R. R. Watts and R. W. Storherr, *J. Assoc. Off. Anal. Chem.*, **52**, 513 (1969).

34. E. S. Windham, *J. Assoc. Off. Anal. Chem.*, **52**, 1237 (1969).

35. M. C. Bowman and M. Beroza, *J. Assoc. Off. Anal. Chem.*, **53**, 499 (1970).

36. J. F. Thompson, J. B. Mann, A. O. Apodaca, and E. Kantor, *J. Assoc. Off. Anal. Chem.*, **58**, 5 (1975).

37. N. F. Ives and L. Giuffrida, *J. Assoc. Off. Anal. Chem.*, **53**, 973 (1970).

38. K. E. Elgar, *Adv. Chem. Ser.*, **104**, (1971).

SELECTION OF GAS CHROMATOGRAPHIC DETECTORS FOR PESTICIDE RESIDUE ANALYSIS

P. T. HOLLAND and R. GREENHALGH

Chemistry Biology Research Institute, Agriculture Canada,
Ottawa, Ontario, Canada

CONTENTS

2.1.   INTRODUCTION

The analysis of pesticides in soils, animal and plant tissues, the establishment of residue tolerances in food, and environmental studies involving trace contaminants are all dependent on trace analytical methods employing gas chromatography (GC). Advances in this technique are closely related to the development of the more sophisticated analytical procedures employed today. Investigations in the early 1950s with gas phase separation techniques revealed unparalleled resolution of complex mixtures of compounds such as those resulting from extracts of natural matrices. However, at that time, the measuring devices used to monitor the eluting components were relatively insensitive. These early detectors measured changes in thermal conductivity or gas density of the effluent, using hot wire filaments, and were limited to the detection of only 0.01 mole% in the effluent. Exploitation of ionization techniques in detector designs in the late 1950s resulted in lower detection limits and greatly extended the applications of gas chromatography.

The argon β-ionization detector (1) and the hydrogen flame ionization detector (FID) (2–4) both respond and provide increased sensitivity to most organic compounds. In the case of the FID the response is proportional to the carbon content of the compound, and this is particularly useful for general analysis of organic mixtures. However, the detector is not selective, and the presence of coelutants necessitates a high degree of sample purification before the full sensitivity of the detector can be utilized. Mainly because of its lack of selectivity the FID has found little use in residue analysis.

The invention and improvement of the electron capture detector (ECD) by Lovelock (5–7) provided the first important tool for residue analytical techniques. Utilizing the electron capture properties of compounds, it provides a selectivity that the earlier detectors lacked. In addition, the ECD is extremely sensitive for those compounds having strong electron capturing properties such as halogenated molecules. The development of this detector coincided with a period of rapid increase in the use of diverse organochlorine compounds, such as the insecticide DDT for agricultural purposes and the industrial high dielectric liquids based on polychlorinated biphenyls (PCBs). The ECD played, and continues to play, a key role in the trace analysis of these compounds. In fact, it is difficult to see how our present extensive knowledge of environmental problems caused by organochlorine compounds could have been attained without the simple, reliable, and sensitive ECD.

Problems associated with the long-term persistence of the organochlorine compounds and their accumulation in various matrices led to their replacement in the 1960s by the more readily degradable organophosphorus and carbamate insecticides. These compounds have a relatively high mammalian toxicity, which necessitates careful regulation of their use and monitoring of crop residues. In recent years the use of herbicides has increased and they now constitute important agrochemicals. They differ from the insecticides in that chemically their structures are very diverse. Because these new insecticides and herbicides often do not have strong electron-capturing properties, GC detector systems other than the ECD are required for analysis of their residues. Several new detectors based on different physical principles have been found to be suitable.

A hydrogen flame can be modified by the addition of an alkali salt vapor so that it gives selective detection of phosphorus, nitrogen, and sulfur compounds (8). The sensitivity to phosphorus compounds is enhanced by a factor of 10^2 over that for the normal FID. Another flame technique employed is the sensitive and selective flame photometric detector (FPD), which responds to both phosphorus- and/or sulfur-containing compounds by measuring characteristic emission bands from hydrogen-rich flames (9). It immediately gained wide acceptance for the analysis of organophosphorus insecticide residues. The selective detection of other heteroatoms, such as the halogens, nitrogen, and sulfur, can be obtained by the oxidation or reduction (pyrolytic or catalytic) of compounds, followed by transfer of the reaction products into an ionizing solvent, generally water, for conductometric detection (10, 11). These three techniques involving element selective detectors, together with electron capture, have dominated the design of GC detectors used for pesticide residue analysis over the last two decades. During this period they have undergone considerable refinement in both the design and the operating conditions required for their optimum performance.

Mass spectrometry (MS) is the most versatile technique available to the residue analyst for selective detection, since the presence of a specific moiety or heteroatom in the molecule is not required. The development of combined GC/MS proceeded independently of residue analysis, although it is now generally accepted as an ancillary tool. In a qualitative role mass spectra have proved invaluable for the identification of unknown compounds and helped in the elucidation of metabolic pathways of pesticides. A more recent development in GC/MS is the selected ion monitoring mode (SIM), which is more sensitive, and can be quantitated by simply monitoring m/z values that are characteristic of the compound. The potential of this technique was demonstrated in some recent work by Harless and Oswald (12). They quantitated and characterized 2,3,7,8-tectrachlorodioxin (TCDD) in human milk at the 1 in 10^{12} level (ppt) using a capillary GC, interfaced to a high-resolution mass spectrometer.

GC detectors have been described in some detail by David (13) and Sevcik (14). These monographs have concentrated on the history of the various detector designs and the theory of their operation. Adlard has also reviewed the development of both general (15) and selective (16) detectors and the practical aspects of GC detectors used in pesticide analysis has been discussed by Aue (17, 18). A useful overview of the subject can be found in the papers presented at a symposium on "Selective Detectors," which includes detailed descriptions of several new designs (19). A recent review of selective nitrogen detectors has also appeared (20). In addition, there are many papers comparing the performance of several selective detectors relevant to the analysis of pesticide residue analysis (21, 22).

This review covers the more recent developments in GC detectors. Emphasis is placed on the practical aspects concerned with the selection and operation of the various types. The objective parameters for sensitivity, selectivity, linearity, and detection limits are used to provide an overall framework for the discussion following the example of Sevcik (14). A complete historical coverage has not been attempted, and discussion of detection mechanisms is minimal. The selection of GC detectors for use in the analysis of residues of various pesticides and pollutants is related to the physiochemical characteristics of the compounds and by comparison of different techniques used for their analysis. The review does not attempt to give a complete coverage of the literature for methods associated with each class of compounds, but instead provides examples illustrating the advantages of various approaches.

Finally, methods of evaluating detector response and the possibilities for anomalous results are described, as these are considered to be important practical problems associated with the selection, optimization, and maintenance of GC detectors for use in the quantitation of compounds at the residue level.

2.2. DEFINITIONS

2.2.1. Detector Response

If a compound, X, elutes from a GC column with an ideal Gaussian profile then the concentration (C_X) in the carrier gas at the column exit at time t is given by

$$C_X = \frac{M_X}{v t_R} \sqrt{\frac{n}{2\pi}} \exp\left[\frac{-n}{2}\left(1 - \frac{t}{t_R}\right)^2\right] \qquad (2.1)$$

where M_X = moles of compound
$\quad n$ = column theoretical plates
$\quad t_R$ = peak retention time
$\quad v$ = carrier gas flow rate

The maximum concentration will be when $t = t_R$ and equation 2.1 then reduces to

$$C_{X,\,max} = \frac{M_X}{vt_R} \sqrt{\frac{n}{2\pi}} \qquad (2.2)$$

The signal (S) produced from a detector by the eluting substance is related to its concentration by

$$S_X = ka_X C_X \qquad (2.3)$$

and the peak signal at $C_{X,\,max}$ is

$$S_{X,\,max} = \frac{ka_X M_X}{vt_R} \sqrt{\frac{n}{2\pi}} \qquad (2.4)$$

where a_X = property of the substance measured by the detector, e.g., electron capture coefficient

k = constant for the experimental factor, which includes amplification factors

For S to be proportional to C_X according to equation 2.3, a_X must be independent of sample concentration, that is, the detector must have a linear response characteristic. The time constant (τ) of the detection system, which is dependent on the detector volume and electronic bandwidth, must also be less than the peak width at half-height ($w_{1/2}$) in seconds. If $\tau \leq w_{1/2}$, then the distortion in peak height and retention time will be less than 5% (23).

In the case of detectors that decompose the eluate, for example, the FID, the detector signal is determined by the quantity of substance eluting per unit time, rather than concentration. Thus incorporating the column flow rate into equations 2.3 and 2.4 gives

$$S_X = k'a_X C_X v \qquad (2.5)$$

$$S_{X,\,max} = \frac{k'a_X M_X}{vt_R} \sqrt{\frac{n}{2\pi}} \qquad (2.6)$$

where k' = experimental constant.

Most of the detectors used for residue analysis are sample flow rate dependent, and equation 2.5 is therefore applicable. The ECD is the exception, since it usually exhibits the properties of a concentration-dependent detector, where the peak signal is inversely proportional to the carrier flow rate (equation 2.4). For the following discussion, however, it is more convenient to class its output along with those of other selective detectors. This assumption is valid as a first approximation, since the total gas flow rate through an ECD is not normally changed by large amounts.

The response (R) is defined as the integral of the signal over the peak elution period and is directly related to measured peak areas.

$$R_X = \int_{t_1}^{t_2} S_X dt \tag{2.7}$$

which on substituting 2.1 into 2.5 and integrating becomes

$$R_X = k' a_X M_X \tag{2.8}$$

substituting into equation 2.6 and using the formula for the number of theoretical plates gives

$$R_X = 1.064 w_{1/2} S_{X,\,max} A \text{ sec} \tag{2.9}$$

where $w_{1/2}$ = measured peak width at half-height.

Therefore, the response is ideally dependent only on two factors, the amount of compound injected and the property of the compound measured by the given detector (a_X).

2.2.2. Sensitivity

The sensitivity or response factor (RF) of a detector is defined as the response per unit of substance i.e.,

$$RF = \frac{\Delta R}{\Delta M_X}$$

$$= k a_X A \text{ sec/mole} \quad (C/\text{mole}) \tag{2.10}$$

Normally RF for ionization dectors are expressed in C/g units.

Sensitivity measurements are of little value unless the detector noise level is also specified, since the most important parameter in any application of a GC detector is the signal/noise ratio. It is also necessary to express the response in absolute units (C/g) to compare different detectors, since the amplification factor in the constant k_1 may be different even for the same class of detector. Sensitivity can, however, still be employed for comparison purposes when studying the response of a number of compounds in the same detector.

The concept of efficiency can be introduced where chemical processes, such as ionization or atomization, form the basis of detection. It represents the degree of conversion of a compound into measurable species. For ionization detectors the efficiency (E) is given by

$$E = \frac{R}{M_x} \frac{100}{F} \%$$ (2.11)

where R/M_x = sensitivity in C/mole.
$\quad\quad$ F = Faraday (96,487 C/mole)

While efficiency is still not useful for describing the practical merits of detectors, it does define the limits of a particular technique. For example, the efficiency of an ECD can approach 100% for highly halogenated compounds. In contrast, an FID is only about 0.01% efficient in producing ions from organic molecules.

2.2.3. Detection Limit

By taking noise into account, a more realistic measure of detector capability is obtained. Following Hartmann (24), the detection limit is defined as the weight of substance required to give a signal of twice the peak(+)–peak(−) noise. The value is normalized by dividing by the peak half-width. This takes into account the fact that the peak area for a particular compound is constant for different retention times, whereas the peak height is not. Thus, narrow peaks have a better detectability than broad ones. It gives the detection limit (L) effectively as that sample flow rate that will give rise to a signal/noise ratio of 2.

$$L = \frac{2nm_x}{R} \text{ g/sec}$$ (2.12)

where n = noise
$\quad\quad m$ = amount of compound injected
$\quad\quad R$ = area response

This is a general definition of detector performance, since it is independent of the detection mechanism or amplification factors. It is also a very practical parameter that relates directly to the measurement process. An objective determination of the detection limit requires that the peak(+)–peak(−) noise include both high- and low-frequency perturbance over the time interval of several peak widths. For the noise estimate to be statistically valid it must also be based on several observations over the time interval (25, 26, 14). Moderate baseline drift or solvent tailing are not generally included. The degree of electronic filtering directly affects the noise and hence the detection limit. Modern detector amplifiers generally have time constants of about 0.1 sec. Further filtering is usually possible, except with capillary columns, and is often present in the recorder, integrator, or data system (23). Detection limits are usually based on injections made at amplification settings where the noise is clearly visible and peaks 10–20× the noise level are obtained.

2.2.4. Minimum Detectable Quantity

The detection limit can be related to analytical measurement limits by defining a minimum detectable quantity (MDQ) for a compound. This is that amount of sample that will produce a peak signal 2× the noise limit.

$$\text{MDQ} = LW_{1/2} \qquad (2.13)$$

$$= \frac{2S_n m}{S_{max}} \text{ g}$$

where S_n = noise signal
S_{max} = peak signal for amount m

Equation 2.13 can be applied to a compound other than that used to define L, if the relative sensitivities of the compounds are taken into account.

For an actual analysis, the MDQ may be larger than that given by equation 2.13 due to adsorption losses, interference from coeluting components or increased baseline noise due to detector contamination. Kaiser (27) has discussed in depth the importance of adequately measuring blank or control samples for use in calculating the detection limit of an analytical method.

Since it is also influenced by the loss of a compound during extraction and cleanup, it is necessary that the analytical limit for a particular procedure be determined experimentally rather than calculated from the MDQ. This analytical limit should be expressed as the concentration in the original sample, e.g., mg/kg, that will lead to a GC peak significantly greater than the control.

2.2.5. Linearity

The linear range of a detector is the range of sample flow rates over which the sensitivity or response factor remains constant. It is expressed as the ratio of the sample flow rate at which the sensitivity has changed by more than 5% to the sample flow rate representing the detection limit.

Log-log plots of response vs amount are often used for linearity evaluation, although these are notoriously insensitive to small changes. For this reason, it is preferable to plot sensitivity against sample flow rate or amount injected. A good linear specification should not be taken for granted. Accurate calibration curves from standards which span the sample concentration range should always be used as there are many factors which can modify an ideal linear response.

Dynamic range is a related parameter where the ratio covers the complete range of usable detector responses going beyond linearity towards detector saturation. The limit may be taken as that sample flow rate where a doubling leads to less than a 20% increase in response.

2.2.6. Selectivity

The selectivity of a detector is expressed as the ratio of the detector sensitivities to two compounds, i.e., RF_1/RF_2, which reduces to the ratio of the compound specific factors a_1/a_2 (cf. equation 2.10). Where these ratios are small for a wide range of compounds, as in the case of the FID, the detector is not selective. For ratios > 10, the selective detection of particular compounds is possible. The term specific detector should be avoided, since in practice no detector is completely specific in its response, element selective is preferred.

As far as the detection mechanism depends on certain atomic species being present, so sensitivity can be expressed in terms of the response to these specific atoms and hence selectivity can also be expressed in elemental terms e.g., selectivity of the FPD to phosphorus with respect to carbon.

2.2.7. Calculation of Detector Performance

The use of the preceding equations are illustrated with some results obtained with an AFID. Figure 2.1A shows the chromatogram for an injection of 10 pg of chlorpyrifos (CP) and 1.2 μg of hexadecane (HD). The area responses for these two components are

$$R_{CP} = 0.25 \times 10 \times 10^{-12} \text{ A sec} \qquad R_{HD} = 0.75 \times 8 \times 10^{-12} \text{ A sec}$$

The response factors to these compounds are

$$RF_{CP} = \frac{R_{CP}}{10^{-11}} \text{ g} \qquad RF_{HD} = \frac{R_{HD}}{1.2 \times 10^{-6}} \text{ g}$$

$$= 0.25 \text{ C/g} \qquad = 5 \times 10^{-6} \text{ C/g}$$

$$= 2.8 \text{ C/gP} \qquad = 6 \times 10^{-6} \text{ C/gC}$$

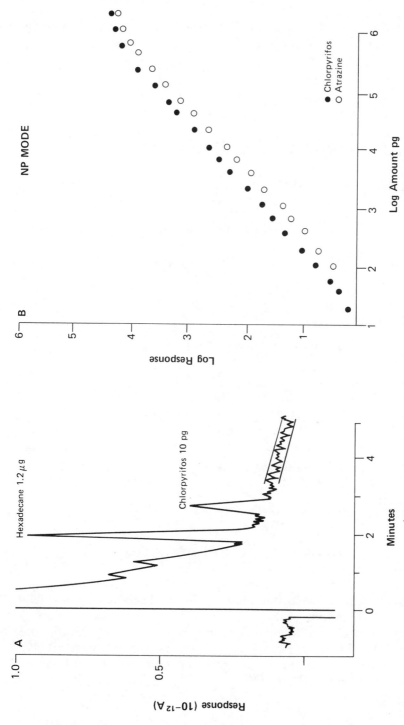

Figure 2.1 Alkali flame ionization detector response in the NP-mode to carbon, nitrogen, and phosphorus compounds. (*A*) Gas chromatogram of chlorpyrifos (10 pg) and hexadecane (1.2 µg). (*B*) Linear range for atrazine (N) and chlorpyrifos (P). GC conditions: column, 1 m × 6 mm, 3% OV-17/Gas Chrom Q; column temperature 215°; column flow 30 ml/min helium; detector flows, 100 ml/min air, 2 ml/min hydrogen; standing current 3 pA.

The estimated noise level (Figures 2.1A) is 4×10^{-14} A and therefore the detection limit for phosphorus is

$$L = \frac{2 \times 4 \times 10^{-14}}{2.8}$$

$$= 0.03 \text{ pg P/sec}$$

The minimum detectable quantity is

$$\text{MDQ} = L \times W_{1/2}$$

$$= 0.3 \text{ pg P}$$

$$= 3.3 \text{ pg chlorpyrifos}$$

The selectivity of the detector to phosphorus against carbon is:

$$\frac{RF_P}{RF_C} = 4.7 \times 10^5$$

The efficiency of the detector for phosphorus is

$$E_P = \frac{2.8 \times 31 \times 100}{96,475}$$

$$= 0.09\%$$

Figure 2.1B shows a log–log plot of signal (peak height) against the amount of chlorpyrifos injected for this AFID. The plot is linear from the MDQ to a sample size of 340 ng chlorpyrifos with a slope of 0.96, giving a linear range of 10^5.

2.3. FLAME IONIZATION DETECTOR (FID)

The response is dependent on the formation of ionic species when carbon compounds are burnt in a hydrogen flame. The mechanism of ion formation is still not well understood, but evidence indicates that pyrolytic cracking and radical reactions in the hydrogen rich part of the flame lead to ions directly (28) and by chemiionization, via reactions with oxygen atoms (29). Proton transfer to water followed by clustering, forms the principal positive charge carriers, (H_3O^+) $(H_2O)_n$ (30).

In the normal detector configuration, the carrier gas is premixed with hydrogen and burnt in a chamber through which excess air is flowing. A cylindrical electrode mounted a few millimeters above the jet tip provides the most efficient form of ion collection (31). A jet polarization of -150 to -300 V is sufficient to collect the more mobile positively charged flame species in a plateau region. The currents produced are in the range 10^{-12} to 10^{-9} A and are measured with electrometer amplifiers. The noise due to the flame is 2×10^{-14} to 10^{-13} A, but pure hydrogen and air supplies are required to attain the lower value. Optimum hydrogen and airflow rates depend on individual detector geometry, type of carrier gas and its flow rates, but are generally in the ranges 30–50 and 200–300 ml/min, respectively. The response is determined principally by carbon content. The sensitivity of an FID is of the order of 0.01 to 0.02 C/g carbon, which results in a detection limit for most organic molecules of about 5 pg/sec and a MDQ of 50 pg for a peak of 10 sec half width. Carbon-heteroatom bonds such as, C—Cl, C—O, and C—N, are less effective in producing ions as compared to the C—H bond. The response for a compound can be calculated from the molecular formula using effective carbon numbers (13, 32). A linear relation is obtained between molar response and carbon number for a homologous series (33). In general, the CH content of a compound determines its response and compounds with no C—H bonds give very poor FID responses. The degree of accuracy from relative molar response calculations is generally insufficient for quantitative purposes and standards are required. Operation of the FID in a hydrogen-rich mode with oxygen for combustion provides an even more universal response (34, 35). For a series of alkanes, chloroalkanes, and chlorobenzenes the carbon response is constant within 3% at 0.31 C/mole carbon.

The linearity of the FID is about 10^7 with nonlinear response becoming apparent above a sample flow rate of 50 μg/sec. The linear range is more affected by changes in flame gas composition, jet size, and collector geometry than the sensitivity (33), and these parameters must be optimized to obtain the maximum values.

Simplicity of construction and operation, high sensitivity, stability, and wide dynamic range have made the FID the most popular detector for the analysis of organic compounds in general. It is quite sensitive to pesticides, except the highly halogenated compounds, but its lack of specificity restricts its use in residue analysis. Coextractives will normally mask the response of pesticide residues unless they are present at high levels (>10 ppm) or extensive sample cleanup has been carried out. Solvent peak tailing may also increase the MDQ of an FID operated at high sensitivity. The use of carbon disulfide as a solvent can minimize this factor, since its FID sensitivity is about 10^{-3} times that of other common organic solvents, although some quenching of sample ionization has been noted. For juvenile hormones, insect pheromones, and natural pyrethroids that lack specific response properties, the FID may be the only choice apart from the much more expensive and complex GC/MS.

The FID has been widely used for pesticide formulation analysis. A collaborative study on simazine in aqueous formations showed a response variation of 1.2% using GC/FID with an internal standard (36).

Another important use of the FID is the estimation of the coextractive levels in relation to possible interferences to other more specific detection methods. When developing new analytical procedures this information is of use in evaluating the degree of sample cleanup required and determining the presence of any major impurities coeluting in the region of the pesticide. Effluent splitter systems have been advocated for this purpose to obtain dual FID and selective detector chromatograms (37, 38), but for routine work simultaneous acquisition is not necessary. The carbon response of the FID implies that the response to lipid coextractives will be approximately the same on a weight basis for a wide range of compounds. Therefore, estimates of coextractive levels can be calculated from direct comparison of peak areas to a hydrocarbon standard.

2.4. ELECTRON CAPTURE DETECTOR (ECD)

This detector has undergone continuous development since its invention by Lovelock (5). Extensive data now exist on detection mechanisms, design, and operational parameters, and these topics have been comprehensively reviewed up to 1974 by Pellizzari (39). The ECD, although simple in design and operation, is actually a complex chemical reactor, and many processes are involved in the production of a signal.

The ECD is based on the formation of a plasma of thermalized electrons inside a cell by irradiation of the carrier gas. These electrons are collected by applying a small voltage to produce a standing current. Compounds entering the cell may capture the thermalized electrons, and the reduced mobility of the negative species formed results in their neutralization before they can be collected. This reduction in the electron concentrations is measured as a decrease in the standing current (7) by applying either a steady voltage (DC mode) or voltage pulses at a constant frequency (pulse mode). Alternatively, the frequency of applying the voltage pulses can be varied to maintain a constant electron current (pulse modulated or constant current mode) (40).

2.4.1. ECD Construction

Parallel plate designs have been largely superceded by concentric cylinder arrangements because these allow a smaller detector volume. An ECD with a cell volume of only 0.15 ml has recently been described for use with capillary GC (41). The features of a concentric cylinder ECD are illustrated in Figure 2.2. The collector is offset from the electron source to reduce space charges and minimize collection of electrons under field-free conditions, which would constitute a background current in the constant current mode (42). When a potential is applied, the mobile electrons can readily diffuse against the carrier gas flow, but electron capture products are neutralized before collection. Argon moderated with 5–10% methane has the optimal gas properties for electron-capture detectors. Electron mobility is high in argon which facilitates their collection. Methane is added to more effectively thermalize the electrons and to reduce the production of long-lived argon metastables, which can directly ionize solute molecules (7). Electron mobility is about a factor of 10 less in nitrogen, and its use in early designs resulted in very limited linear ranges. The development of the

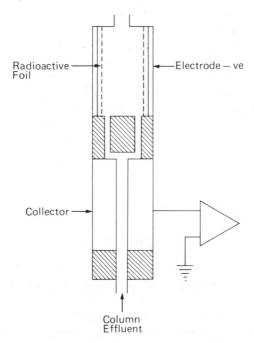

Figure 2.2 Schematic of a concentric cylinder electron capture detector cell.

constant current operating mode and optimization of other operating parameters such as the pulse duration have improved the ECD performance with nitrogen (42), which is a more desirable carrier gas both from cost and versatility points of view.

The usual source of electrons is β-particles from a metal foil containing ^{3}H or ^{63}Ni. Foils with the ^{3}H imbedded in titanium have a temperature limit of about 225° before loss of radioactive material becomes excessive (43). This low temperature limit renders them very prone to contamination. The use of scandium tritide foils has been reported (44, 41) with a temperature limit of 325°. However, this type of radiation source appears to have lost favor with commercial detector manufacturers. It is possible that the low energy of ^{3}H β particles (18 keV) makes tritium sources very sensitive to subtle changes in surface structure. The source most commonly in use today is ^{63}Ni (67 keV). It is a low-energy β emitter and has the advantage of a long half-life and also the detector can be operated at temperatures up to 400° (45). Its disadvantages are a low activity and high cost of commercial foils. Whereas ^{3}H sources can be obtained with 1 Ci of activity, ^{63}Ni sources rarely exceed 15 mCi. In detectors of similar design these activity differences are directly reflected in the sensitivity and dynamic range. In practice, recent ^{63}Ni designs with carefully optimized electrode geometries have matched or exceeded the routine performance of earlier designs using ^{3}H sources (42, 46, 47), which usually lose a good percentage of their initial sensitivity. Other ionization sources have been considered including ^{241}Am (48), ^{55}Fe (49), and photoionization (50); however, only the last has been developed commercially. The standing current of a ECD is an extremely important parameter that determines the upper limit of the dynamic range, which occurs when the eluate has removed most of the electrons. Cell contamination, column bleed and carrier gas impurities such as chlorinated solvents and oxygen all reduce the standing current and, therefore, affect dynamic range.

2.4.2. DC and Constant Frequency Operation

In the DC mode, a constant polarizing voltage of 20 to 50 V is applied and compounds are detected by a decrease in the standing current. This method is simple and sensitive but prone to instability and anomalous response due to space charges and direct ionization of sample (7). Contact potentials, caused by temporary condensation of sample are also sometimes noted as negative tailing of eluting peaks (7). These problems are overcome by pulsed operation (7) where collection of electrons only takes place during a brief period, typically 0.5 μsec for nitrogen and 0.1 μsec for argon/methane, when the cell is polarized with a 50-V pulse. Between pulses the thermalized electron concentration builds up and reaches a steady state level of 4×10^6–6×10^7 electrons/ml in ^{63}Ni cells (48, 42). Eluting compounds can react with this electron plasma, reducing the current collected when the pulse is applied. Optimum sensitivity is obtained with longer pulse periods (0.1–1 μsec), but shorter pulse periods are useful to reduce sensitivity and shift the linear range (7). For both the DC and pulsed modes of ECD operation the cell current is related to the sample concentration by an equation of the following form (51):

$$\frac{\Delta S}{S_b - \Delta S} = KaC \qquad (2.14)$$

where ΔS = reduction in current induced by compound
$\quad\quad S_b$ = standing current
$\quad\quad a$ = rate constant for electron attachment to compound
$\quad\quad C$ = concentration of eluate in carrier gas
$\quad\quad K$ = experimental constant

As can be seen, ΔS is only a linear function of C for values of ΔS that are small relative to the standing current. This results in a small linear range, which is illustrated by the following calculation. The short term noise of an ECD of about 1×10^{-12} A (7) is determined by the statistical variation in the radioactive emission giving rise to the standing current. Thus, the detection limit is 2×10^{-12} A. The upper limit to the linear range can be taken as that concentration of eluate that reduces the standing current by 5% (equation 2.14). This reduction represents 2×10^{-10} A for a 4×10^{-9} A standing current and leads directly to a linear range of 10^2. This is the order of linearity reported for an ECD operated in the DC mode (45). The upper limit of the dynamic range is taken as an 85% reduction in standing current, which leads to a dynamic range of about 10^4. Contamination of an ECD greatly reduces detection limits and linearity because of increased noise and reduced standing current. Rigorous cleaning is necessary to remove baked-on column bleed from detector foils. Sonicating tritium foils in methanolic detergent (3:1) and passing hydrogen through ^{63}Ni cells at 400° have been recommended.

2.4.3. Constant Current Operation

Although analog circuitry has been designed to linearize the output described by equation 2.14 (52), constant current operation has dominated recent approaches to improve the linear range of the ECD. Maggs et al. (40)

have shown that if the output current of the ECD is fed back to the pulse circuit so that the frequency of pulses, f, is changed to keep the cell current constant then,

$$f - f_0 = \frac{a}{D} C_x \qquad (2.15)$$

where f_0 = pulse frequency with carrier gas only
$\quad\;\; D$ = experimental constant dependent on the cell current
$\quad\;\; a$ = rate constant for electron attachment

Thus, an increase in sample concentration is directly related to the increased frequency of pulses required to maintain a constant cell current. This relation holds until the electron concentration is reduced to 0.5% of the initial compared to the 95% tolerable for linear DC operation (42). This extends the linear range of an ECD by a factor of 200 to about 2×10^4. Detection limits are often improved over DC operation, which has been ascribed to the higher concentration of electrons in the pulsed mode (42). The current to which the output is referenced is usually 10–50% of the cell standing current with a baseline frequency of 1–5 kHz (40, 47). The output voltage from this type of detector circuit is now proportional to the pulse frequency rather than the cell current. Sullivan and Burgett (47, 53) have derived useful equations describing constant operation for various electron capture mechanisms. Contamination of the detector results in a reduction of linear range, but sensitivity is not as badly affected as in the constant frequency modes.

2.4.4. Compound Response

Under certain conditions, strongly electron-capturing materials can undergo essentially complete electron attachment, and the detector exhibits a coulometric mass flow-dependent response (54, 55). In this case the detection limit is given by the solute flow required to produce a change in signal of twice the noise level. For a noise of 1×10^{-12} A, it is

$$DL = 2 \times 10^{-12}/1.6 \times 10^{-19} \text{ A sec/molecules}$$

$$= 1.2 \times 10^7 \text{ molecules/sec}$$

Organochlorine pesticides are strongly electron capturing and the ECD is extremely sensitive to them, particularly aldrin. Assuming complete ionization, the detection limit for aldrin is predicted to be about 7×10^{-14} g/sec with a MDQ of 0.35 pg for a half-peak width of 5 sec, which can be obtained in practice. The responses of other organochlorine pesticides, including lindane, dieldrin, heptachlor and Mirex are similar and independent of detector flow rate, which is indicative of a coulometric response (55). Hypercoulometric responses have also been observed for some halo- or nitro- compounds (56). It is possible that ions for nondissociative electron capture can capture further electrons after neutralization (57). Coulometric responses depend on specific detector parameters, e.g., DC mode at low flow rates, and are not always observed. Constant current ECDs with small cell volumes and moderate source activity generally do not exhibit coulometric responses and behave as concentration-dependent detectors with the response proportional to detector flow (40, 42). The detection limit of these detectors is not markedly different from those exhibiting coulometric responses. When coulometric responses are observed in constant current ECDs, they affect the linearity characteristics by reducing the apparent response factor at low sample flow rates (53).

The relationship between compound structure and ECD response has been extensively studied (39). The probability for electron attachment depends on several factors, including electron affinity and electron capture cross section. These parameters are not readily calculable from first principles, and empirical rules have been formulated to predict relative response based on the compound type and functional groups. Table 2.1 lists the

TABLE 2.1. Relative Molar Response Factors for Electron Capture Detection

Chemical Class	Relative Response[a]
Hydrocarbons	10^{-2}
Aliphatic ethers and esters	10^{-1}
Aliphatic alcohols, ketones, aldehydes, amines, nitrites, monochloro and -fluoro compounds	1
Enols, α, β diesters, monobromo, dichloro, hexafluoro compounds	10
Trichloro compounds, acyl chlorides, anhydrides, carbamates, triazines, alkyl leads, organophosphates	10^2
Monoiodo, dibromo, trichloro and mononitro compounds, di- and trisulfides	10^3
1,2-diketones, conjugated esters, quinones, diiodo tribromo, polychloro and dinitro compounds, and organomercurials	10^4

[a] With respect to chlorobenzene.

approximate relative molar responses for a variety of chemical classes expressed relative to chlorobenzene (6, 46, 58). Increasing substitution of halogen or keto groups has a synergistic effect and allylic substitution enhances electron attachment by factors of up to 100 (6). The configuration of sulfur in organophosphorus esters also affects the ECD response. Cook et al. (59) showed that the sensitivity decreased in the following order: phosphorodithioate > phosphorothiolate > phosphorothionate > phosphate.

The reaction of amino, hydroxy or carboxy compounds with derivatives containing halogen or nitro moieties is a means to increase ECD response (60, 61), as well as improving volatility and thermal stability. Derivatives containing more than one chloro, bromo, or nitro group are not always practical due to their longer retention times. The perfluoro derivatives are of interest in that heptafluorobutrates often have shorter retention times than trifluoroacetates, also the pentafluorobenzyl derivatives are up to 150 times more sensitive than the heptafluorobutryl (62). Derivatization reactions are discussed in detail in a book edited by Blau and King (63).

Detector temperature can also effect the detection limits. Nondissociative electron capture response is reduced by increased temperature, whereas dissociative electron capture shows a positive temperature dependence (64, 65). For the effects to be apparent the detector must be functioning below the coulometric limit. Dieldrin shows a 10-fold increase in response for a change in detector temperature from 80 to 300° (40). Since most moderately to strongly electron-attaching compounds undergo dissociative electron attachment, it is advantageous to operate the detector at as high a temperature as possible without causing undue thermal degradation of samples. A temperature of about 300° is a good compromise for ^{63}Ni detectors, which also helps minimize detector contamination. For reproducible quantitative measurements, the cell temperature should be controlled to at least 0.5°.

2.4.5. ECD Performance

An interlaboratory study was recently conducted of ECD performance in 27 laboratories (66) and included a variety of commercial detectors. Chlorpyrifos was used as a standard compound for the evaluation. The linear ranges for DC, pulsed or constant current ECDs were found to be similar to the theoretical estimates (4.2, 4.3) for most models as were the detection limits. The ECD is 10 times more sensitive to aldrin than to chlorpyrifos. The improved linear range of the constant current detectors was confirmed under routine conditions, although most required argon/methane for optimum performance.

2.4.6. Summary

The use of the ECD for pesticide residue analysis is well established, especially for organochlorine pesticides and pollutants. Its high sensitivity is an advantage, but selectivity is also very important in determining the practical limits for an analysis. It is apparent from the relative response factors (Table 2.1) that organophosphorus and carbamate insecticides have much less favorable responses. Although the ECD is of sufficient sensitivity to measure these compounds at the residue level, the selectivity relative to possible coextractives is only 10^2–10^3. The fact that both organophosphorus and carbamates are more polar than the organochlorine compounds is a further complications, since extracts, even after cleanup, are liable to contain some moderately polar contaminants which have electron-capturing capability. In addition, analyses of these insecticides are liable to interference from small amounts of the more strongly electron-capturing compounds, such as the ubiquitous PCBs and phthalate esters. This broad spectrum of possible ECD responses can make interpretation of ECD chromatograms difficult and necessitates confirmation for identification. For these reasons, more selective detectors are preferred for the analysis of pesticides having only moderate electron attachment probability. However, for ultimate sensitivity, where a compound has, or can be made, by derivatization, to have a high electron capture ability, the ECD is the detector of choice.

2.5. FLAME PHOTOMETRIC DETECTOR (FPD)

The invention of the FPD by Brody and Chaney (9), based on earlier sulfur emission work (67), has provided gas chromatography with the most reliable and selective detector for the analysis of phosphorus and sulfur compounds. Development of this technique has been reviewed up to 1971 by Selucky (68). There have been remarkably few changes in the original design or improvements in performance since that period.

2.5.1. FPD Design

The FPD is based on the element specific chemiluminesence produced when compounds are burnt in a hydrogen-rich flame. The sulfur and phosphorus emission bands are derived from excited S_2 and HPO species, respectively, and are formed by reactions with hydrogen atoms in the upper part of the flame (69, 70). Carbon also gives flame emission bands, which are formed in the lower region of the flame. The emission spectra of HPO, S_2, CH_3, and C_2 radicals have been reported (70). Carbon radicals can quench S_2 and HPO emission (71–73).

In the original FPD design (9), the column effluent was mixed with air and combusted in a burner of a similar design to an FID. The flame burns in a stream of H_2 that is present in excess of stoichiometric requirements. Normally, only the upper part of the flame is monitored by a horizontal optical system, which focuses the emission onto a photomultiplier. Filters are employed to improve selectivity with transmission maxima at 394 nm for sulfur and 526 nm for phosphorus detection. A dual channel model was introduced to simultaneously monitor these two bands (74). A reduction in the noise level of the photomultiplier by a factor of 3 has been reported using water cooling (75). Optical fiber light pipes are also used to transmit light and at the same time provide thermal insulation of the photomultiplier (21, 38).

The original detector design (9) is prone to flameout due to the solvent peak reducing the oxidant flow. This can be corrected by the use of relight devices or solvent venting. A more satisfactory solution is to premix the carrier gas with hydrogen instead of air, i.e., reversing the gas lines to give the normal FID configuration (76). Sensitivity is not markedly affected, but selectivity to carbon may be somewhat reduced, since oxidation now occurs higher in the flame. A jet design where the carrier gas is premixed with hydrogen before introduction via a concentric tube surrounding an air jet also provides improved resistance to flameout (77, 78). A more recent development is a dual burner detector, where the sample is combusted in the first flame, recombusted and the emission measured in the second larger flame (70). If flameout occurs, the second flame will relight the first. The jet, gas flows and geometries for these various FPD designs are illustrated in Figure 2.3.

2.5.2. FPD Optimization and Performance

The critical dependence of emission on the combustion process means that the flow rates must be optimized for each design. The optimum flow rates for several detectors are summarized in Table 2.2. The important parameters are the O_2/H_2 ratio and the total oxygen and carrier flow rate (78, 79). They are generally different for phosphorus and sulfur compounds, the latter requiring a higher O_2/H_2 ratio. Once optimized, a given detector should exhibit a stable response provided gases are under precise flow control. In practice it is recommended to use separate ON/OFF valves for shutdown rather than the flow controllers. The performance of the various

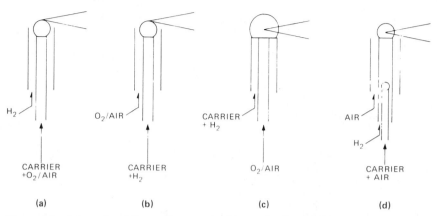

H_2 O_2/AIR CARRIER $+ H_2$ AIR H_2

CARRIER $+O_2/AIR$ CARRIER $+H_2$ O_2/AIR CARRIER $+ AIR$

(a)   (b)   (c)   (d)

Figure 2.3 Schematic of flame photometric detector jet flow configuration and viewing geometry. (a) Normal (Tracor, Bendix). (b) Reverse (Tracor, Hewlett-Packard, Perkin-Elmer). (c) Reverse double jet (Pye). (d) Double burner (Varian).

FPDs are compared in Table 2.3. Based on the detection limit and the selectivity, jet configuration 3c (Figure 2.3) appears to give the best performance as well as being the most economical in flame gases. The detection limit for phosphorus of the Bendix FPD, (jet configuration 3a) (80), is also low, but the sulfur limit is higher. This results in a poor S/P selectivity so that, in the S mode at low sample flows, the phosphorus response can be actually greater than that for sulfur.

The response of the FPD to phosphorus is linear, and for sulfur it should be a square function of sample flow rate due to the formation of the species S_2. However, in practice the exponent of log–log plots of the sulfur response can range from 1.6 to 2.0 (79) and is dependent on the oxygen/hydrogen ratio. The sensitivity, noise, linear range, and response exponent are also very dependent on the jet design and gas flows.

In an interlaboratory study the linearity and sensitivity of a number FPDs were compared using chlorpyrifos as a standard (66). Three operating conditions were studied: (1) optimum gas flows for configuration 3a (Figure 2.3) (80), (2) operator chosen flows for configuration 3a, and (3) type 3a with reversed air and hydrogen gas lines i.e., configuration 3b. The data are expressed in Table 2.4.

The results obtained are in accord with the values reported in the literature for type 3a and 3b detectors. The detection limit and linearity for phosphorus were rather better with operator chosen conditions than the preselected flows. This suggests that individual detectors of the same type differ in their response. The reversed configuration was only marginally inferior for sul-

TABLE 2.2. Optimal Gas Flow Rates (mL/min) for FPD in the P and S mode

Gas	Tracor (83) P	S	Bendix (80) P	S	Tracor (76) P	S	Pye (77, 78) P	S	Varian (70) P	S
Hydrogen	170	170	200	100	200	50	30	30	140	140
Carrier	80	80	40	40	60	60	60	60	30	30
Air 1	100	100	100	125	—	—	—	—	80	80
Air 2	—	—	—	—	50	50	20	24	170	170
Oxygen	20	20	—	—	20	10	—	—	—	—
Oxygen + carrier	120	120	140	165	130	120	80	84	280	280
Oxygen/hydrogen ratio	0.24	0.24	0.1	0.25	0.15	0.40	0.13	0.16	0.27	0.27

TABLE 2.3. Performance of FPDs to Phosphorus and Sulfur Compounds[a]

Detector	Tracor (83) 3a		Bendix (80) 3a		Tracor (76) 3b		Pye (77, 78) 3c		Varian (70) 3d	
Jet Configuration[b]	P	S	P	S	P	S	P	S	P	S
Response exponent	1.01	1.84	1.04	2.1	1.01	1.87	0.98	1.74	1.0	2.0
Detection limit (pg/sec)	1.7	35	0.2	50	2.4	20	0.2	4	0.5	50
Linear Range	3×10^4	5×10^2	5×10^4	2×10^2	10^4	10^3	5×10^4	5×10^7	5×10^4	5×10^2
Selectivity[c]										
X/C	$>10^5$	3000	—	—	—	—	10^6	10^4	$>5 \times 10^5$	10^3
P/S	500	—	1000	—	500	—	500	—	5000	—
S/P	—	50	—	0.1	—	50	—	5	—	10

[a] Low sample flows.
[b] See Figure 2.3.
[c] X = P or S.

75

TABLE 2.4.   Interlaboratory Survey of FPD Sensitivity and Linearity to Chlorpyrifos (66)

Conditions	Detection Limit[a] (pg X/sec)[b]		Linearity	
	P	S	P	S
Specifed flows	0.6	6.7	2500	100
Operator chosen flows	0.4	10.5	3500	50
Reverse configuration specified flows	0.75	3.3	1300	150

[a] Mean values $N = 6$.
[b] X = P or S.

fur compounds but offers the best performance for phosphorus. Taking into account its superior resistance to flameout, it is the preferred mode of operation.

2.5.3.   Sulfur Response

The interpretation of FPD response in the S mode is complex because of the following characteristics which result from the squared response function (77).

1.   Halving the amount injected or a doubling of the retention time decreases the peak height by a ¼ rather than the usual ½ for other detectors. This effect can be useful for confirming a response as being due to S. Sensitivity is greatest at short retention times.

2.   Areas of peaks must be evaluated using the peak width at ¼ peak height rather than half peak height. Linearization can be accomplished electronically (79) but requires careful calibration and does not change the other anomolous effects.

3.   In the S mode, the selectivity ratios, S/P and S/C, increase linearly with sample flow rate. Hence the relatively poor selectivities shown in Table 2.3 are improved by a factor of 100 at high sample flow rates, i.e., in the ng S/sec region.

4.   In the P mode, the P/S selectivity decreases in inverse ratio to the sulfur flow rate and the responses are equal at about 10 ng S and P/sec. Because of these selectivity differences, interference from S in the P mode is more likely to occur than from P in the S mode, unless working at S levels

near the detection limit. Positive responses in the P mode should always be checked in the S mode in case sulfur is interfering (74).

5. Presence of a constant low level of sulfur, as a result of contamination or by design, enhances sulfur detection limits and gives a moderate true linear range. Addition of 6 ng S/sec, as SO_2, to the flame improved the sensitivity and detection limits for aldicarb and malathion by a factor of 8 (81). A linear response was obtained for 1 to 100 ng of a sulfur analyte by using a CS_2 doped FPD (82). It was emphasized that internal standards should be employed with doped flames, since small changes in S background affect the sensitivity.

6. Sulfur compounds are readily adsorbed on active surfaces and can be quite difficult to chromatograph, resulting in nonlinearity at low levels.

2.5.4. Selectivity and Quenching

Direct interference from hydrocarbons due to CH and C_2 emissions is not a major problem in the P mode due to its high selectivity, although hydrocarbon fragments can directly quench HPO and S_2 emissions. It is manifested as negative peaks due to eluting hydrocarbons quenching a constant sulfur or phosphorus background (77). Sugiyama et al. (72) studied this problem and have shown that introduction of a variety of compounds containing carbon, hydrogen, and oxygen into the flame caused an exponential reduction in S_2 emission for a linear increase in carbon flow rate. At a sulfur flow of 1.25 ng S/sec, they found that a flow of 500 ng C/sec caused a 25% reduction in S_2 emission. A similar magnitude was predicted for the effect on P emission. Although quenching is a well-known phenomenon, this work indicates that it can occur when only microgram quantities of coextractives, or solvent tail, coelute with pesticide residues. The importance of adequate column resolution has been noted (83). The extent of quenching can be estimated by the use of fortified controls.

The double burner FPD was designed to reduce quenching effects (73). Because the sample has been oxidized in the first flame, the principal carbon species in the second flame is CO_2, which has a much lower quenching ability. It can be estimated from the work of Patterson (73) that a flow of 5 μg C/sec would be required to cause 25% quenching. Direct analyses of petroleum products for sulfur containing compounds are possible with the twin burner that give totally misleading results with a single burner system (73). Another advantage of the twin burner is that it provides uniformity of response for phosphorus and sulfur compounds of different structures. The response of FPDs with a single burner design does not accurately reflect their true phosphorus and sulfur content (74, 83). This may be due to self-quenching by the hydrocarbon moiety or different combustion characteristics. Although separation of the combustion and excitation flame processes minimizes these effects, it is also possible to modify a single burner FPD to produce a selective and uniform response (84).

2.5.5. FPD Response to Other Elements

Aue and Hastings (85) reported on the effects of removing the filters from a type 3a FPD (Figure 2.3). It produced an increase in both sensitivity and noise, which resulted in no improvement in detection limit. However, it did allow extremely sensitive detection of tin compounds (82) and, to a lesser extent, other elements such as Fe, As, Sb, Bi, and Pb. The detection of selenium and tellurium has also been reported using a 484-nm filter with a detection limit of 2 pg Se/sec and gave a squared response similar to sulfur (82, 86). The improved sulfur detection limits stemming from an increased sulfur background, is also applicable to selenium and tellurium (82). In addition, the quenching effects of added methane were noted (86). The FPD has also been made selective to chlorine. A type 3d burner (Figure 2.3) was used together with an electrically heated indium pellet between the two flames. The emission of InCl in the upper flame was monitored at 360 nm (87). The detection limit was 11 pg aldrin/sec (6 pg Cl/sec) and selectivity was much improved over a ^{3}H ECD in the analysis of crop extracts.

2.5.6. Summary

The high selectivity and reproducible operation of the FPD have made it the preferred detector for the analysis of organophosphorus pesticides as well as other P- or S-containing compounds. Provided the nature of the sulfur response and the possibility of quenching are kept in mind, it can give excellent results. The two-burner detector is an advance in technology that extends the usefulness of the FPD by reducing hydrocarbon quenching. Unfortunately, its sensitivity is not as high as other designs. The flow rate/response data (70) for the current version, indicate that improvements may be possible by further optimizing flame parameters.

The overall collection efficiency of the FPD is low, approximately only 1 photon is collected for every 10^8 atoms of P combusted based on a photomultiplier gain of 10^5. It is only the high, relatively noisefree, gain of the photomultiplier that gives the FPD sufficient sensitivity for residue analysis. It is possible that this efficiency can be further improved by the use of different excitation methods such as microwave radiation.

2.6. ALKALI FLAME IONIZATION DETECTOR (AFID)

2.6.1. AFID Design and Operation

This detector has been widely adopted by residue chemists for the detection of organophosphorus compounds and, more recently, for nitrogen-containing pesticides such as carbamates and triazines. Many variants of the original design of Karmen and Giuffrida (8) have been investigated and the early approaches have been reviewed by David (13) and Brazhnikov (88). The basic feature of all AFIDs is the introduction of an alkali salt into a relatively cool hydrogen flame or plasma. This results in a greatly increased sensitivity to phosphorus with suppression of the normal FID hydrocarbon response. Nitrogen, halogen, and sulfur-containing compounds also respond at levels 1–3 orders of magnitude less than do those containing phosphorus.

The mechanism of operation of the AFID is still poorly understood (89). Of the many mechanisms proposed, the most plausible involves the ionization of alkali atoms by combustion products with high electron affinities (90). The phosphorus response is postulated to be due to the PO_2 radical, while the nitrogen response is thought to be due to the CN radical. A different mechanism may be necessary to explain the weaker halogen and sulfur responses, which may be positive or negative (7, 91).

The many variables influencing AFID response include jet and collector geometry, means of introduction of alkali, type of salt, and flame gas flow rates. In order to obtain good sensitivity and reproducibility it is necessary to provide a flame environment containing a stable alkali concentration. Figure 2.4 schematically outlines the geometries of four widely used designs including the early modified FIDs with a salt-tipped jet or collector (Figure 2.4a,b. The more sophisticated 3-electrode (92) and heated bead (90) designs (Figure 2.4c,d) offer the advantages of reduced noise and higher reproducibility. Higher reproducibility is especially noted for the heated bead detector. The performance of the latter two configurations and that of the salt-tipped jet is summarized in Table 2.5. These data indicating the relative performance of the three designs is based on the literature values and those obtained in our laboratory. The chemical structure of the detected compound is known to affect the response. This fact, together with the lack of absolute response and noise data in much of the early literature, makes exact comparisons difficult.

The AFIDs can be subdivided into two basic types. Those where the flame performs the dual function of heating the alkali salt to provide alkali atoms in the flame, and combusts the sample to give the reactive species. Optimization of the detector geometry and gas flows may be a compromise

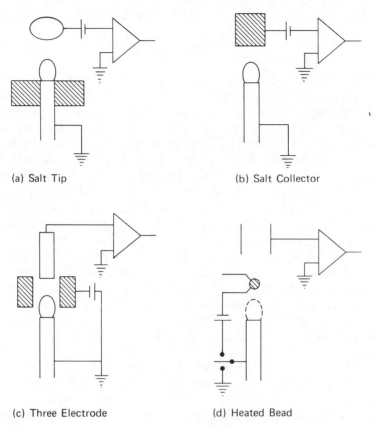

(a) Salt Tip   (b) Salt Collector

(c) Three Electrode   (d) Heated Bead

Figure 2.4   Schematic of alkali flame ionization detector configuraton. (*a*) Salt tip. (*b*) Salt collector. (*c*) Three electrode (Pye). (*d*) Heated bead (Perkin-Elmer, Hewett-Packard, Varian).

between these two requirements. More recently, these two functions have been separated by the use of an electrically heated bead.

2.6.2. Salt-Tipped Jet FID

This design has the advantage of simplicity and can be readily made by modifying standard FIDs (93, 94). Aue reviewed some of the factors influencing response with a salt-tipped jet FID (17). The hydrogen flow rate is the principal means of controlling the response of this type of detector. Increasing the hydrogen flow rate results in a corresponding increase in the sensitivity to phosphorus together with higher background current and noise levels. The collector height and physical condition of the salt surface also affects the response but to a lesser degree. Rubidium salts have gained wide

TABLE 2.5. Alkali Flame Ionization Detector Performance

Detector Source	Salt-Tipped Jet Rb_2SO_4 Pellet	3-Electrode RbCl Annulus	Heated Bead[a] Rb Silicate Glass
Background current (pA)	5×10^3	10^3	10
Noise (pA)	10	1	8
Sensitivity (Coul./g)	—	—	—
P	200	100	5
N	2	1	0.5
S	0.2(−)	0.5	n.d.
Cl	0.5(−)	(−)	0.002
Selectivity (P/C)	2×10^4	10^5	5×10^5
Detection limits (pg/sec)			
P	0.1	0.02	0.02
N	10	2	0.2
S	100	4	n.d.
Linearity	10^2–10^3	10^4	10^4–10^5

[a] Operating in the N/P mode.

81

acceptance, although cesium or potassium salts have also been used (95). The effect of the salt type is difficult to determine because the true alkali effects may be masked by the purity and physical nature of the salt.

Background currents are generally in the range to 10 nA, with noise levels of 10 to 100 pA. Although very high sensitivities (Table 2.5) can be obtained, the high noise level restricts the detection limits to 0.1 pg P/sec. In turn, this leads to MDQs for organophosphorus insecticides of the order of 1 pg (96) and for the nitrogen-containing triazine herbicides of about 100 pg (17). The detection limits are somewhat lower for commercial designs, where attention was paid to reducing the noise level (24, 97). The main problem associated with the salt-tipped design, even with well-regulated hydrogen flows, is that the salt tip deteriorates relatively rapidly, leading to poor reproducibility and high noise levels. The high salt consumption also necessitates frequent cleaning of the detector. The low hydrogen flows required for optimum nitrogen response, can result in the solvent peak extinguishing the flame.

2.6.3. Three-Electrode AFID

Three-electrode designs greatly improve AFID performance. Although the Pye design (Figure 2.4c) is the best documented, other geometries have been reported (98, 99). The alkali source is in the form of an annulus positioned above the jet and is negatively polarized.

Control of the rate of volatilization of the alkali salt is achieved either by varying the distance between the flame and the alkali source or changing the temperature of the flame. This results in lower and more reproducible concentrations of alkali halide in the flame, which reduces the background current and noise level. Hydrocarbon cations are collected more efficiently and this leads to higher selectivity (92). The negative ions from the alkali atom ionization reactions in the flame are collected on the rod electrode to provide the measured signal. In later models, to improve the response to nitrogen, the CsBr salt has been changed to RbCl and the jet bore enlarged (100). The effect of changing the salt seems mainly to result in a reduction of the noise level (98).

Several groups have examined the optimization of the Pye 3-electrode design (101–104). Optimization is a complex process with the various interactions of the parameters. It was established that different conditions are required for the maximum response to phosphorus, nitrogen, and sulfur (104). The highest sensitivity is obtained with the probe assembly (salt annulus electrode plus collector probe) as close to the jet as the design permits, together with high airflow ($\approx$350 ml/min) and the hydrogen flow optimized for each element. Phosphorus and sulfur generally require a hydrogen flow of about 34 ml/min; nitrogen requires a cooler flame obtained with flows of about 30 ml H_2/min. Operating the detector at high sensitivities, results in rapid evaporation of the salt annulus, which in turn leads to a reduced lifetime of the annulus and contamination of the detector. Adequate phosphorus sensitivity for residue analytical purposes can be obtained by operating at reduced hydrogen flows leading to lower background current and longer salt life. Variation of the probe assembly height is another approach to optimizing this detector, especially for nitrogen (104, 105). Increases in P/C, N/P, or S/P selectivity have been obtained by careful adjustment of the probe height and gas flows (104).

It can be seen in Table 2.5 that the sensitivity of the 3-electrode design is lower than that of the salt-tipped jet; however, a lower noise level results in better detection limits. The overwhelming advantages of this type of detector are greater stability of operation and ease of adjustment.

The stability of operation still does not compare with that of the FID. The absolute sensitivity of an AFID to nitrosamines varied by 20–30% over a 4-week period as compared with only 4% for an FID (106). In practice, by frequent analysis of standards, the reproducibility should approach that obtained on an FID. In the analysis of cruformate in a crude blood extract, a relative standard deviation of 6.2% was obtained operating at the nanogram level, compared to 6.8% for standards (95). The selectivity of the 3-electrode AFID against carbon response is dependent on concentration (92, 104) and detector optimization. When operated at maximum sensitivity to phosphorus, the solvent peak is negative followed by a positive tail. Microgram amounts of hydrocarbons give positive responses with a P/C selectivity of 10^4–10^5.

The sulfur response is also very dependent on concentration and detector optimization. This is reflected in a narrow linear range, at best 5×10^2, and a tendency to give split peaks with microgram quantities of sulfur compounds (104).

2.6.4.   Heated Alkali Source AFID

Some of the early AFID designs separated the combustion and salt vapour production aspects of the detector (107). However, the first commercial design to gain wide acceptance employing this advantageous concept was produced by Perkin-Elmer in 1974 (108, 109) (Figure 2.4d). The alkali source is a 1–2 mm diam glass bead incorporating rubidium, formed on a platinum wire, which can be heated electrically. The bead is normally positioned about 2 mm above the jet. In the so-called N/P mode, which provides maximum sensitivity to nitrogen and phosphorus compounds, the jet, bead, and detector housing are polarized at −180 V and a hydrogen flow rate of 2–4 ml/min is used. In the P mode, which is slightly less sensitive but more selective to phorphorus, the jet is grounded and higher hydrogen flows are used. Several other manufacturers now offer AFIDs with heated bead alkali sources (110–112) that give performances similar to the Perkin–Elmer design.

The Hewlett-Packard design differs somewhat in that the bead is attached to a negatively polarized collector (110), and therefore positively rather than negatively charged products are detected. The following discussion will concentrate on data for the Perkin-Elmer model, as little information is currently available on the optimization of the other models.

The two important parameters involved in operation of the heated bead AFID are the bead temperature and the hydrogen flow rate. The bead temperature is governed by the heat provided by the electrical current through the bead, together with heat from the hydrogen plasma. The background or bead current represents ionization of rubidium released from the bead, and indirectly is a measure of the bead temperature. In the N/P mode the negatively polarized bead and jet repel the negatively charged particles from the flame which are then measured at the collector electrode. Salt loss is minimal due to the return of Rb ions to the bead.

Response to nitrogen and phosphorus is a linear function of the bead current over the range 1–300 pA in both modes (109, 113, 114). At high bead currents the noise increases rapidly and the lifetime of the bead is reduced so that bead currents < 10 pA are generally employed. The best signal/noise ratio for nitrogen was obtained at 5–8 pA in the N/P mode (109).

In the P mode alkali emission is caused by the hydrogen flame burning at the grounded jet with a hydrogen flow of 30–35 ml/min. It is thus operating in a fashion similar to the 3-electrode AFID and bead currents are 50–100 pA (113). Optimum signal/noise for phosphorus using nitrogen or helium carrier gas occurs at 25–25 and 35–40 ml/min hydrogen flow rates, respectively. Sensitivities and detection limits are about a factor of 3 lower for phosphorus and 30 lower for nitrogen over the N/P mode (Table 2.5). The phosphorus response is more affected by the flame gas composition than is the nitrogen response and increased selectivity to P is obtained with higher hydrogen and lower air flows, although they increase the detection limit (113). It was observed that organochlorine compounds give positive responses at the nanogram level, in contrast to the salt-tipped jet and 3-electrode designs which give negative peaks. The P/C selectivity is better than 10^6 and linearity is 10^4 (108).

The effect of hydrogen flow rate on response in the N/P mode is complex and controversial. At the low flow rates used, a flame is not sustained at the jet but a hydrogen plasma is formed around the hot bead. The sensitivity for phosphorus compounds increases with hydrogen flow rate, but data from both the manufacturer and our laboratory show that increasing flow in the range 1–7 ml/min cause a reduction in nitrogen response with an optimum signal/noise ratio at 2–3 ml/min. The cooler flame favors nitrogen response but not phosphorus. In contrast, a study employing nitrogen carrier gas rather than helium indicated the nitrogen response increasing with hydrogen flow rate (114). As the sensitivities to nitrogen and phosphorus in the latter work were about 5 times lower than manufacturer's specifications, it is not clear whether the discrepancy in the N response is due to the type of carrier gas or to some other factor. The airflow rate at which the detector background is maximized for a given bead current and hydrogen flow rate is much lower using nitrogen rather than helium carrier gas (115), which was attributed to dilution and cooling effects. At optimal signal/noise, the ratio of sensitivities (N/P) is about 0.1 (Table 2.5). This is much higher than for other types of AFID, and combined with the very low noise level of this detector (approaching that for the FID) results in very good detection limits for nitrogen. Gas chromatograms obtained in this laboratory with a Perkin-Elmer heated bead detector are shown in Figure 2.5 for some organophosphorus and triazine compounds. The linearity of the detector to both nitrogen and phosphorus is better than 10^4 in the N/P mode (109).

The effect of bead composition has been the subject of two studies (116, 117). Low sodium and high rubidium content results in low noise, high sensitivity and stability characteristics. Cesium salts are not as effective as those of rubidium (16). The performance of 2.4–19.6% Rb, as the hydroxide, in quartz beads (117) showed that additives to lower the melting point of the bead are not necessary and may in fact be a cause of noise. Although the height of the bead above the jet would be expected to influence both the detector response and selectivity, no data have been reported on the effect of this parameter.

The bead current, and hence response, after an initial rapid drop, decreases only slowly with time. Excellent reproducibility can be obtained by increasing the heating current to maintain the bead current constant to 0.1 pA. Over a 6-hr period, the relative standard deviation for the response to 11.8 ng azobenzene was less than 1% (114). In the P mode, the reproducibility for 8 ng malathion injections over a 7-hr period was 4.6% (113).

2.6.5. Structure/Response Correlations

For compounds containing both nitrogen and phosphorus, the AFID response is mainly attributed to the phosphorus because of its greater sensitivity. However, the response of organophosphates is structure

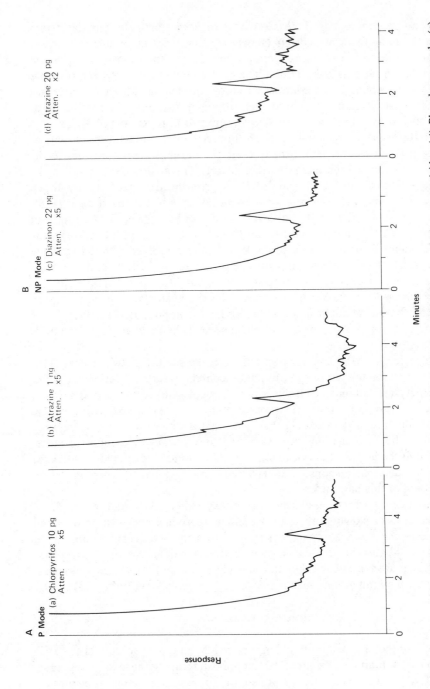

Figure 2.5 Response of heated bead alkali flame ionization detector to organophosphorus and nitrogen pesticides. (*A*) Phosphorus mode: (*a*) chlorpyrifos 10 pg (Atten × 5), (*b*) atrazine 1 ng (Atten × 5). (*B*) NP mode: (*a*) diazinon 22 pg (Atten × 5), (*b*) atrazine 20 pg (Atten × 2). GC conditions: column, 1 m × 6 mm, 4% SE-30/6% QF-1/Gas chrom Q; column flow 30 ml/min helium. Perkin-Elmer NPD.

dependent, and is particularly affected by substitution at the phosphorus atom. Data indicate the following trends (118):

1.   Oxygen analogues have relative response factors 2–4 times less than the corresponding phosphorothioates. Similar ratios are also observed with the FPD in the P mode.

2.   Phosphorodithioate response factors are 10–20% higher than the corresponding phosphorothioate.

3.   Heteroatom substitution in the ring of aryl phosphorothioates increases the relative response. Thus, chlorpyrifos and parathion give slightly higher molar responses than diethyl phenyl phosphorothioate.

4.   Phosphoramidates, with the nitrogen directly attached to phosphorus, have reduced response factors compared to the alkyl analogs.

For nitrogen compounds, the chemical structure has an even larger effect on response, which has been attributed to the CN radical mechanism (90). Nitrogen bonded to oxygen or to $C{=}O$ is relatively ineffective. For example, the amide propanil shows only about 50% the response on the AFID of the triazine herbicides (118) and vinylbarbital has only 5% the response of the tertiary amine nicotine (90). The following relative response order can be made from available data.

$$CH_3N(R)_2 > CH_3N{=}CH \cdot R > CH_3N(\overset{\overset{\displaystyle O}{\|}}{C}R)_2 >$$

$$Ar \cdot NO_2 > NH_2\overset{\overset{\displaystyle O}{\|}}{C} \cdot R > NH(\overset{\overset{\displaystyle O}{\|}}{C} \cdot R)_2$$

It is probable that relative molar responses are also sensitive to detector operating parameters. The difficulty of predicting responses on the AFID means standards are essential for detector calibration.

2.6.6.   Summary

The long-term stability of AFIDs is not as great as other flame detectors and certain precautions must be taken to obtain best performance. Gas flows to the detector should be carefully controlled with individual pressure regulation for each instrument. A flame stabilization period of about 1 hr should be sufficient for the 3-electrode and less for the bead-type detectors providing that the alkali source is in good condition. Overnight operation is not advocated as it shortens the life of the alkali source and increases detector contamination. Column bleeds, while not giving directly observable signals, may cause silica buildup on the alkali source. Only carefully conditioned materials should be used. Injections of chlorinated solvents or silylating reagents are not recommended with most AFIDs as the alkali source can be readily contaminated resulting in baseline instability. A recent AFID detector design reputedly overcomes this problem by switching off the heat to the bead during solvent elution (112). Rapid restabilization of the bead temperature is achieved by control of the current via the bead resistance rather than by voltage.

High noise levels are generally due to a dirty or depleted alkali source and/or buildup of salt deposits around the collector and its connections. The latter problem is a major disadvantage of the AFID as salts can cause leakage at the input to the high impedance electrometer amplifiers. In this respect, the low salt consumption of the bead designs is a great advantage. However, the operation of the bead detectors at subnanogram levels requires the use of the most sensitive range of the amplifiers, which is more liable to noise and drift problems.

Detector parts, except for the alkali source, can be cleaned with distilled water followed by acetone or isopropanol. Alkali sources can sometimes be rejuvenated by careful burnishing with dry cotton swabs to remove silica layers.

The AFID has a larger number of variables affecting performance than other detectors. With careful initial optimization using the S/N data to establish the best conditions, day-to-day variations can be accommodated by minor changes to the hydrogen flow or the current to the bead. The short term stability of the modern designs is now more than adequate for automation and they can be used with capillary columns (115).

The reproducibility of heated bead AFIDs is illustrated by two studies on the analysis of amphetamines (115, 110). Using internal standards, relative standards deviations for nanogram levels were 1.5% and 5%, respectively.

2.7. ELECTROCHEMICAL DETECTORS

Electrolytic conductivity and microcoulometric titration are two techniques that have gained acceptance for pesticide residue analysis. Selucky has reviewed their development up to 1972 (119), and more recently they have been discussed by Hall (20). The attractive features of these detectors are that nitrogen, sulfur, or halogens may be detected with a high degree of specificity. Under suitable conditions the response of these detectors is independent of structure. In the 1960s Coulson developed detectors for pesticide analysis based on electrolytic conductivity (CCD 10, 120) and coulometric (MCD 121, 122) principles. In 1976 Hall described a conductivity detector (HCD) based on a smaller pyrolyzer and a new cell design (123). Improved versions of these three electrochemical detectors are produced commercially.

The first stage of these detectors involves the controlled pyrolysis of components to produce specific products for electrochemical detection. The effluent gases may be passed through a trap (scrubber) to remove interferences and provide the desired selectivity.

For conductometric detection, the pyrolysis gases are swept into a gas–liquid separator where they contact a thin film of conductivity solvent. The solvent then flows through a conductivity cell containing 2 electrodes where changes in conductance produced by ionizable components are measured. Solutes are removed from solution by passage through an ion-exchange column. In the coulometric detector, the pyrolysis gases interact with an electrolyte and a charge generating system compensates electrochemically for any reaction, the quantity of charge required being a direct measure of the pyrolysis product.

2.7.1. Pyrolysis Conditions

The pyrolysis step is a common feature of the CCD, HCD, and MCD. The modes in which the pyrolysis reactor may be operated to form selected products are listed in Table 2.6. Reagent and purge gases are mixed with the column effluent prior to introduction into the pyrolysis reactor. The reductive mode uses hydrogen as reagent gas, while the oxidative mode uses oxygen. Pyrolysis in the absence of reagent gas is a further mode of operation. The reactor temperature, dimensions, and gas flow rates are generally a compromise between maximizing product formation and minimizing adsorption losses, tailing, and interferences.

Selective detection of nitrogen compounds is achieved in the reductive mode. A nickel catalyst is necessary for uniform production of ammonia, although some amines give good responses without catalyst (124, 125). Helium or hydrogen carrier gas is recommended, since nitrogen can give rise to some background ammonia production. The CCD reactor is a 4 mm i.d. $\times$ 280 mm quartz tube packed with fine-stranded nickel wire (126). The HCD uses a shorter pyrolyzer with either a nickel tube or a wire-packed quartz tube. Activation of the catalyst is important to achieve good efficiency. One method proposed involves oxidation in a flame followed by acid etching with final heat activation under hydrogen (127). A minimum reagent gas flow of 50 ml/min hydrogen is necessary (128), although a total

TABLE 2.6. Electrochemical Detector Reactor Products in the Reductive and Oxidative Modes

Element	Product	Comments
Reductive mode		
Halogens	HX	NH_3 may be removed with an acid scrubber
Sulfur	H_2S	Low conductivity
Nitrogen	NH_3	Catalyst required, H_2S and HX may be removed with an alkaline scrubber
Oxidative mode		
Halogens	HX	SO_2, SO_3 may be removed with a CaO scrubber
Sulfur	SO_2, SO_3	HX may be removed by an Ag scrubber
Carbon and nitrogen	SO_2, NO_2	Do not transfer well to solvent

reactor flow (carrier, reagent, and purge) of 150 ml/min is generally used to minimize peak tailing. The conversion efficiency increases with reactor temperature, an upper limit of 800–850°C being advised to reduce deterioration of the quartz. An alkaline scrubber is used to remove acidic halogen and sulfur compounds. Both the CCD and HCD generally use Sr $(OH)_2$ on quartz wool (126), while BaO on perlite is favored in the MCD furnace due to the higher stability of its acid salts (129). The scrubber is maintained at 350–450°C to minimize adsorption of ammonia and to decompose any NH_4Cl formed from compounds containing both nitrogen and chlorine. The efficiency of NH_3 formation in the catalytic reductive mode has been reported as 70–80% (126).

The conversion of halogen compounds to HX proceeds without catalysis under reducing, oxidative, or pyrolytic conditions (Table 2.6). The optimal conditions for the determination of pesticides with a CCD in the reductive Cl mode are furnace 850°C, hydrogen 10 ml/min, and a 0.5 min i.d. × 280 mm quartz reactor tube (130). Replacement of stainless steel and glass fittings in the transfer line to the conductivity cell by a Teflon insert showed that HCl, equivalent to 0.5 ng chlorinated pesticide was previously adsorbed (128). Teflon transfer lines also reduce adsorption losses of ammonia in the catalytic-reductive mode (126). At temperatures above 250–275°C Teflon decomposes and product loss by diffusion is also observed. An acidic scrubber may be used to remove interfering ammonia (131). Ionization from hydrogen sulfide and ammonia can also be suppressed with a nonaqueous electrolyte (132). The temperature dependence of hydrogen chloride formation is much greater for aromatic than for aliphatic chlorine and by maintaining the reactor at 710°C, a selectivity of 10^3 was obtained for organochlorine pesticides relative to polychlorinated biphenyls (130).

Oxidative reaction conditions have been investigated mainly for sulfur detection, although halogens and to a lesser degree nitrogen also respond. For the selective detection of sulfur, removal of hydrogen chloride with silver has been advocated (120), although silver scrubbers lead to some loss of sulfur compounds (133). Synthetic aluminosilicate (Al_2SiO_5) can also selectively remove HCl if kept at 650°C (119). The oxidative mode does not offer any advantages for halogen detection over the reductive mode, and it is subject to sulfur and carbon interference. Sulfur compounds can be removed by a CaO scrubber, but the oxygen should be saturated with water vapour to prevent $CaCl_2$ formation.

The oxidative or pyrolytic modes without scrubbers have been shown to be of use for the general screening of both organochlorine and phosphorothioates insecticide residues (133, 134). A furnace temperature of 850°C was judged optimum for furnace life with a reasonable sensitivity. Oxygen flows of 40 ml/min or above gave maximum response for sulfur compounds, which is a 20% increase over response in the pyrolytic mode. An increase in response up to 300% and greatly improved reproducibility was obtained for organochlorine compounds under oxidative conditions. The responses to chlorine and sulfur were affected by structure of compounds, including the presence of nitrogen. For compounds containing only nitrogen, the response is also very structure dependent. For example, prometryn showed the same sensitivity as malathion and was unaffected by oxygen flow (0–100 ml/min), whereas the sensitivity of azobenzene was reduced from 30 to 9 times by increasing oxygen flow (134).

2.7.2. Coulson Conductivity Detector (CCD)

The Coulson cell (Figure 2.6) is a combined separator and conductometric cell made of glass and using platinum electrodes. Conductivity is determined with a simple DC bridge system, generally at an applied voltage of 30 V (122). The conductivity (C) of a solution entering the cell at a given cell temperature and applied voltage is given by:

$$C = \frac{C_{sp}A}{d} \qquad (2.16)$$

where C_{sp} = specific conductivity
    A = electrode area
    d = electrode separation

The specific conductivity depends on the concentration of ions present and their equivalent ionic conductances. For a given element and cell, the detector response can be described by:

$$R_{CCD} = \frac{KME}{f} \qquad (2.17)$$

where K = proportionality constant

M = mass of element in compound eluting from column

E = efficiency factor for conversion in pyrolyser and transfer to solvent

f = solvent flow rate

The following solution reactions can occur for ammonia, hydrogen chloride, and sulfur oxides depending on the pH of the conductivity solvent:

$$NH_3 + H_3O^+ \rightarrow NH_4^+ + H_2O \qquad (2.18)$$

$$NH_3 + H_2O \rightarrow NH_4^+ + OH^- \qquad pK_b = 4.76 \qquad (2.19)$$

$$HCl + H_2O \rightarrow H_3O^+ + Cl^- \qquad (2.20)$$

$$SO_x + 3H_2O \rightarrow SO_{x+1}^{-2} + 2H_3O^+ \qquad (2.21)$$

The drop in conductivity caused by the neutralization reaction (2.18) is greater than the increase caused by the weak dissociation reaction (2.19) mainly due to the high equivalent ionic conductance of H_3O^+. Acidic solvents have been recommended for maximum sensitivity of nitrogen detection (123, 127). However, only a few ppm of acid can be tolerated in the solvent without producing excessive background. Thus, the dynamic range of the neutralization reaction is strictly limited and when exceeded, the response changes from positive to negative. The dissociation reaction is more reliable and is ensured by maintaining the solvent slightly alkaline (pH 7.5–8.0) with a regenerating bed consisting of basic resin (IRN-78) followed by mixed resin (IRN-150). In the absence of a scrubber, acidic products produce negative neutralization peaks. For the detection of hydrogen chloride and sulfur oxides by the dissociation reactions (2.20 and 2.21), the solvent should be maintained slightly acid (pH 6.0–6.5), by use of acidic resin (IRN-77) in front of the mixed resin.

The solubility of the reaction gases and the specific conductivity are affected by temperature. Maintaining the cell temperature constant at $<20°C$ is reported to improve sensitivity and stability (135, 136). In addi-

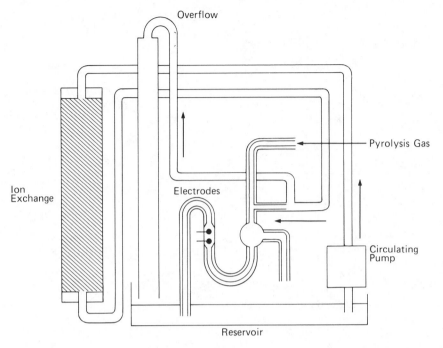

Figure 2.6   Schematic diagram of Coulson electrolytic conductivity detector.

tion, increasing the proportion of the liquid from the gas/liquid separator that is in contact with the electrode improves sensitivity. Decreasing the cell flow rate has the same effect but at the risk of increased tailing (135, 137). Higher bridge voltage of 60 V (131) or 100 V (138) also improves sensitivity, but the noise is also increased.

2.7.3.   Hall Conductivity Detector (HCD)

The salient features described by Hall (123) are a small cell volume and AC rather than DC conductivity measurements which overcome polarization problems. The conductivity cell is optimized according to equation 2.16 for maximum conductivity by increasing the electrode area and decreasing the electrode separation. The cell consists of two concentric electrodes ca. 1.7 mm i.d. separated by 0.2 mm. It is of Teflon and stainless steel construction and incorporates the gas–liquid separator. A more recent version of this cell (Figure 2.7) uses a second reference cell and allows differential bipolar pulse conductivity measurements (139). Circulation and regeneration of the solvent is similar to the CCD.

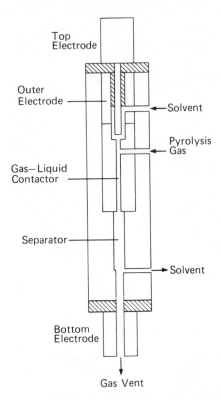

Figure 2.7 Schematic diagram of Hall electrolytic conductivity detector.

Work with the HCD has led to the adoption of alcohol based electrolyte solvents, which are claimed to result in reduced noise and higher selectivity. The solubility of hydrogen sulfide, ammonia and carbon dioxide are much less in alcohol as compared to water. Ethanol (123) or *n*-butanol (132) have been recommended for the reductive determination of chlorine and the oxidative determination of sulfur. However, alcohol reduces the sensitivity by a factor of 4 for HCl (123). An isopropanol:water mixture (15:85 v/v) has also been recommended for nitrogen determination in the catalytic reductive mode. It is claimed to be optimal for reducing noise and spikes, without too great a reduction in sensitivity (132). The effect of the solvent flow rate was studied for nitrogen determinations (140); both signal and noise increased steeply below 0.5 ml/min and the optimal signal/noise was at 0.5–0.7 ml/min. The HCD was 7 times more sensitive for nitrogen than the CCD, and 4 times more sensitive for the chlorine response in the oxidative mode.

2.7.4. Microcoulometric Detector (MCD)

The design and operation of the MCD has been reviewed (141). Its application to organonitrogen pesticides, in particular carbofuran, has been described in depth (129). Three types of cells are available depending on the elements to be detected. Each consists of a glass reservoir containing electrolyte into which the reactor effluent is bubbled. One reference, one sensing, and two generating electrodes form the electrochemical basis of the cell. Nitrogen analysis involves the hydrogen cell which contains an equilibrium concentration of H^+ for reaction with the base. The protons are replenished by the platinum black electrodes. The hydrogen cell is affected by CO_2, which must be efficiently scrubbed to obtain high selectivity. The halogen cell detects HX by precipitation with Ag^+. The sulfur cell uses oxidation of SO_2 by I_3^-, with NaN_3 added to counteract halides and strong oxidants. The MCD response is not affected by electrode geometry, unlike the conductivity detector, and the electrical charge generated is equivalent to the amount of reactant entering the cell.

Osmotic cells have also been investigated as electrochemical GC detectors and have been applied to the detection of traces of alcohols, aromatic hydrocarbons, and chlorinated solvents (142).

2.7.5. Electrochemical Detector Performance and Maintenance

Sensitivity data in the literature is often presented as the amount required to produce a half-scale deflection but without any indication of peak width or noise levels. This makes it difficult to compare these detectors on an absolute basis using the concept of detection limit. Unlike the ionization and flame detectors, losses in the reactor system must be taken into account, as well as baseline noise. Sensitivity, data linearity, and selectivity are presented in Table 2.7 for clean systems incorporating all significant modifications. Absorption losses in the reactor and transfer lines and noise resulting from contamination by crude samples made these limits (Table 2.7) difficult to maintain in practice. Detection of nitrogen is particularly affected by contaminated or oxidized catalyst and acidic residues in the pyrolyzer or transfer lines. For reproducible results these detectors require a minimum of 10 ng N per peak, i.e., approximately 30 ng atrazine. The linear range of electrochemical detectors is also largely determined by the reactor system and nitrogen is the most readily affected. The adsorption of traces of ammonia and incomplete conversion of large samples leads to sigmoid calibration curves. For chlorine or sulfur, the compact HCD performs rather better than the CCD or MCD (Table 2.7). The small reactor gives problems in the catalytic reductive mode, when early failure of the catalyst results in irreproducible reduction of nitrogen to ammonia. This compromises the higher ultimate sensitivity of the Hall cell and has led many laboratories to prefer the older CCD with its larger reactor.

On the limited data available the selectivity of the HCD is generally better than the CCD due to the use of alcohol based solvents (Table 2.7). The

TABLE 2.7. Electrochemical Detector Performance

Detector Electrolyte	Coulson Conductivity (water)	Hall Conductivity (alcohol based)	Micro- coulo- metric
Detection limits (pg/sec)			
N[a]	2	1	50
Cl[b]	5	0.5	50
S[c]	5	1	50
Linearity			
N	10^3	10^2–10^3	10^2
Cl	10^4	10^5	10^3
S	10^2	10^3	10^2
Selectivity			
N/C	10^5	5×10^5	
N/Cl	2×10^4	2×10^4	
Cl/C	$> 10^6$	$> 10^6$	
Cl/N	10–10^3	10^4	
S/C	10^2	10^4	
S/N	10–10^3	10	

[a] Atrazine in catalytic reduction mode ($Sr(OH)_2$ scrubber).
[b] Lindane in reductive mode.
[c] Parathion in oxidative mode.

selectivity against carbon compounds is very high in the reductive modes, but not as good in the oxidative mode, especially with aqueous conductivity solvents. The CCD is rather more sensitive to nitrogen interference in the reductive halogen or oxidative sulfur modes than the HCD.

Use of the MCD as a selective GC detector appears to be decreasing compared to that of the conductivity detectors. The detection limits are at least a factor of 10 higher and the cell and associated electronics are more complex. The large cell volume results in a 5-sec time constant, which makes close GC peaks difficult to monitor.

None of the electrochemical detectors are suitable for capillary column GC because of their relatively long time constants. They also require solvent venting because large samples cause coking of the reactor tube and the acids produced can rapidly overload the scrubber and electrolyte. Postcolumn in-line vent valves mounted in the column oven have been recommended (143), but they can lead to sample absorption and leakage problems. It is preferable to ensure sufficient detector back pressure, so that a simple T to a valve vented to the atmosphere may be used (144). This provides minimum dead volume, although the detector baseline will be affected during venting. Some hydrogen must continue to flow over the catalyst during venting to prevent its oxidation.

Electrochemical detectors are more liable to contamination and other problems than the simple flame detectors. They are best dedicated to routine applications where high sensitivity is not required and where they can receive regular preventative maintenance (145). The following points apply to obtaining best performance from the CCD or HCD in the catalytic-reductive mode, but many are also relevant to the MCD and to other furnace modes:

1. Every 2 to 4 months clean the reactor and cell, reactivate the catalyst, and replace the scrubber and mixed resin bed.

2. Avoid fluorinated stationary phases such as QF-1 or OV-210 and chlorinated solvents. Halogenated bleed and unvented solvent tail can quickly saturate the alkaline scrubber.

3. Avoid cyano-silicone stationary phases such as OV-225, which cause elevated ammonia background.

4. Vent as long as possible after injection and also while on standby to minimize reactor contamination by solvent, coextractives, and column bleed.

5. Use high-grade gases and ensure a leak free GC system. Allow sufficient time for reactor cooling before disassembly as the hot catalyst is rapidly deactivated by oxidation.

6. Use high quality deionized water when replenishing the electrolyte.

7. Soxhlet extract the regenerating resins with methanol followed by water before use.

8. Run the detector continuously for highest stability.

2.8. MASS SPECTROMETRY (GC/MS)

The role of GC/MS in pesticide residue analysis is mainly qualitative, such as confirming the identification of pesticide residues and their degradation products. The ability of GC/MS to provide selective detection at subnanogram levels of any compound that can be chromatographed is a very attractive one. The cost of instrumentation is quite high and the use of GC/MS for routine residue analysis is not usually justified. However, in some cases the technique can be applied to obtain quantitative results more rapidly than by other methods.

The GC/MS literature is vast and only the main features of the technique as applied to residue confirmation and identification will be covered.

Several reviews (146–149) and books (150–152) provide useful information on the technology and applications of GC/MS to residue analysis. A book concerned with the analysis of water pollutants contains several interesting papers showing advanced applications of the technique to organic residue analysis (153). The proceedings have been published in an ACS symposium on MS and NMR in pesticide chemistry (154) and two short reviews on pesticide GC/MS have appeared (155, 156). The mass spectra of a variety of model organophosphorus compounds have been interpreted in some detail (157).

2.8.1. Ion Source

The principal components of GC/MS systems are outlined in Table 2.8. Jet separators have proved to be the most satisfactory interface for packed columns, while direct coupling of capillary columns is rapidly gaining favor because of high resolution and maximum sample transfer with minimum sample loss or degradation (158, 159). Electron impact (EI) is the favored ionization method as it usually provides the most structural information. Large libraries of standard spectra are available with the NIH/EPA collection of 32,000 checked spectra being the most important (160). Three other collections specialize in spectra of pesticides and pollutants (161–163). A useful summary of the principal ions in spectra of a variety of pesticide and pollutant compound classes has been prepared (164).

Chemical ionization (CI) has recently been reviewed (165). It is a valuable ancillary technique, as many compounds do not give molecular ions under EI ionization. Strong MH (M + H) ions are generally observed with isobutane reagent gas, which are useful for quantitative purposes. Some fragmentation occurs, especially with the more energetic methane reagent gas. The CI (isobutane) spectra for 23 organophosphorus compounds have been reported (166).

Negative ion chemical ionization shows some promise for pollutant detection as many electron capturing compounds, such as the organochlorines give higher sensitivities than in the positive CI or EI modes (167, 168). Several mechanisms of ionization appear to be involved. Under some conditions chlorinated compounds are ionized by charge transfer from Cl^-. This reaction, being second order with respect to the sample, may be of limited quantitative value. Another related technique is atmospheric pressure ionization (API) where the effluent from a ^{63}Ni electron capture cell, or a Townsend discharge, is directed into a quadrupole analyzer (169, 170). Both negative and positive ions can be examined. Very high sensitivities can be obtained and application to environmental problems are now appearing in the literature (148). The negative ion mass spectra of some pesticides has

TABLE 2.8. GC/MS Configurations

Component	Type	Comments
Interface	Separator	
	Jet	Rugged, low dead volume, good yield
	Silicone membrane	High enrichment, fragile
	Fritted glass effusion	Reliable
	Direct coupling of capillary columns	Requires >200 l/sec pumping at source
Ion source	Electron impact 20–70 eV electrons	Extensive fragmentation
		at 1 torr MH$^+$ usually predominates
	Chemical ionization	Requires >200 l/sec pumping at source
	Ion/molecule reaction using ions produced from EI of reagent gas (CH$_4$ or i-C$_4$H$_{10}$	
Chromatogram monitoring	Total ion current (TIC)	Affected by fields and carrier gas ionization
	Preanalyzer electrode	Difficult to implement on magnetic instruments
	Fast integration of output over selected mass range	Recommended, but TIC gives more information on interface and MS performance
	Split to FID	
Mass separation	Magnetic sector	3–8 kV Accelerating voltage
		Resolution to >10,000 with electric sector
	Quadrupole	10–20 V accelerating voltage
		Unit mass resolution, fast scan rate
Ion detector	Electron multiplier	Gain 10^5–10^6
Output recording	Potentiometric recorder	Limited to chromatogram monitoring by slow response
	Oscillographic recorder	Direct spectral output, mass marking optional
	Daya system	Data acquisition using analog to digital conversion or ion counting
		Maximum versatility for spectral storage, retrieval, background subtraction, and file searching.

been reported (171) and the spectra of 16 organophosphorus insecticides obtained with different ionization methods has been compared (172).

Quadrupole mass analyzers are widely used for GC/MS. The reproducible mass scale allows ready focusing on chosen masses, but ion transmission is variable and often results in poor sensitivity at high mass. The high scan rates possible are useful for capillary GC/MS and selected ion monitoring. Data systems control spectral scanning by stepping the quadrupole through the mass range rather than in a sweep. Magnetic sector instruments cannot be scanned as fast but provide more reliable transmission and higher resolution. Mass focusing is not reproducible due to magnet hystersis, making computer control difficult unless the field is controlled via a Hall probe (173).

2.8.2. Operation Modes

There are several modes in which a GC/MS may be employed (147). For the identification of a few components present at 0.1–1 mg/ml, the spectrometer may be scanned over the entire mass range to give individual spectra as each component emerges, using the total ion current (TIC) as a monitor. Background spectra can be subtracted from the sample spectra with a data system to remove peaks due to column bleed and other contaminants. The TIC is a nonselective monitor which produces a chromatogram similar to an FID. This may result in difficulties when attempting to match peaks with those observed in a gas chromatogram from a selective detector. The postcolumn "plumbing" of a GC/MS is also more extensive than a normal GC, which may cause problems of sample adsorption and decomposition at ng levels.

For complex mixtures or where the component concentration is <0.1 mg/ml, it is preferable to acquire sequential data from repetitive scans using a data system. The scan rate should be sufficient to obtain several spectra during the elution time of each GC peak. The stored data can then be processed in a number of ways. A reconstructed TIC chromatogram can be produced by plotting the sum of intensities of all peaks vs scan number. Figure 2.8 shows such a plot for a GC/MS analysis of a freshwater extract estimated to contain 46 μg/liter fenitrothion. It is seen that the reconstructed TIC plot is not a selective or sensitive method of presenting the data. Plots of the abundance of specific ions vs. scan number (mass chromatograms) are invaluable for assessing where in a chromatogram particular components are eluting. Plots for m/z 109 and m/z 125 are also shown in Figure 2.8 (b and c, respectively). These ions are important fragments in dimethyl phosphorothioates spectra and show clearly where the fenitrothion elutes. A reasonable spectrum for fenitrothion was obtained after background subtraction even though only a few ng were injected.

Water Extract
Progmd: 175–200 Deg (Se 30/OV210)

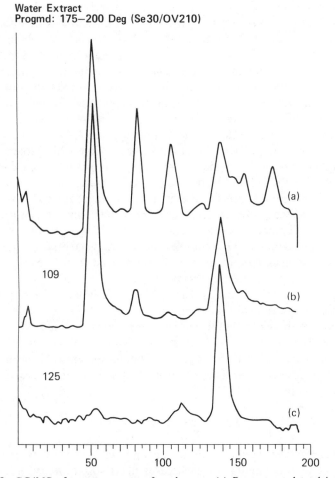

Figure 2.8 GC/MS of a water extract after clean-up. (*a*) Reconstructed total ion current chromatogram. (*b*) Mass chromatogram *m/z* 109. (*c*) Mass chromatogram *m/z* 125.

Mass chromatograms of isotope peaks allow determination of the number of hetero atoms, e.g., chlorine, present in the compound providing an even more positive assignment. A plot of the sum of intensities for a set of characteristic ions (when all are present) is even more specific (166). Peak areas from such plots can be useful for quantitative purposes, as other contributions from coeluting components or background are greatly reduced.

Another mode of acquiring mass-spectral data is by selected ion monitoring (SIM) where the ion current is measured at selected masses. Although the information content is reduced, there is a large gain in sensitivity. A scanned mass spectral peak typically is 1 msec wide and an amplifier bandwidth of ca. 3 kHz must be used. If only one peak is monitored then the bandwidth can be reduced to that which will adequately follow the chromatographic profile (ca. 3 Hz). A stepped quadrupole can similarly use an increased peak integration time. This increases the signal/noise by factors of up to 1000 depending on the number of peaks monitored. For example, analysis of chlorinated biphenyls gave a 50-fold increase in signal/noise by changing from a complete 40–400 amu scan at 11 msec/peak to a 7 mass SIM scan at 540 msec/peak (156).

The real time output from a SIM run has a time scale rather than the spectrum number scale of mass chromatograms plotted from continuous scan data. The GC/MS-SIM peak areas are very suitable for quantitative analysis. The linear range for the analysis of chlordane by SIM was 10^4 (155). SIM is the preferred technique for quantitating or confirming the presence of components present at less than 10 ng/injection, due to the improved signal/noise. For complex mixtures, sequential scanning with storage of complete spectra is preferable, provided there is enough sample, as this provides the greatest information.

Mass spectrometry allows the use of isotopically labeled internal standards which are chemically identical to the sample but separable by mass. This is an advantage for labile compounds as extraction and GC losses can be accurately compensated. Deuterium labeled compounds are widely used in the biomedical field (149). The most important use of labeled internal standards in pesticide chemistry so far is ^{37}Cl for the quantitative determination of chlorodioxin (TCDD) (12).

2.8.3.   GC/MS Sensitivity

Quantitative residue analysis at the ng/kg level extends existing analytical techniques to their limits. In such cases mass spectrometry becomes the method of choice. It is, therefore, useful to discuss the parameters of sensitivity and selectivity as applied to MS. Sensitivity or response factor (RF) is defined as the total charge/gram of substance that would be collected if all ion masses were monitored simultaneously. In practice, it is measured by taking the integrated ion current at the electron multiplier for a specific ion mass following the introduction into the ion source of a known quantity of reference compound. This is transformed into the total ionization per gram by using the known abundance of the measured ion in the spectrum of the compound. EI is a nonselective ionization method, hence the RF is relatively independent of compound type. To date, most manufacturers and users do not employ this parameter to describe the performance of their instruments. Values for RF of 0.03 C/g (methyl stearate) are readily obtainable on modern instruments at low resolution.

The following calculation shows how the RF may be used to determine response under various operating conditions. The charge, Cm, collected at a given ion mass m in the spectrum is determined by the sample flow rate, the relative abundance of the ion and the time spent on that mass.

$$Cm = \frac{fYRFTmt}{100} \qquad (2.22)$$

where f = column sample flow rate
     Y = separator yield
    Tm = abundance of m as % of total ionization
     t = half peak width for each ion peak or integration time

Thus, for a sample flow from the GC of 0.1 ng/sec and a separator yield of 33% with a scan rate of 3 msec/peak

$$Cm = 3 \times 10^{-17} Tm \text{ C} \qquad (2.23)$$

$$= 180 Tm \text{ ions}$$

Generally, large peaks in EI spectra are approximately 10% of the total ionization, whereas small peaks may be only about 0.1%. The sensitivity of CI is about 20% of EI, but the large peaks carry more than 50% of the total ionization. Thus, in the above example with a column, sample flow of 0.1 ng/sec would give EI spectra with large and small peaks containing 1800 and 18 ions, respectively. The major peaks in the CI spectra would also contain 1800 ions.

The observed signal for MS peaks is the result of amplification by the electron multiplier whose gain (10^5–10^7) is sufficient such that the noise on the peaks is determined by ion statistics rather than the baseline electrometer noise. Thus, the concept of detection limit is different from other detectors, as well as being dependent on which peak in the spectrum is used for the definition. Counting statistics give $1/\sqrt{Cm}$ as the coefficient of variation for the mean intensity of a peak containing Cm ions. The actual noise measured on a peak depends on the bandwidth of the amplifying system but $1/\sqrt{Cm} \approx 0.25$ will approximate to the definition for detection limit as a signal of twice the noise. Hence, the MS detection limit is 10–20 ions. Identification from a scanned spectrum, whether by file searching or by manual interpretation, requires small peaks to be significant and large peaks to have 5% relative intensity accuracy. In the example above the peaks at 0.1% of the total ionization contain 18 ions, and therefore a sample flow rate of 0.1 ng/sec is the detection limit. This represents a GC/MS MDQ of 1 ng for a 10 sec GC peak half-width. Peaks of 10% relative abundance will have a 2.5% coefficient of variation. Capillary column GC/MS requires no separator and for 2 sec GC peaks scanned at 1 msec/amu the MDQ becomes 0.2 ng. Commercial systems will yield spectra at these limits, but column and ion-source contamination can make it difficult to distinguish the spectra from background.

In the SIM mode with large peaks monitored at 500 msec integration time, the detection limit is reduced to 0.006 pg/sec or a MDQ of 0.06 pg on a packed column in the above calculation. These limits have been approached in practice using standards (175). Prime requirements for ultimate GC/MS sensitivity are clean GC and vacuum systems, but SIM limits are routinely 10–100 × higher due to background at the monitored masses contributing to the noise. Thus GC/MS detection limits in the SIM mode are similar to those for the ECD or the AFID, but analysis is not limited to those compounds having strong electron capture properties or containing phosphorus.

2.8.4. GC/MS Selectivity

The versatility of the mass-spectrometer as a detector is based on the nonspecific ionization methods. Selectivity derives from the mass-separation process and the degree of selectivity depends on the contribution from contaminants to the ionization for the major ion masses of the sample. The selectivity in GC/MS is enchanced by the additional resolving power of the chromatographic system. Contamination in GC/MS arises from coeluting extractives as well as the more constant instrument background and column bleed. As with other detectors, coextractives are generally the factor limiting the sensitivity of an assay and determine the degree of sample cleanup required to lower the analytical detection limit. Capillary columns assist by improving the separation of components and increasing sample flow rate to the source.

Prominent peaks in the spectra of common contaminants are listed in Table 2.9. The ion-masses and their isotope peaks should be avoided in any SIM measurements. Stationary phases other than OV-1, OV-17, Dexsil, or their equivalents generally have too great a bleed to be satisfactory for GC/MS above 200°C unless reduced detection limits can be tolerated. The choice of even numbered, high mass peaks for monitoring reduces the chance of interference. Heptafluorobutyrates are preferable to acetates as derivatives for alcohols or amines due to their higher mass. The selectivity of low resolution GC/MS in the SIM mode is generally inferior to the FPD, and column cleanup is necessary to analyze natural matrices for residues present below about 1 mg/kg.

Magnetic sector instruments can be operated at increased resolution for improved selectivity. The sensitivity is reduced, but a lowering of the

TABLE 2.9. Mass Spectra of Common Major Contaminants[a]

Compounds	Source	Peaks[b]
Silicones	Column and septum	73, 147, 207, 221, 281, 295, 351, 367, 429, 503
	TMS reagent polymers	
Poly phenyl ethers	Diffusion pump oil	77, 115, 139, 141, 149, 168, 260, 354, 446
Hydrocarbons	Ubiquitous	C_nH_{2n-1} and C_nH_{2n+1} ion series
Dibutyladipate	Plasticizer	100, 101, 111, 129, 185, 203
Phthalates	Plasticizers, pump oil	149, 167, 279
Tributyl phosphate	Plasticizer	99, 155, 211
Fatty acid methyl compounds	Methylated natural extracts	74, 87, $(59 + C_nH_{2n})$ ion series

[a] Masses of major ions only.
[b] In order of mass.

analytical limit usually results from the rejection of contaminant ions. A resolution of 2000–3000 is sufficient to resolve mass positive carbon/ hydrogen/oxygen ions from mass deficient chlorinated hydrocarbon or silicone ions. Similarly, isobaric ions containing different numbers of chlorine atoms can often be resolved. In the analysis of tetrachlorodioxin in tissue and milk, a resolution of 12,000 provided sufficient sensitivity and selectivity for assays at 100 ng/kg in the presence of μg-mg/kg levels of DDE and PCBs (176). Lack of stability and sensitivity at high resolution was overcome by the use of narrow mass scans ($\pm$1–2 amu) combined with signal averaging techniques.

2.8.5. Confirmation of Residues

Mass spectrometry has been cited as the ideal technique for providing confirmation of residues detected by other methods (177). Adequate confirmation requires that at least 3 ions should be measured, either acquired via scanning GC/MS and presented as mass chromatograms, or via SIM measurements (178). The ions should have the same retention time and intensity ratios (to $\pm$5%) as a standard sample. For compounds with less than 3 characteristic ions, CI in both positive and negative ion modes are recommended alternatives. Sphon points out the dangers in improving marginal spectra by data system manipulations such as background subtraction, where background peaks are much stronger than those due to sample (178). Confirmation via spectral file searches give objective matches but poor correlations sometimes arise when sample and file spectra were taken on different instruments.

Mass spectrometry is currently a very active field of research and it is expected that future advances will further increase its usefulness for the analysis of low levels of pesticides and pollutants. API and negative ion CI techniques show great promise for improvements in the sensitivity and selectivity of GC/MS analysis for certain compound classes.

2.9. MISCELLANEOUS SELECTIVE DETECTORS

2.9.1. Infrared Spectroscopy (IR)

IR has been used as an adjunct to GC for many years, but stopped flow or trapping systems were necessary to obtain reasonable sensitivity. The recent development of special purpose ir detectors for GC with heated microcells and of simple construction, has led to a renewed interest in the technique for selective detection (179, 180). The wavelength region to be monitored is selected by the choice of the appropriate monochromator or filter. Carbonyl compounds are detected around 1720 cm^{-1} and hydroxyl compounds around 3600 cm^{-1}. The MDQ for the GC/ir of carbonyl compounds is about 5–10 ng (172).

The use of Fourier transform (FT) techniques has led to sophisticated FT-ir instruments incorporating data systems (181, 182). These are capable of acquiring complete spectra in less than 1 sec at 4 cm^{-1} resolution and simultaneously monitoring the absorption over several wavelength ranges. Facilities, such as averaging and background subtraction give versatility similar to GC/MS. Less than 100 ng is required for a complete spectrum of small molecules (182). The detection limit for low molecular weight carbonyl compounds monitored at 1760–1680 cm^{-1} is about 10 ng. Mirex and derivatives gave adequate spectra from a few micrograms of each component (182). The environmental applications of FT-ir have been reviewed (183).

As the sensitivity of GC/ir is at least an order of magnitude less than for GC/MS, it is unlikely to replace GC/MS as a primary identification technique for residues or metabolites. However, the group characteristics of ir spectra provide complementary information to MS, which is of value when analysing unknown samples. Inexpensive fixed wavelength ir detectors may be useful for determining biologically active compounds such as the long chain acetate pheromones that are difficult to selectively detect except by SIM-GC/MS.

2.9.2.   Microwave Plasma Emission Detector (MPED)

This technique is similar in principle to the FPD, but can be applied to almost any element. A helium or argon microwave discharge is used to decompose compounds and excite the component atoms (184–186). A grating spectrometer is employed to separate the atomic emission lines and the signal is amplified with a photomultipler. A commercial version is available that monitors several wavelengths simultaneously.

Helium is the preferred plasma gas as all elements exhibit line spectra, whereas in argon some of the lighter elements only give diatomic band spectra. Early designs could only sustain a microwave discharge in helium at reduced pressures (5–50 torr), which reduced sensitivity and complicated the detector design. Recent improvements in microwave resonant cavity design have resulted in MPEDs which operate satisfactorily with helium at atmospheric pressure (187–189). These detectors also have the advantage of axial viewing, which avoids the problem of reduced transmission of the quartz walls encountered in transverse geometries. Wall contamination can be minimized, but not eliminated, by the addition of about 1% of oxygen or nitrogen to the plasma gas (185). The selectivity of the MPED is dependent on the degree of separation of the measured line from carbon emission lines and is determined by the relative positions of the lines and resolution of the spectrometer. The selectivity to lead was improved by the use of wavelength modulation combined with phase-sensitive detection techniques (190).

MPED data for some of the elements is presented in Table 2.10 as the wavelength, detection limit, and selectivity to carbon. The linearity for most elements is 10^4 (189, 191). The detection limits for the halogens are higher than those for the ECD or HCD but the MPED can differentiate between the various halogens atoms. Similarly, the MPED is not as sensitive as the FPD for phosphorus and sulfur compounds. Bache and Lisk demonstrated the potential of the MPED when they analyzed plants and soil for phorate, disulfoton, bromoxynil and endosulfan at the 0.2 mg/kg level (186). The detection limits for nitrogen and oxygen with the more recent models are likely to be lower than those reported in Table 2.10 (189). The principal limitation to the use of the MPED in the residue analysis of nitrogen compounds is the difficulty of eliminating traces of atmospheric nitrogen from the GC system. The high sensitivity and selectivity of the MPED toward some of the heavier elements has led to its use in GC methods for volatile compounds of Mn (190), Hg (192), Se (191), As (193), and Pb (194). The atomic response is independent of molecular structure and can be used to calculate atomic ratios in compounds (185).

The MPED is the most versatile of the GC detectors although opinions differ on the ease of operation and its reliability. A wide range of elements can be measured at subnanogram levels with high selectivity and

TABLE 2.10. Performance of the Microwave Plasma Emission Detector

Element	Wavelength (nm)	Detection Limit (pg/sec)	Selectivity[a]	Ref.
H	486.1	30	n.d.[b]	186
D	656.1	90	n.d.	186
C	247.9	80	—	186
N	746.9	2900	n.d.	186
O	777.2	3000	n.d.	186
P	253.6	2	2×10^4	190
S	545.4	63	250	190
F	685.6	9	3500	190
Cl	481.0	16	2400	190
Br	470.5	10	1400	190
I	206.1	31	1100	190
Si	251.6	29	3900	190
Se	204.0	2	10^4	192
As	228.8	1	2×10^6	194
Mn	257.6	0.25	2×10^6	190
Hg	253.7	0.05	10^4	193
Pb	283.3	0.50	10^6	190

[a] Selectivity to carbon as a hydrocarbon.
[b] Not determined.

simultaneous determination of several elements is possible on the more complex instruments. Applications to residue analysis are likely to increase, particularly for the less expensive single channel instruments. The MPED is particularly well suited to the analysis of the heavier elements.

2.9.3.   Atomic Absorption Spectroscopy

The standard atomic absorption spectrometer (AA) can be used as a selective GC detector either with flame or carbon furnace excitation. It gives nanogram sensitivity to many elements under suitable conditions and is of extremely high selectivity. Methods using GC/AA have been developed for tetraalkyl lead compounds (195) and volatile As, Se, and Sn compounds (196). Various metals such as Cr, Cu, and Co can also be derivatized and determined as volatile chelates (197). The principal problem for metal analysis is in the formation of complexes that will chromatograph satisfactorily at low levels (63).

2.9.4. Photoionization Detector (PID)

Although the principles of the technique are well established, it required the development of a sealed uv source (198) to overcome the practical problems of the earlier flow discharge designs. Photons ionize the GC effluent and the current produced between electrodes is measured as in the FID. With a 10.2 eV uv source, most molecules are ionized with the exception of the permanent gases, C_1–C_4 paraffins, methanol, acetonitrile, and the chloromethanes. The sensitivity for benzene is 0.3 C/g, about 30 times that for the FID and as the noise level is similar (4×10^{-14} A), the detection limit is lower and approaches that of the ECD. The linearity of the PID is 10^7 (19). The lack of a solvent response for the above compounds should help detection limits in practical analyses but, as with the FID, the almost universal response to larger molecules is likely to cause problems from other interferences. Temperature-programmed separations are more affected by column bleed as it has a high response to silicon compounds, unlike the FID.

Chromatograms of nanogram amounts of sulfur-based pesticides have been published (199) for pure standards. It is difficult to assess the value of the PID to residue analysis of natural matrices. The work of Oyler et al. (200) illustrates a possible role for the PID. Polynuclear aromatic hydrocarbons (PAHs) were concentrated and separated on a reversed phase HPLC packing and CH_3CN/H_2O fractions analyzed directly by GC/PID with a MDQ of 0.1 ng PAH. The separating power of HPLC and the lack of solvent response of the PID combined well. With a 9.5 eV uv source, a degree of selectivity is obtained as compounds with ionization potentials below about 9.7 eV are not ionized (199). The response to hydrocarbons is eliminated whilst that to aromatics, mercaptans and amines remains high.

2.9.5. Chemiluminescent Detector (Thermal Energy Analyzer)

The TEA detector was originally developed to provide sensitive detection of nitric oxide (201), but it has been applied in the residue field for the detection of nitrosamines (202, 203). A catalytic pyrolyzer is used to cleave the N—NO bond and the NO is swept into a cell containing ozone. The reaction of NO with the O_3 produces excited NO_2, which emits light in the near-infrared for detection. The light is measured by a photomultiplier after passing through a filter. The several stages involved in the detection process result in a very specific response. The selectivity over most amines and nitrocompounds is $> 10^3$ (202). However, ethylene glycol dinitrate and C-nitroso compounds respond and false positive results from these and other sources must be eliminated by confirmatory methods. The sensitivity is high with a MDQ of 0.5 ng for the lower alkyl nitrosamines in food extracts corresponding to residues in the sample of 0.1 μg/kg (204, 205). The linearity is greater than 10^3. The short term reproducibility of response for 6 nitrosamines was 3.4–10.9% (204). Selectivity for the analysis of fish extracts was found to be superior to the AFID or CCD (204).

2.10. SELECTION OF DETECTORS FOR PESTICIDE RESIDUE ANALYSIS

The performance of the ECD, FPD, AFID, and electrolytic conductivity detector is summarized in Table 2.11. These four selective detectors are readily available and cover the requirements for the GC analysis of the majority of pesticides and pollutants. The data presented are based on the best current technology, e.g., constant current linearized ECD and heated bead AFID. No one detector can fulfill all residue requirements and often there is a choice between several detectors for a given application. The following sections discuss the choice of detectors based on the compound class and mention some of the derivatives that can be used to change selectivity.

2.10.1. Organochlorine Insecticides and Polyhalogenated Aromatic Compounds

The ECD remains the prime detector of choice for the GC analysis of chlorinated insecticides of the DDT, cyclodiene, lindane and toxaphene classes. Routine column chromatographic cleanup procedures are available that readily extend the analytical limits to 1 μg/kg in plant and animal tissues or soil (206). The principal interferences arise from environmental pollutants such as the PCBs or polybrominated biphenyls (PBBs), which are usually present as a complex mixture (207). Column chromatography, capillary column GC (41, 208) and perchlorination (209) are practical approaches to separating PCBs from insecticides, but more selective detection is often still required. Operation of the HCD in the chlorine mode with a reduced pyrolyzer temperature (130) is a useful method of discriminating against PCBs and has found application in the analysis of mirex residues (210). GC/MS is the only other detector which can provide selective detection amongst polyhalogenated compounds (155). Chemical transformations, monitored using GC/ECD, are useful as confirmatory tests for some compounds (211). Analysis of the actual electron capture products from pesticides using tandem GC/ECDs has been proposed as an approach to residue confirmation at the pg level (212, 213). Laboratory contamination by

TABLE 2.11. Summary of the Performance of Selective GC Dectors

Detector	Element or Function Group	Detection Limit (pg/sec)	Selectivity[a]	Linearity	Pesticide Analysis Application
Electron capture[b]	Trichloro, nitro	0.5^d	10^5	2×10^4	2,4-D and 2,4,5-T esters, polynuclear aromatics, nitro herbicides
	Polyhalogenated	0.05^e	10^6	2×10^4	Organochlorine insecticides, PCBs and PBBs, dioxins, pentachlorophenol
Flame photometric	P	0.2^d	10^6	5×10^4	Organophosphorus insecticides
	S	4	10^4	5×10^{2g}	Carbamates containing S, ethylene thiourea
Alkali flame[c] ionization	P	0.02^d	5×10^5	2×10^4	Organophosphorus insecticides
	N	0.2^f	5×10^4	2×10^4	Carbamate insecticides, triazines, urea and nitro herbicides, nitrosamines
Hall electrolytic conductivity	Cl	0.5^e	$>10^6$	10^5	Organochlorine insecticides, phenoxy herbicides
	N	1^f	5×10^5	10^3	Carbamate insecticides, triazines, urea and nitro herbicides, nitrosamines

[a] With respect to carbon (hydrocarbon).
[b] Constant current.
[c] Heated bead.
[d] Parathion.
[e] Aldrin.
[f] Atrazine.
[g] Quadratic response.

111

phthalate plasticizers has been implicated in false positive DDT and PCB analyses by ECD (214).

2.10.2. Phenoxy Herbicides, Picloram and Dicamba

The analysis of 2,4-D, 2,4,5-T, and related di- and trichlorinated acidic herbicides in water and soils is readily accomplished to the 10 μg/kg level using GC/ECD of the methyl esters (215). The ECD sensitivity for 2,4-D is about a tenth that of 2,4,5-T, and monochloro compounds such as MCPA give only weak responses. Esterification with reagents such as 2-chloro-ethanol or pentafluorobenzyl bromide (PFB) (216) enhances the ECD response. Although PFB esters show a uniformly high response, ECD detection limits are compromised by high blanks even following column clean-up to remove excess reagent (217). Enhanced responses of derivatized extractives is another problem with ECD detection. Capillary columns provided sufficient resolution to enable an FID to be used in the analysis of MCPA in well water. Less than 100 μg/kg of MCPA was detected as the methyl ester (217).

GC/MS in the SIM mode employing a deuterated internal standard (D$_3$-methyl ester) achieved 2% reproducibility for 2,4-D in urine at the 10 μg/kg level (218). A study of chlorophenoxy esters levels in air were compared by GC/MS and ECD detection (219). The MDQ for 2,4-D butyl ester was less than 1 pg for GC/MS (SIM) and about 50 pg for ECD. The results of air sample analyses by GC/MS and by constant current [63]Ni ECD showed good correspondence. However, a ^{3}H DC-ECD detector gave consistently high results that were not explained.

2.10.3. Phenols

The wood preservative pentachlorophenol is a common environmental contaminant. It can be analyzed by GC/ECD either as the parent phenol or its methyl ether. Nitrophenol herbicides, such as DNOC, are similarly well suited to GC/ECD analysis. Other phenols of interest are those arising from hydrolysis of organophosphorus or carbamate insecticides. They may be produced chemically as part of a multiresidue method or confirmatory test and may be found in natural matrices as degradation products of insecticides. Because of the diversity of the structure of the phenols, derivatization is usually employed to ensure a uniform response to a selective detector. The trichloroacetates (220), 2,4-dinitrophenyl ethers (221) and penta-fluorobenzyl ethers (222) have been shown to be suitable for ECD determinations. To fully exploit the high sensitivity of the ECD, a postreaction column cleanup must be used for these derivatives. Extractive pentafluorobenzylation of phenols and carboxylic acids has been studied. The phenols were derivitized in high yield using pentafluorobenzyl bromide in $5N$ NaOH plus methylene chloride, but carboxylic acids were not derivitized unless a phase transfer catalyst, such as tetrahexylammonium hydroxide was present (223).

The high sensitivity and selectivity of the FPD or AFID to phosphorus has led to the use of phosphoryl derivatives. Carbamate phenols were determined by FPD or ECD following reaction with $(CH_3O)_2P(S)Cl$. Carbofuran has been analyzed in silage at 0.04 mg/kg levels following hydrolysis and derivatization (224). Steroidal alcohols and phenols from blood plasma were reacted with $(CH_3)_2P(S)Cl$ forming phosphinyl derivatives of which 10–100 pg gave good peaks on an AFID (225). This reagent should be of general utility in improving the detectability of alcohols and phenols. It is more reactive than the thiophosphoryl reagent and the derivatives are more stable and volatile.

2.10.4.   Chlorinated Dibenzo-p-dioxins and Furans

These extremely toxic compounds occur as manufacturing impurities in polychlorinated aromatic compounds. Their GC analysis in commercial formulations of 2,4,5-T (226), PCBs (227), and pentachlorophenol (228) has been carried out using GC/ECD or low resolution GC/MS (SIM) at a detection limit of 10 μg/kg. A detailed study of the EI mass-spectra of a variety of dioxins (1–8 Cl) has shown that isomers for a given chlorine content can be distinguished (239) using the low mass region ($m/z < 150$). This information has been used to identify the dioxin isomers present in PCB pyrolosates and fly ash samples from incinerators (229, 230).

The environmental persistence of the above commercial chemicals raises questions regarding the fate of the associated dioxins and dibenzofurans. In foodstuffs, ng/kg concentrations may be significant and existing analytical techniques must be used at their limits for this sensitivity. The criteria for GC/MS detection in low level analyses where no suitable control exists have been discussed (177). For tetrachlorodioxin (TCDD) these criteria are: (1) signal/noise ratio of at least 2.5:1, (2) a clearly defined valley for TCDD as a shoulder on a contaminating peak, (3) duplicate analysis if the signal/noise is between 2.5:1 and 10:1 or the ratio of m/z 320 and 322 peak heights is not in the correct isotopic proportion, (d) confirmation of all positive results between 2.5 and 10:1 signal/noise by another measurement method. The analytical limits for the low to medium resolution GC/MS (SIM) determination of TCDD in cleaned-up extracts of a wide variety of tissues were in the range 1–20 ng/kg (176, 231). TCDD was not detected at a limit of 1 ng/kg in bovine milk from animals grazing range land that had been sprayed with 2,4,5-T (232). Neither hexa- nor octachlorodioxin were detected in bovine milk at a 50 ng/kg limit using GC/ECD (233).

2.10.5.   Organophosphorus Insecticides

Most organophosphorus insecticides can be satisfactorily detected by an ECD, especially phosphorothioates containing a halogen or nitro moiety. However, the higher selectivity and uniform response of the FPD and AFID have made them the detectors of choice for routine analysis. Chromatograms for the extracts of a high organic sandy loam soil containing 0.04 mg/kg fensulfothion together with an untreated control were obtained using an FID, heated bead AFID, and FPD in our laboratory (Figure 2.9A,B). In spite of a charcoal column cleanup step, the FID response shows a complex pattern of coextractives which completely masks the chlorpyrifos.

In contrast, the AFID (P mode), (see Section 2.6.4) shows a relatively simple chromatogram and the fensulfothion (0.8 ng) can be readily determined. The AFID in the N/P mode was not as selective as shown in Figure 2.10. Several large peaks due to coextractives are observed, including one of which overlaps the fensulfothion peak. The interferences were presumably due to nitrogen containing compounds whose response is greatly attenuated in the P mode. The FPD gave a similar chromatogram to the AFID (P mode), except that the signal/noise ratio is not as high.

Although the FPD has a higher detection limit than the AFID, it is often preferred because of its greater reliability and long-term reproducibility. The greater sensitivity of the AFID is more beneficial when dealing with samples whose size is limited. For general residue work, however, either detector is satisfactory.

The CCD may be used in the oxidative mode to detect S from phosphorothioates (133) but it offers no advantages for P detection and oxygen analogues and other organophosphorus compounds lacking sulfur do not respond. The phenols resulting from aryl organophosphate hydrolysis can be derivatized for electron-capture detection (see Section 2.10.3) and this technique is the basis for confirmatory tests (211, 222). Triaryl phosphate flame retardants are usually detected following hydrolysis to the phenols (234), although direct GC at high temperatures with heated bead AFID detection has recently been reported (235).

2.10.6.   N-Methyl Carbamate Insecticides

The analysis of carbamates is not as well established as for organophosphates due to problems related to thermal stability and the lack of sensitive

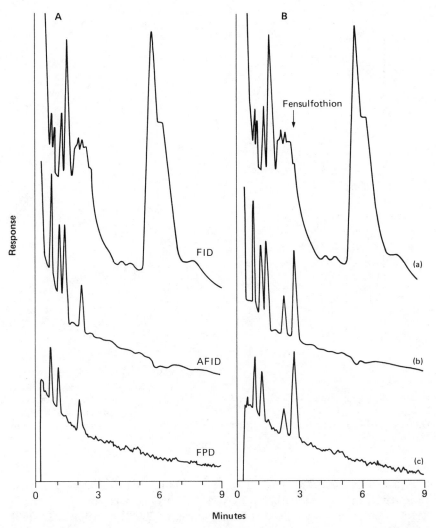

Figure 2.9 Response of the flame ionization, alkali flame ionization, and flame photometric detectors to a high organic sandy loam soil extract containing fensulfothion. (*A*) Chromatogram of untreated soil extracts: (*a*) flame ionization detector, (*b*) alkali flame ionization detector (P mode), (*c*) flame photometric detector (P mode). (*B*) Chromatograms of treated soil extracts: (*a*), (*b*), and (*c*) as above. GC conditions: column, 1 m × 3 mm, 4% SE-30/6% QF-1/Gas Chrom Q; column flow, 50 ml/min nitrogen; column temperature 205°; detector flows; FID/AFID; 35 ml/min hydrogen, 240 ml/min air, FPD; 200 ml/min hydrogen, 50 ml/min air, 20 ml/min oxygen. Perkin-Elmer NPD, Bendix SPED.

115

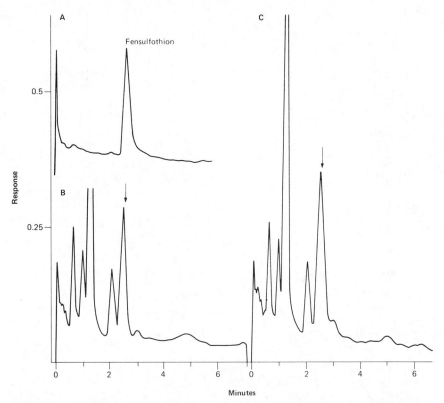

Figure 2.10   Response of the heated bead alkali flame ionization detector (N/P mode) to high organic sandy loam soil extract containing fensulfothion. (*a*) Fensulfothion (0.8 ng). (*b*) Untreated soil extraction. (*c*) Soil extract containing fensulfothion. GC conditions as in Figure 2.9.

nitrogen detectors. The GC of nitrogen containing pesticides has been reviewed (236). Most carbamates can be chromatographed, if carefully deactivated packings are used at temperatures of 200°C or below (237–239). The AFID (239) and electrolytic conductivity detectors (240) have been used for their analysis. The heated bead AFID in particular has improved stability and sensitivity to nitrogen over earlier versions. A chromatogram for 8 ng of underivatized carbofuran and 4 ng each of the 3-keto and 3-hydroxy metabolites using such a detector is shown in Figure 2.11*A* (this laboratory). A chromatogram of the cleanup extract from a peat soil field treated with 2.2 kg/ha carbofuran shows the presence of only carbofuran and the 3-keto analog at estimated levels of 0.76 and 0.14 mg/kg, respectively (Figure 2.11*B*). The MDQ for these compounds was of the order of

0.5–1 ng. Similar limits have been reported for a HCD (240), but the long term reliability of this detector in the catalytic reductive mode has yet to be proven (see Section 2.7.5). Interference from nitrogenous coextractives of soil and plant samples can be particularly troublesome for the analysis of carbamates using nitrogen selective detectors.

As a result of the instability and poor detector response of carbamates, a large number of analytical methods have been developed based on a wide range of derivatives (241, 242). The response of carbamates to the ECD can be improved by acylating the N—H group using trifluoroacetic- or pentafluoropropionic anhydride, with (243) or without catalyst (244). Unfortunately, these methods also greatly enhance the ECD response to the blank and coextractives, especially in the catalyzed reaction, and are best applied

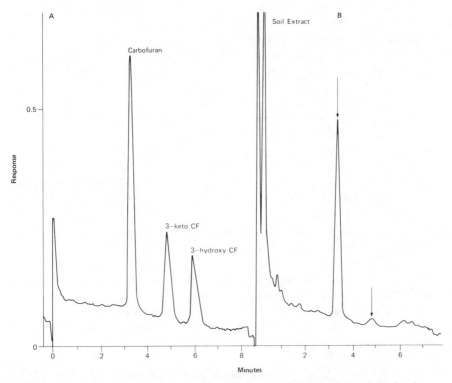

Figure 2.11 Analysis of carbofuran and its 3-hydroxy and 3-keto analogs in soil using a Perkin-Elmer NPD (AFID N/P mode). (*a*) Standards: carbofuran (8 ng); 3-keto carbofuran (4 ng), 3-hydroxy carbofuran (4 ng). (*b*) cleaned up extract of high organic peat soil, field treated showing the presence of carbofuran and its 3-keto analog. GC conditions: column 1.8 m × 6 mm, 4% SE-30/6%-QF-1/Gas Chrom Q; column temperature 205°; column flow 30 ml/helium; detector flows 100 ml/min air, 2 ml/min hydrogen; standing current 3 pA.

to cleaned-up samples. It is important that excess reagent be removed. The heptafluorobutryl (HFB) derivatives have been analyzed both by using an ECD or a CCD in the halogen mode (245, 246). The latter method offers no advantage over the ECD as the sensitivity is lower and coextractive response is similarly magnified.

The FPD is suitable for the analysis of sulfur-containing carbamates such as Mesurol and aldicarb. This detector is more selective than the AFID or the CCD in the oxidative mode (80) when 10–100 ng amounts of pesticide are analyzed due to its quadratic response (see Section 2.5). The stabilization of sulfoxide moieties by the Pummer rearrangement as induced by trifluoroacetic anhydride (TFA) has been studied (247). Mesurol, its sulfoxide, sulfone and phenol analogs were derivatized with TFA in ethyl acetate (1 hr, 100°C) and analyzed using the FPD in the S mode (248) (Figure 2.12). The sensitivity of this method for Mesurol and its metabolites on blueberries was 0.3 mg/kg, without any cleanup of the crude extracts (249).

A temperature programmed chromatogram of the TFA derivatives of Mesurol, its oxidation products and phenol analogs is shown in Figure 2.12A. This insecticide is used as a bird repellant on blueberries and the chromatogram in Figure 2.12B of the derivatized crude blueberry extract after only methylene chloride partition demonstrates the usefulness of selective element detectors in residue analysis. This particular sample contained Mesurol and its sulfoxide and sulfone oxidation products at estimated levels of 5, 0.9, and 0.03 mg/kg, respectively.

Alkylation also improves the thermal stability of carbamates and other pesticides with NH groups (239). On-column transesterification of carbamates with methanol forms N-methyl methylcarbamate, which can be analyzed on a Porapak column using AFID detection (250, 251). Hydrolysis of carbamates to their phenols followed by derivitization for selective detection has often been employed (see Section 2.10.3).

The CI (CH$_4$) mass spectra of 25 carbamates have been reported (252). The base peak of the HFB derivatives of all carbamates is protonated methylamine heptafluorobutyrate (m/z 228) and this has been used for a GC/MS (SIM) analysis of carbofuran and metabolites in crops to 0.02 ng/kg with minimal cleanup (253). This study illustrated the selectivity of the GC/MS technique.

2.10.7.   Ureas and N-Aryl Carbamates

These compounds pose similar GC analysis problems as those of the N-methyl carbamates, as they are often thermally unstable. A method involving hydrolysis, diazotization, and iodination to form the iodo analog of the substituted aniline moiety, which can be detected by the ECD with high

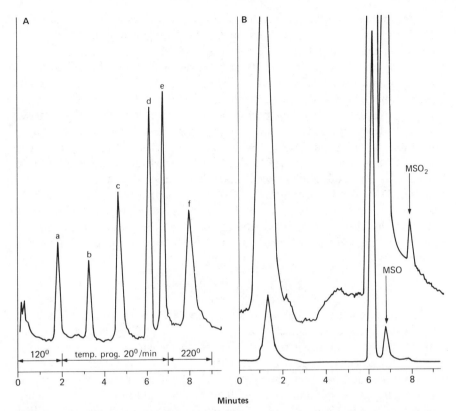

Figure 2.12 Analysis of the trifluoroacetyl derivatives of Mesurol, its sulfoxide and sulfone oxidation products and phenol analogs in blueberries using a Bendix SPED (FPD-S mode). (*A*) Standards: (*a*) Mesurol phenol TFA (3 ng), (*b*) Mesurol sulfoxide phenol di TFA (8 ng), (*c*) Mesurol sulfone phenol TFA (6 ng), (*d*) Mesurol TFA (5 ng), (*e*) Mesurol sulfoxide di TFA (12 ng), (*f*) Mesurol sulfone TFA (12 ng). (*B*) Extract of field treated blueberries after partion and derivatization showing the presence of Mesurol, Mesurol sulfoxide, and sulfone. GC conditions: column, 1 m × 6 mm, 3% OV-225/Gas Chrom Q; column temperature 175; column flow 60 ml/min nitrogen; detector flow 50 ml/min air, 10 ml/min oxygen, 50 ml/min hydrogen.

sensitivity has been reported (254). However, analysis of the intact compound is preferable and methylation of the amide NH appears the most effective method of improving stability (239). A CH_3I/NaH/DMSO methylation is the basis of a routine analysis for herbicide residues in foodstuffs to 0.1 mg/kg using CCD detection (255). The derivatives are also suitable for the ECD or AFID, but column cleanup of the methylated extract is required before the higher sensitivity of these detectors can be used. Metabolites can also be analyzed following alkylation but methylation of desmethyl compounds results in formation of the parent herbicide. GC/MS has been suggested as a means of distinguishing metabolites and parent by the use of CD_3I in the methylation (256).

Insect growth regulators of the diflubenzuron type are also suited to the methylation procedure and GC/ECD analysis (257). However, this and many other phenyl-substituted pesticides that are difficult to GC, are more readily analyzed using high performance liquid chromatography (HPLC) with uv detection. This technique may supercede GC for routine analysis of these compounds (MDQ 5–10 ng). Lawrence (258) has compared GC/EC and CCD with HPLC/uv for the routine analysis of several herbicides in food and showed that their sensitivity was in the order GC/ECD > HPLC/uv > GC/CCD. Coextractive response of the ECD was high in many cases. The AFID has not received much attention as a detector for these compounds, although it should be suitable, especially for the N-methylated derivatives which will have optimal responses (see Section 2.6.5).

2.10.8. Triazine Herbicides

Although chlorotriazines, such as atrazine will give a moderate response to the ECD, nitrogen selective detectors are generally preferred to give a more uniform response across the class, independent of substitution and to give better selectivity against coextractives. The CCD is extensively used for this purpose and has been shown to be of value for quantitation of residues to 0.01 ng/kg in soil and water without cleanup and in corn with cleanup, whereas the ECD was only of qualitative use for corn samples even following cleanup (259). Three methods of determining atrazine, ametryne, and terbutryne in water have been compared (260). GC/AFID could detect 1 μg/kg, whereas uv or polarographic detection were limited to 10 μg/kg. The 3-electrode-AFID and CCD have been compared for detection of several nitrogen pesticides (118). The CCD gave a more selective and uniform response, but the heated bead AFID may be more suitable for triazine analysis because of its greater sensitivity and stability. Hydroxyatrazine, a metabolite of atrazine, may be analyzed by GC/AFID following alkylation or trimethysilylation (261). The direct injection of crude silylation mixtures is not recommended with the AFID as the salt source is affected. A 40% loss of sensitivity after 10 injections occurred with a 3-electrode AFID, whereas FID response was unaffected (this laboratory).

2.10.9. Miscellaneous Herbicides

The dinitroaniline herbicides have excellent ECD characteristics and either this or an N selective detector (CCD or AFID) is used in their GC analysis. Trifluralin and benefin are very difficult to separate by GC. GC/MS with monitoring of a characteristic fragment from each has been used to quantitate mixed residues in soil down to 0.01 mg/kg (262).

The pyridinium salt herbicides such as paraquat present problems for GC analysis as they are difficult to concentrate and decompose on injection. Catalytic reduction of the quaternary centres to form the tertiary amine followed by organic extraction, concentration, and GC/AFID has been used for the analysis of diquat, paraquat, and cyperquat in soil and plant (263, 264). A recent review of paraquat analysis (265) does not include the excellent work by Draffan et al. (266). They analyzed paraquat in human bile and plasma down to 0.025 mg/kg using $NaBH_4$ reduction followed by GC/AFID. The limit was extended to 5 μg/kg by using GC/MS (SIM) detection with either paraquat-d_6 or the diethyl analog as the internal standard. High sensitivity is necessary in this analysis as low concentrations of paraquat are physiologically damaging.

The growth regulator ethephon (β-chloroethylphosphonic acid) can be very sensitively detected in apple extracts by FPD by forming the dimethyl ester. Esterification proceeded readily with diazomethane, but could not be achieved using BF_3/MeOH, HCl/MeOH or methyl fluorosulfonate (267). A simpler procedure decomposes ethephon in situ to ethylene at high pH in a sealed flask (268). The ethylene is detected by GC/FID using direct injection of headspace gas and gives sensitivity to 0.1 mg ethephon/kg. The herbicide glyphosate [N-(phosphonomethyl) glycine] can be determined in water and soil by GC/FPD following dual derivatization involving methylation and acylation. The very polar nature of the above two phosphonic acids and the dealkyl degradation products of phosphorothioates and phosphates makes extraction and the concentration step difficult. An HPLC method using post column formation of a fluorescent derivative is a superior approach (269).

2.10.10. Pyrethroid Insecticides

Use of these compounds is rapidly increasing as more active and photochemically stable forms become commercially available. In general, methods for the analysis of natural and some synthetic pyrethroids are limited because of the lack of heteroatoms suitable for selective detection. GC/FID has been used to determine resmethrin residues in cereals to 0.2 mg/kg (270) and GC/MS with CI was used to determine the 6 natural pyrethrin esters to 0.04 mg/kg on silage (271). The newer highly active analogs containing halogen and/or cyano groups are more amenable to GC/ECD analysis. Permethrin was analyzed in potato to 0.01 mg/kg following column cleanup (272) and sheep wool (this laboratory). Permethrin, cypermethrin, and two other synthetic pyrethroids gave MDQs of 10 pg using GC/ECD (273). Fenvalerate was detected at 0.01 mg/kg on crops using GC/ECD (274).

A general method for permethrin residues in soil, animal, and soil matrices using either GC/ECD or GC/CCD (halogen mode) following gel-permeation/Florisil column cleanup has been reported (275). Both detectors were satisfactory for residue determinations below 0.05 mg/kg, but the lower sensitivity of the CCD required injection of 100 mg equivalents of initial sample vs <1 mg for the ECD. It was also more difficult to maintain the cis–trans isomer separation with the CCD because of its greater detector volume.

2.10.11. ETU and Ethylenebisdithiocarbamate Fungicides

The principles of detector selection for most fungicide analyses follow those applied to insecticides and herbicides in the previous sections. The ethylenebisdithiocarbamate (EBDC) fungicides, such as Maneb and Zineb, are widely used agricultural chemicals. Approaches to their analysis, together with that of ethylenethiourea (ETU), a carcinogenic degradation product, form useful examples of selective detection and derivatization.

The EBDC metal complexes cannot be chromatographed intact. Heating with H_2SO_4 and $SnCl_2$ results in the formation of CS_2 and ethylenediamine. The CS_2 evolved is generally measured colorimetrically (276), but a GC method has also been proposed using a thermal conductivity detector to measure CS_2 (277). As CS_2 also results from the decomposition of dimethyl-dithiocarbamates, e.g., thiram and plant sulfur compounds, a more specific analysis procedure for EBDCs has been developed which detects the ethylenediamine (278). A TFA derivative is formed and EBDC residues could be determined on fruits and crops to 0.1 mg/kg using GC/ECD.

The analysis of ETU in fungicide formulations, crops and processed foods has been the subject of a recent report (279). ETU can be chromatographed intact in nanogram quantities both on packed columns (280) and glass capillary columns (W. D. Marshall, private communication). However, standard GC methods use various derivatives to improve stability and increase sensitivity. Alkylation of the thiocarbonyl group with butyl bromide (281) or m-trifluoromethylbenzyl bromide (282) forms stable derivatives which can be detected by AFID (283), ECD (282) or FPD in the S mode (281, 282). The N-trifluoroacetyl or N-pentafluorobenzyl derivatives have been used in conjunction with S-benzyl or S-(O-chlorobenzyl) derivatives (284, 285). The FPD gave better selectivity than the ECD, but since it is often operating near the detection limit for sulfur, quenching can cause problems (see Section 2.5). The trifluoromethylbenzyl derivative was suitable for determination of ETU on crops to 0.01 mg/kg (FPD) or 0.002

mg/kg (ECD) and is preferable to dual derivatives which require an extra column cleanup (285). However, the retention time shift and sensitivity increase when S-trifluoroethylbenzyl ETU is trifluoroacetylated can be used for confirmation.

2.10.12.   Organometallic Compounds

Although the main sources of mercury and tin are inorganic, their methyl analogs are of great environmental concern. Both organomercury and tin compounds are used in agriculture as fungicides. The dialkyl and the mono-alkylhalo derivatives of mercury have good GC properties. Methods for methyl mercury in tissues and foodstuffs using GC/ECD of the chloride or bromide can measure 0.1 mg/kg with minimal cleanup (286) or below 0.01 mg/kg with extensive cleanup (287, 288). Inorganic mercury can be alkylated with $(CH_3)_4Sn$ and converted to CH_3HgBr for GC/ECD determination (288). The microwave plasma emission detector is very sensitive and selective for mercury (Section 2.9.4) and has been employed to determine CH_3HgCl in samples (0.05–0.5 g) of various tissue to 1 μg/kg (190). Dimethyl and monomethyl mercury halides are detected with equal sensitivity by the MPED, whereas the ECD is rather less sensitive to the former. The high selectivity of the MPED also allows minimal cleanup of extracts and permits the use of chlorinated solvents.

Residues of the miticide, tricyclohexyltin hydroxide, and its degradation products may be determined to 0.1 mg/kg on fruit by GC of their bromides and use of the CCD in the halogen mode (289). Detectors specific to tin would appear to be more suitable as demonstrated by the use of a heated furnace AA which gave very selective detection of Sn(Me)$_4$ down to 10 ng (196). No doubt, the MPED could also be tuned to Sn, although no applications have yet been reported. The most interesting development in the analysis of organotin compounds stems from work by Aue and Flinn (82), who showed that a modified FPD is extremely sensitive to this element. A special constricted quartz tube was used as a flame shield and the interference filters removed. This configuration was about 2000 times more sensitive to tin than to phosphorus and picogram quantities of a variety of alkyl tin compounds were readily determined.

Sodium and ammonium salts of monomethylarsinic and dimethylarsinic acids are employed as contact herbicides. The analysis of arsenicals has recently been reviewed (290). These compounds may be chromatographed as the alkyl arsines following reduction with NaBH$_4$ (191). An analytical method using this procedure has been applied to formulations or residues with the MPED, and gave a MDQ of 20 pg for As. Levels of 0.5–1 μg/kg of the herbicides could be determined in fresh or salt water (191). The heated furnace AA can also be used to determine alkyl arsines with detection limits of about 5 ng (196).

Selenium is an important trace element that may accumulate at toxic levels in foodstuffs. Several GC methods are available that use the volatile piazselenol complexes.

The ECD response of piazselenols has been examined, and the sensitivity of the 4,6-dibromopiazselenol was shown to be three times that of the 5-nitro complex (291). The 5-nitropiazselenol can be detected by the ECD with a MDQ of 1 pg Se and has been used to analyze a variety of plant and animal tissues (0.2–1.0 g) at the 0.1 mg/kg level (292). Similarly 4,5-dichloropiazselenol has been used as the basis for a very sensitive method for selenium at levels below 0.01 mg/kg (275). It was also noted that the FPD and heated bead AFID gave responses to the complex. Aue and coworkers have studied the FPD response to Se in some detail (82, 86) (see Section 2.5.5). Although the FPD detection limit of selenium is higher than of the ECD, the higher selectivity of the FPD should result in some useful applications. The MDQ for 5-nitropiazselenol with this detector was 40 pg Se and the analytical limits were 0.1 μg/kg in water and 15 μg/kg for solid samples (192). The column cleanup required for ECD methods (293) was unnecessary with this detector. Plants and animals can convert inorganic Se to volatile methylated compounds which are amenable to GC. Flameless AA has been used to detect these compounds down to 0.1 ng as Me_2Se or Me_2Se_2 (294).

2.11. SAMPLE CONSTRAINTS GOVERNING DETECTOR SENSITIVITY AND SELECTIVITY

Typical routine extraction methods increase the residue concentration by factors of 1–10 going from mg/kg of sample to mg/liter of extract, although larger factors of 100–1000 are often obtained. This concentration factor may be limited by initial sample size since a minimum final volume 0.5–1.0 ml is required for quantitative work employing external standards. This limit can be lowered to 10 μl if an internal standard is used. High concentration factors may also give levels of coextractives that will precipitate or cause deterioration of the GC column performance. A concentration factor of 1 means a 5-μl injection makes available to the detector 5 ng of a residue for each mg/kg in the sample, while a factor of 1000 results in 5 ng of residue per μg/kg sample. This amount is of the order required to give a good response on most selective detectors.

Coextractives are also important in relation to interference with the detector response to residues. It is difficult to be specific about the level of co-extractives tolerable for a detector because the degree of interference from each component depends on whether it coelutes with the residue. In general, the greater the selectivity of the detector the better the discrimination against coextractives and the less sample cleanup required. As the majority of volatile coextractives from animal, plant or soil matrices are likely to be carbon compounds, the selectivity needs to be high against this group. In practice the selectivity against carbon as a hydrocarbon is used in detector evaluation but other compounds may cause more serious interferences, e.g., in the analysis of phosphorus compounds, sulfur compounds interfere on the FPD and nitrogen compounds on the AFID.

Concentration and interference problems will generally necessitate fractionation procedures to remove coextractives (cleanup) on extracts of plant or animal samples being analyzed for pesticide residues below ca 1 mg/kg.

2.12. CONCLUSION

In order to utilize the technique of gas chromatography to the best advantage, an understanding of the principles governing the operation of each GC detector is important for the selection of a detector and optimization of its performance. The parameters of sensitivity, selectivity, linearity, and detection limits should be used objectively as a means of detector evaluation. The sensitivities of currently available GC detectors are generally adequate for the analysis of 0.1–1 ng of a compound, which provides pesticide residue levels of 0.01–0.1 mg/kg when a 10 g sample is used. The type of sample and analyte together with the extraction and cleanup method used will determine the selectivity required in order to overcome any interferences.

For phosphorus compounds the various designs of the AFID and FPD provide very sensitive and selective detectors. The heated bead AFID also offers improved analysis of nitrogen compounds over earlier designs because of its greater sensitivity and stability. The electrolytic conductivity detectors have improved steadily over the years and offer high element selectivity which can be made class dependent, e.g., organochlorine insecticides and nitrosamines. However, their reliable operation requires more extensive preventative maintenance than for the flame based detectors. Spectroscopic techniques such as MS, IR and microwave plasma emission offer variable selectivity. The latter detector could be used for many elements, including some that are not readily determined by other detectors. A need still exists for a sensitive and selective detector with a linear response for sulfur compounds.

Chemical derivatization offers a simple means of increasing both the sensitivity and selectivity of analytical methods. For derivatives with high electron capturing properties, the ECD is an extremely sensitive detector in the constant current mode. However, the relatively poor selectivity necessitates more extensive cleanup. The use of perfluoracetyl and phosphinyl derivatives coupled with the use of element selective and sensitive detectors such as the ECD, CCD, FPD, and the AFID would appear to be the logical choice for future development, together with on-column reactions.

The use of capillary column GC enhances the selectivity of analytical methods by increasing the resolution of components in a mixture. The narrow peak widths also lead to improved peak detectability. However, their low sample capacity necessitates the use of sensitive GC detectors in order to avoid overloading the column with coextractives. The detectors are also required to have a short-time constant in order to retain high resolutions. The ECD, FID, and AFID with small volume cells will fulfill these requirements.

Gas chromatography is expected to continue to play an important role in the analysis of pesticide residues even though the number of compounds suitable for HPLC analysis is steadily increasing. This is mainly due to the wide variety of sensitive and element selective detectors that are available for routine GC analysis and the use of GC/MS for confirmation purposes.

REFERENCES

1. J. E. Lovelock, *J. Chromatogr.*, **1,** 35 (1958).
2. S. R. Lipsky, R. A. Landowne, and J. E. Lovelock, *Anal. Chem.*, **31,** 852 (1959).
3. I. G. McWilliams and R. G. Dewar, *Nature*, **181,** 760 (1958).
4. J. Harley, W. Nel and V. Pretorious, *Nature*, **181,** 177 (1958).
5. J. E. Lovelock and S. R. Lipsky, *J. Am. Chem. Soc.*, **82,** 431 (1961).
6. J. E. Lovelock, *Nature*, **187,** 49 (1960).
7. J. E. Lovelock, *Anal. Chem.*, **35,** 474 (1963).
8. A. Karmen and L. Giuffrida, *Nature*, **201,** 1204 (1964).
9. S. S. Brody and J. E. Chaney, *J. Gas Chromatogr.*, **4,** 42 (1966).
10. D. M. Coulson, *J. Gas Chromatogr.*, **3,** 134 (1965).
11. O. Piringer and M. Pascalau, *J. Chromatogr.*, **8,** 410 (1962).
12. R. L. Harless, E. O. Oswald, and M. K. Wilkinson, presented at 28th Annual Conference, American Society for Mass Spectrometry, St. Louis, May 1978.
13. D. J. David, *Gas Chromatographic Detectors*, Wiley-Interscience, New York, 1974.
14. J. Sevcik, *Detectors in Gas Chromatography*, Elsevier, Amsterdam, 1976.
15. E. R. Adlard, *C. R. C. Crit. Rev. Anal. Chem.*, **5,** 1 (1975).

16. E. R. Adlard, *C. R. C. Crit. Rev. Anal. Chem.*, **5**, 13 (1975).
17. W. A. Aue, *Adv. Chem.*, **104**, 39 (1971).
18. W. A. Aue, *J. Chromatogr. Sci.*, **13**, 329 (1975).
19. A. C. S. Symposium on Substance Selective Detectors, San Francisco, Sept., 1976, *J. Chromatogr.*, **134**, 1 (1977).
20. R. C. Hall, *C. R. C. Crit. Rev. Anal. Chem.*, **8**, 323 (1978).
21. W. P. Cochrane and R. Greenhalgh, *Chromatographia*, **9**, 255 (1976).
22. L. S. Ettre, *J. Chromatogr.*, **112**, 1 (1975).
23. H. C. Smit and H. L. Walg, *Chromatographia*, **8**, 311 (1975).
24. C. H. Hartmann, *J. Chromatogr. Sci.*, **7**, 163 (1969).
25. H. W. Johnson and F. H. Stross, *Anal. Chem.*, **31**, 1206 (1959).
26. I. G. Young, *2nd International Gas Chromatography Symposium ISA, Pittsburgh, 1959*, p. 75.
27. H. Kaiser, *International Congress in Analytical Chemistry (IUPAC) Kyoto, Japan* (M. Senda, Ed.), Butterworths, London, 1977, p. 35.
28. J. Sevcik and M. Klima, *Chromatographia*, **9**, 69 (1976).
29. A. T. Blades, *J. Chromatogr. Sci.*, **14**, 45 (1976).
30. T. McAllister, A. J. C. Nicholson, and D. L. Swingler, *Int. J. Mass Spec. Ion Phys.*, **27**, 43 (1978).
31. J. M. Gill and C. H. Hartmann, *J. Gas Chromatogr.*, **5**, 605 (1967).
32. J. C. Sternberg, W. S. Gallaway, and D. T. C. Jones, *International Gas Symposium, Lansing, Michigan, 1961*, Academic Press, New York, 1962.
33. R. G. Ackman, *J. Gas Chromatogr.*, **6**, 497 (1968).
34. P. Russer, T. A. Gough, and C. J. Woollam, *J. Chromatogr.*, **119**, 461 (1976).
35. T. A. Gough, M. A. Pringner, and C. J. Woollam, *J. Chromatogr.*, **150**, 533 (1978).
36. A. H. Hofberg, L. C. Heinrichs, and G. A. Gentry, *J. Assoc. Off. Anal. Chem.*, **59**, 758 (1976).
37. M. Riva and A. Carisano, *J. Chromatogr.*, **36**, 269 (1968).
38. R. Pigliucci, W. Averill, J. E. Purcell, and L. S. Ettre, *Chromatographia*, **8**, 165 (1975).
39. E. D. Pellizzari, *J. Chromatogr. (Chromatogr. Rev.)*, **98**, 323 (1974).
40. R. J. Maggs, P. L. Joynes, A. J. Davies, and J. E. Lovelock, *Anal. Chem.*, **43**, 1966 (1971).
41. B. Brechbuhler, L. Gay, and H. Jaeger, *Chromatographia*, **10**, 478 (1977).
42. P. L. Patterson, *J. Chromatogr.*, **134**, 25 (1977).
43. G. R. Shoemaker, J. E. Lovelock, and A. Zlatkis, *J. Chromatogr.*, **12**, 314 (1963).
44. C. H. Hartmann, *Anal. Chem.*, **45**, 733 (1973).
45. P. G. Symonds, D. C. Fenimore, B. C. Petit, J. E. Lovelock, and A. Zlatkis, *Anal. Chem.*, **39**, 1428 (1967).

46. P. Devaux and G. Guiochon, *J. Chromatogr. Sci.*, **8**, 502 (1970).

47. J. J. Sullivan, *J. Chromatogr.*, **87**, 9 (1973).

48. J. E. Lovelock, R. J. Maggs, and E. R. Adlard, *Anal. Chem.*, **43**, 1962 (1971).

49. D. J. Dwight, E. A. Lorch and J. E. Lovelock, *J. Chromatogr.*, **116**, 257 (1976).

50. W. E. Wentworth, A. Tishbee, C. F. Batten, and A. Zlatkis, *J. Chromatogr.*, **112**, 229 (1975).

51. W. E. Wentworth, E. Chen, and J. E. Lovelock, *J. Phys. Chem.*, **70**, 445 (1966).

52. D. C. Fenimore and C. M. Davis, *J. Chromatogr. Sci.*, **8**, 519 (1970).

53. J. J. Sullivan and C. A. Burgett, *Chromatographia*, **8**, 176 (1975).

54. J. E. Lovelock, *J. Chromatogr.*, **99**, 3 (1974).

55. W. A. Aue and S. Kapila, *J. Chromatogr.*, **11**, 225 (1973).

56. W. A. Aue and S. Kapila, *J. Chromatogr.*, **112**, 247 (1975).

57. J. Rosiek, I. Sliwka, and J. Lasa, *J. Chromatogr.*, **137**, 245 (1977).

58. M. Satouchi and T. Kojima, *Anal. Lett.*, **5**, 931 (1972).

59. C. E. Cook, C. W. Stanley, and J. E. Burney, *Anal. Chem.*, **36**, 1560 (1964).

60. W. P. Cochrane, *J. Chromatogr. Sci.*, **13**, 246 (1975).

61. C. F. Poole, *Chem. Ind.*, **1976**, 479.

62. S. B. Martin and M. Rowland, *J. Pharm. Sci.*, **61**, 1237 (1972).

63. K. Blau and G. S. King, *Handbook of Derivatives* for *Chromatography*, Heyden, London, 1977.

64. W. E. Wentworth and E. Chen, *J. Gas Chromatogr.*, **5**, 170 (1967).

65. C. F. Poole, *J. Chromatogr.*, **118**, 280 (1976).

66. W. P. Cochrane, R. B. Maybury, and R. Greenhalgh, *J. Environ. Health Sci.*, **B14**, 197 (1979).

67. M. Drager and B. Drager, Ger. Pat. 1,133,818 (1962).

68. M. L. Selucky, *Chromatographia*, **4**, 425 (1971).

69. P. T. Gilbert, *Nonmetals in Analytical Flame Spectroscopy*, (R. Mavrodineau, Ed.), Macmillan, London, 1970, Chapter 5.

70. P. L. Patterson, R. L. Howe, and A. Abu-Shumays, *Anal. Chem.*, **50**, 339 (1978).

71. W. L. Crider and R. W. Slater, *Anal. Chem.*, **41**, 531 (1969).

72. T. Sugiyama, Y. Suzuki, and T. Takeuchi, *J. Chromatogr.*, **80**, 61 (1973).

73. P. L. Patterson, *Anal. Chem.*, **50**, 345 (1978).

74. M. C. Bowman and M. Beroza, *Anal. Chem.*, **40**, 1448 (1968).

75. W. E. Dale and C. C. Hughes, *J. Gas Chromatogr.*, **6**, 603 (1968).

76. C. A. Burgett and L. E. Green, *J. Chromatogr.*, **12**, 356 (1974).

77. A. R. L. Moss, *Scan (Pye-Unicam)*, **4**, 5 (1974).

78. R. Greenhalgh and M. Wilson, *J. Chromatogr.*, **128**, 157 (1978).

79. J. G. Eckhardt, M. B. Denton, and J. L. Moyers, *J. Chromatogr. Sci.*, **13**, 133 (1975).

80. R. Greenhalgh and W. P. Cochrane, *Int. J. Environ. Anal. Chem.*, **3**, 213 (1974).

81. J. M. Zehner and R. A. Simonaitis, *J. Chromatogr. Sci.*, **14**, 348 (1976).

82. W. A. Aue and C. G. Flinn, *J. Chromatogr.*, **142**, 145 (1977).

83. H. W. Grice, M. L. Yates, and D. J. David, *J. Chromatogr. Sci.*, **8**, 90 (1970).

84. J. Sevcik and Nguyen Thi Phuong Thao, *Chromatographia*, **8**, 559 (1975).

85. W. A. Aue and C. R. Hastings, *J. Chromatogr.*, **87**, 232 (1973).

86. C. G. Flinn and W. A. Aue, *J. Chromatogr.*, **153**, 49 (1978).

87. B. Versino and H. Vissers, *Chromatographia*, **8**, 5 (1975).

88. V. V. Brazhnikov, M. V. Gur'ev, and K. I. Sakodynsky, *Chromatogr. Rev.*, **12**, 1 (1970).

89. J. Sevcik, *Chromatographia*, **6**, 139 (1973).

90. B. Kolb, M. Auer, and P. Pospisil, *J. Chromatogr. Sci.*, **15**, 53 (1977).

91. M. Drescher and J. Janak, *J. Chromatogr. Sci.*, **7**, 451 (1969).

92. F. P. Speakman, *Column (Pye-Unicam)*, **3**, 2 (1968).

93. W. H. Stewart, *Anal. Chem.*, **44**, 1547 (1972).

94. B. M. Johnson, B. D. Kaimanad, and R. W. Lambrecht, *Anal. Chem.*, **48**, 1271 (1976).

95. W. A. Aue, C. W. Gehrke, R. C. Tindle, D. L. Stalling, and C. D. Ruyle, *J. Gas Chromatogr.*, **5**, 381 (1967).

96. H. K. DeLoach and D. D. Hemphill, *J. Assoc. Off. Anal. Chem.*, **52**, 533 (1969).

97. R. A. Mees and J. Spaans, *Z. Anal. Chem.*, **247**, 252 (1969).

98. A. Verga and F. Poy, *J. Chromatogr.*, **116**, 17 (1976).

99. V. V. Brazhnikov and E. B. Shmidel, *J. Chromatogr.*, **112**, 527 (1976).

100. D. K. F. Swan, *Column (Pye-Unicam)*, **14**, 9 (1972).

101. R. Greenhalgh and J. Dokladalova, *Column (Pye-Unicam)*, **14**, 4 (1972).

102. R. Greenhalgh and M. Wilson, *Column (Pye-Unicam)*, **15**, 10 (1972).

103. N. Mellor, *J. Chromatogr.*, **123**, 396 (1976).

104. R. A. Hoodless, M. Sargent, and R. D. Teble, *J. Chromatogr.*, **136**, 199 (1977).

105. A. T. Chamberlain, *J. Chromatogr.*, **166**, 180 (1976).

106. T. A. Gough and K. Sugden, *J. Chromatogr.*, **86**, 65 (1973).

107. K. Abel, K. Lannean and R. K. Stevens, *J. Assoc. Off. Anal. Chem.*, **49**, 1022 (1955).

108. B. Kolb and J. Bischoff, *J. Chromatogr. Sci.*, **12**, 625 (1974).

109. B. Kolb, M. Linder and B. Kempken, *Angew, Gas Chromatog.*, *Helf* **1974**, 21.

110. C. A. Burgett, D. H. Smith, and H. B. Bente, *J. Chromatogr.*, **134**, 57 (1977).

111. P. L. Patterson, R. Howe, V. Hornung, and C. H. Hartmann, *J. Chromatogr.* **167,** 381 (1978).

112. R. C. Hall, B. J. Ehrlich, and P. W. Thiede, Abstract 58, 29th Pittsburgh Conference, Feb., 1978.

113. B. P. Semonian, J. A. Lubkowitz, and L. B. Rogers, *J. Chromatogr.*, **151,** 1 (1978).

114. J. A. Lubkowitz, J. L. Glajch, B. P. Semonian, and L. B. Rogers, *J. Chromatogr.*, **133,** 37 (1977).

115. M. J. Hartigan, J. E. Purcell, M. Novotny, M. L. McConnell, and M. L. Lee, *J. Chromatogr.*, **99,** 339 (1974).

116. J. A. Lubkowitz, B. P. Semonian, H. Galobardes, and L. B. Rogers, *Anal. Chem.*, **50,** 672 (1978).

117. R. Greenhalgh, J. Muller, and W. A. Aue, *J. Chromatogr. Sci.*, **16,** 8 (1978).

118. R. Greenhalgh and W. P. Cochrane, *J. Chromatogr.*, **70,** 37 (1972).

119. M. L. Selucky, *Chromatographia*, **5,** 359 (1972).

120. D. M. Coulson, *J. Gas Chromatogr.*, **4,** 285 (1966).

121. D. M. Coulson and L. A. Cavanagh, *Anal. Chem.*, **32,** 1245 (1960).

122. R. L. Martin, *Anal. Chem.*, **38,** 1209 (1960).

123. R. C. Hall, *J. Chromatogr. Sci.*, **12,** 152 (1974).

124. J. W. Rhoades and D. E. Johnson, *J. Chromatogr. Sci.*, **8,** 1616 (1970).

125. B. E. Pape, *Clin. Chem.*, **22,** 739 (1976).

126. G. C. Patchett, *J. Chromatogr. Sci.*, **8,** 155 (1970).

127. P. Jones and G. Nickless, *J. Chromatogr.*, **73,** 19 (1972).

128. J. F. Lawrence, *J. Chromatogr.*, **87,** 333 (1973).

129. C. C. Cassil, R. P. Stanovick, and R. F. Cook, *Res. Rev.*, **26,** 63 (1969).

130. J. W. Dolan and R. C. Hall, *Anal. Chem.*, **45,** 2199 (1973).

131. D. M. Hailey, A. G. Howard, and G. Nickless, *J. Chromatogr.*, **100,** 49 (1974).

132. B. E. Pape, D. H. Rodgers, and T. C. Flynn, *J. Chromatogr.*, **134,** 1 (1974).

133. W. P. Cochrane, B. P. Wilson, and R. Greenhalgh, *J. Chromatogr.*, **75,** 207 (1973).

134. W. P. Cochrane and R. Greenhalgh, *Int. J. Environ. Anal. Chem.*, **3,** 199 (1974).

135. J. F. Lawrence and A. H. Moore, *Anal. Chem.*, **46,** 755 (1974).

136. G. Winnert, *J. Chromatogr. Sci.*, **14,** 255 (1966).

137. J. F. Lawrence and N. P. Sen, *J. Chromatogr.*, **47,** 367 (1975).

138. J. F. Lawrence, *J. Chromatogr.*, **128,** 154 (1976).

139. R. C. Hall, B. J. Ehrlich, and P. W. Thiede, paper 33, 29th Pittsburgh Conference, Feb., 1978.

140. B. P. Wilson and W. P. Cochrane, *J. Chromatogr.*, **106,** 174 (1974).

141. L. A. Beaver, *Analytical Methods for Pesticides and Growth Regulators* (G. Zweig and J. Sherma, Eds.), Academic Press, New York, 1972.

142. A. Berton, *Off. Emball*, **1974,** 82.

143. J. McDonald and J. W. King, *J. Chromatogr.*, **124,** 124 (1976).

144. L. E. St. John and C. A. Bache, *J. Assoc. Off. Anal. Chem.*, **55,** 1152 (1972).

145. L. P. Sarna and G. R. B. Webster, *J. Assoc. Off. Anal. Chem.*, **57,** 1279 (1974).

146. C. J. W. Brooks and B. S. Middleditch, *Mass Spectrometry*, Vol. 4, (R. A. Johnstone, Ed.), Chemical Society, London, 1977.

147. C. Fenselau, *Anal. Chem.*, **49,** 563A (1977).

148. E. C. Horning, D. I. Carroll, I. Dzidic, R. N. Stillwell, and J. P. Thenot, *J. Assoc. Off. Anal. Chem.*, **61,** 1222 (1978).

149. W. D. Lehman and H. R. Schulten, *Angew. Chem. Int. Edit.*, **17,** 221 (1978).

150. W. H. McFadden, *Techniques of Combined GC/MS*, Wiley-Interscience, New York, 1973.

151. S. Safe and O. Hutzinger, *Mass Spectrometry of Pesticides and Pollutants*, C.R.C. Press, Cleveland, 1973.

152. B. J. Millard, *Quantitative Mass-Spectrometry*, Heyden and Sons, London, 1977.

153. *Identification and Analysis of Organic Pollutants in Water*, (L. W. Keith, Ed.), Ann Arbor Science Publishers, Ann Arbor, Michigan, 1976.

154. *Mass-Spectrometry and NMR Spectrometry in Pesticide Chemistry*, (R. Haque and F. J. Biros, Eds.), Plenum Press, New York, 1974.

155. G. Vander-Velde and J. F. Ryan, *J. Chromatogr. Sci.*, **13,** 322 (1975).

156. W. L. Budde and J. W. Eichelberger, *J. Chromatogr.*, **134,** 147 (1977).

157. J. M. Desmarchelier, D. A. Wunster and T. K. Fukuto, *Residue Rev.*, **63,** 77 (1970).

158. D. Henneburg, N. Henrichs, and G. Schomburg, *J. Chromatogr.*, **122,** 343 (1975).

159. H. J. Stan, *Chromatographia*, **10,** 233 (1977).

160. The Chemical Information System Data Base, Office of Standard Reference Data, National Bureau of Standards, Washington, D.C.

161. S. I. M. Skinner and R. Greenhalgh, *Reference Guide to Mass Spectra of Insecticides, Herbicides and Fungicides and their Metabolites*, Agriculture Canada, Ottawa, 1977.

162. T. Cairns and R. Jacobsen, *Mass Spectral Data Compilation of Pesticides and Industrial Chemicals*, U.S. Food and Drug Admin., Los Angeles, 1978.

163. J. Freudenthal and L. G. Gramberg, *Catalogue of the Mass Spectra of Pesticides*, National Institute of Public Health, Bilthoven, Netherlands, 1975.

164. R. Mestres, C. L. Chevallier, C. I. Espinoza, and R. Cornet, *Ann. Fals. Exp. Chim.*, **70,** 177 (1977).

165. W. J. Richter and H. Schwarz, *Angew. Chem. Int. Edit.*, **17**, 424 (1978).

166. H. J. Stan, *Freseenius Z. Anal. Chem.*, **287**, 104 (1977).

167. R. C. Dougherty, J. Dalton, and F. J. Biros, *Org. Mass Spectrom.*, **6**, 1171 (1972).

168. D. F. Hunt, G. G. Stafford, F. W. Crow, and J. W. Russel, *Anal. Chem.*, **48**, 2008 (1976).

169. J. R. Kennings, *Mass-Spectrometry* Vol. 4, (R. A. W. Johnstone, Ed.), Chemical Society, London, 1977.

170. E. C. Horning, D. I. Carroll, I. Dzidic, and R. N. Stillwell, *Pure Appl. Chem.*, **50**, 113 (1978).

171. P. C. Rankin, *J. Assoc. Off. Anal. Chem.*, **54**, 1340 (1971).

172. K. L. Busch, M. M. Bursey, J. R. Hass, and G. W. Sovocool, *Appl. Spectrosc.*, **32**, 388 (1978).

173. S. C. Gates, M. J. Smisko, C. L. Ashendel, N. D. Young, J. F. Holland, and C. C. Sweely, *Anal. Chem.*, **50**, 433 (1978).

174. D. W. Kuehl, *Anal. Chem.*, **49**, 521 (1977).

175. H. T. Cory, P. T. Lascelles, B. J. Millard, W. Snedden, and B. W. Wilson, *Biomed. Mass Spectrom.*, **3**, 117 (1976).

176. L. A. Shadoff and P. A. Hummel, *Biomed. Mass Spectrom.*, **5**, 7 (1978).

177. *Federal Register USA*, **42**, No. 35 (1977).

178. J. A. Sphon, *J. Assoc. Off. Anal. Chem.*, **61**, 1247 (1978).

179. H. H. Hausdorff, *J. Chromatog.*, **134**, 131 (1977).

180. P. Coffey, D. R. Mattson, and J. C. Wright, *Am. Lab.*, **1978**, 77.

181. D. L. Wall and A. W. Mantz, *Appl. Spectr.*, **31**, 552 (1977).

182. P. Coffey, D. R. Mattson, and J. C. Wright, *Am. Lab.*, **1978**, 126.

183. P. R. Griffiths, *Appl. Spectrosc.*, **31**, 497 (1977).

184. A. J. McCormack, S. C. Tong, and W. D. Cooke, *Anal. Chem.*, **37**, 1470 (1965).

185. W. R. McLean, D. L. Stanton, and G. E. Penketh, *Analyst*, **98**, 432 (1973).

186. C. A. Bache and D. J. Lisk, *Anal. Chem.*, **39**, 787 (1967).

187. C. I. M. Beenakker, *Spectrochim. Acta*, (*B*), **31**, 483 (1976).

188. C. I. M. Beenakker, *Spectrochim. Acta*, (*B*), **32**, 173 (1977).

189. B. D. Quimby, P. C. Uden, and R. M. Barnes, *Anal. Chem.*, **50**, 2112 (1978).

190. P. C. Uden, R. M. Barnes, and F. O. DiSanza, *Anal. Chem.*, **50**, 852 (1978).

191. Y. Talmi and A. W. Andren, *Anal. Chem.*, **46**, 2122 (1974).

192. Y. Talmi, *Anal. Chim. Acta*, **74**, 107 (1975).

193. Y. Talmi and D. T. Bostick, *Anal. Chem.*, **47**, 2145 (1975).

194. D. Reamer, W. H. Zollerad, and T. C. O'Haver, *Anal. Chem.*, **50**, 1449 (1978).

195. Y. K. Chan, P. T. Wong, and H. Seitoh, *J. Chromatogr. Sci.*, **14**, 162 (1972).

196. G. E. Parris, W. R. Blair, and F. E. Brinckman, *Anal. Chem.*, **49**, 378 (1977).

197. W. R. Wolf, *J. Chromatogr.*, **134**, 159 (1977).

198. J. N. Driscoll, *J. Chromatogr.*, **134**, 49 (1977).

199. J. N. Driscoll, *Am. Lab.*, **1978**, 137.

200. A. R. Oyler, D. C. Bodenner, K. J. Welch, R. J. Linkkonen, R. M. Carlson, H. L. Kopperman, and R. Caple, *Anal. Chem.*, **50**, 837 (1978).

201. R. K. Stevens and J. A. Hodgeson, *Anal. Chem.*, **45**, 443A (1973).

202. D. H. Fine, F. Rufeh, and D. Lieb, *Anal. Chem.*, **47**, 1188 (1975).

203. D. H. Fine and D. P. Raunbehler, *J. Chromatogr.*, **109**, 271 (1975).

204. T. A. Gough, K. S. Webb, and R. F. Eaton, *J. Chromatogr.*, **137**, 293 (1977).

205. T. A. Gough, K. S. Webb, M. A. Pringuer, and B. J. Wood, *J. Agric. Food Chem.*, **25**, 663 (1977).

206. W. L. Oller and M. F. Crammer, *J. Chromatogr. Sci.*, **13**, 296 (1975).

207. I. S. Krull, *Res. Rev.*, **66**, 185 (1977).

208. P. E. Mattson and S. Nygren, *J. Chromatogr.*, **123**, 396 (1976).

209. H. L. Crist and R. F. Moseman, *J. Assoc. Off. Anal. Chem.*, **60**, 1277 (1977).

210. D. J. Hallett, R. J. Norstrom, F. I. Onuska, M. E. Comba, and R. Sampson, *J. Agric. Food Chem.*, **24**, 1189 (1976).

211. W. P. Cochrane, *J. Chromatogr. Sci.*, **13**, 246 (1975).

212. S. Kapila and W. A. Aue, *J. Chromatogr.*, **148**, 343 (1978).

213. W. A. Aue and S. Kapila, *Anal. Chem.*, **50**, 536 (1978).

214. J. A. Singmaster and D. G. Crosby, *Bull. Environ. Contam. Toxicol.*, **16**, 291 (1976).

215. B. A. Olsen, T. C. Sneath, and N. C. Jain, *J. Agric. Food Chem.*, **26**, 640 (1978).

216. H. Agemian and A. S. Y. Chau, *J. Assoc. Off. Anal. Chem.*, **60**, 1070 (1977).

217. M. L. Hattula and S. Raisanen, *Bull. Environ. Contam. Toxicol.*, **16**, 355 (1976).

218. C. H. Van Peteghem and A. M. Heydrickx, *J. Agric. Food Chem.*, **24**, 635 (1976).

219. S. O. Farwell, F. W. Bowes, and D. F. Adams, *Anal. Chem.*, **48**, 420 (1976).

220. L. I. Butler and L. M. J. McDonough, *Anal. Chem.*, **54**, 1357 (1971).

221. E. R. J. Holden, *J. Assoc. Off. Anal. Chem.*, **56**, 713 (1973).

222. J. A. Coburn and A. S. Y. Chau, *J. Assoc. Off. Anal. Chem.*, **57**, 1272 (1974).

223. J. M. Rosenfeld and J. L. Crocco, *Anal. Chem.*, **50**, 701 (1978).

224. M. C. Bowman and M. Beroza, *J. Assoc. Off. Anal. Chem.*, **50**, 933 (1967).

225. K. Jacob and W. Vogt, *J. Chromatogr.*, **150**, 339 (1978).

226. C. Rappe, H. R. Buser and H. P. Bosshardt, *Chemosphere*, **5**, 431 (1978).

227. G. W. Bowes, M. J. Mulvihill, B. R. T. Simoneit, A. L. Burlingame, and R. W. Riseborough, *Nature*, **256**, 305 (1975).

228. H. R. Buser and H. P. Bosshardt, *J. Assoc. Off. Anal. Chem.*, **59**, 562 (1976).

229. H. R. Buser and C. Rappe, *Chemosphere*, **2**, 199 (1978).

230. H. R. Buser, H. P. Bosshardt, C. Rappe, and R. Lindahl, *Chemosphere*, **5**, 419 (1978).

231. L. A. Shadoff, R. A. Hummel, D. K. Jenson, and N. H. Mahle, *Ann. Chem.*, **67**, 583 (1977).

232. N. H. Mahle, H. S. Higgins, and M. E. Getzendaner, *Bull. Environ. Contam. Toxicol.*, **18**, 123 (1977).

233. L. L. Lamparski, N. H. Mahle, and L. A. Shadoff, *J. Agric. Food Chem.*, **26**, 1113 (1978).

234. R. F. Brady, Jr. and B. C. Pettitt, *J. Chromatogr.*, **93**, 375 (1974).

235. P. Lombardo and I. J. Egry, *J. Assoc. Off. Anal. Chem.*, **62**, 47 (1979).

236. H. Maier, E. Bode, and M. Riedmann, *Res. Rev.*, **54**, 113 (1975).

237. M. Riva and A. Carisano, *J. Chromatogr.*, **42**, 464 (1969).

238. E. J. Lorah and D. D. Hemphill, *J. Assoc. Off. Anal. Chem.*, **57**, 570 (1974).

239. R. Greenhalgh and J. Kovacicova, *J. Agric. Food Chem.*, **23**, 325 (1975).

240. R. C. Hall and D. E. Harris, *J. Chromatogr.* **169**, 245 (1979).

241. H. W. Dorough and J. H. Thorstenson, *J. Chromatogr. Sci.*, **13**, 212 (1975).

242. F. D. Magallona, *Res. Rev.*, **56**, 1 (1975).

243. J. Sherma and T. M. Shafik, *Arch. Environ. Contam. Toxicol.*, **3**, 55 (1975).

244. J. N. Seiber, *J. Agric. Food Chem.*, **20**, 443 (1972).

245. J. Lawrence, *J. Chromatogr.*, **123**, 287 (1976).

246. J. Lawrence, D. A. Lewis, and H. A. McLeod, *J. Chromatogr.*, **138**, 143 (1977).

247. R. Greenhalgh, R. R. King, and W. D. Marshall, *J. Agric. Food Chem.*, **26**, 475 (1978).

248. R. Greenhalgh, W. D. Marshall, and R. R. King, *J. Agric. Food Chem.*, **24**, 266 (1976).

249. R. Greenhalgh, G. W. Wood, and P. A. Pearce, *J. Environ. Sci. Health*, **B12**, 229 (1977).

250. H. A. Moye, *J. Agric. Food Chem.*, **19**, 452 (1971).

251. P. T. Holland, *Pestic. Sci.*, **8**, 354 (1977).

252. R. C. Holmstead and J. E. Casida, *J. Assoc. Off. Anal. Chem.*, **58**, 541 (1975).

253. R. A. Chapman and J. R. Robinson, *J. Chromatogr.*, **140**, 209 (1977).

254. I. Bannok and H. Geissbuhler, *Bull. Environ. Contam. Toxicol.*, **3**, 7 (1968).

255. J. F. Lawrence, *J. Assoc. Off. Anal. Chem.*, **59**, 1061 (1976).

256. A. Buchert and H. Lokke, *J. Chromatogr.*, **115**, 682 (1975).

257. J. F. Lawrence and K. M. S. Sundaram, *J. Assoc. Off. Anal. Chem.*, **59**, 938 (1976).

258. J. F. Lawrence, *J. Chromatogr. Sci.*, **14**, 558 (1976).

259. R. Purkayastha and W. P. Cochrane, *J. Agric. Food Chem.*, **21**, 93 (1973).

260. C. E. McKane, T. H. Bryant and R. J. Hance, *Analyst*, **97**, 653 (1977).

261. S. U. Khan, R. Greenhalgh, and W. P. Cochrane, *J. Agric. Food Chem.*, **23**, 430 (1975).

262. G. B. Downer, M. Hall, and D. N. B. Mallen, *J. Agric. Food Chem.*, **24**, 1223 (1976).

263. R. R. King, *J. Agric. Food Chem.*, **26**, 1460 (1978).

264. S. U. Khan and K. S. Lee, *J. Agric. Food Chem.*, **24**, 684 (1976).

265. P. F. Lott, J. W. Lott, and D. J. Doms, *J. Chromatogr. Sci.*, **16**, 390 (1978).

266. G. H. Draffen, R. A. Clare, D. L. Davies, G. Hawkesworth, S. Murray, and D. S. Davies, *J. Chromatogr.*, **139**, 311 (1977).

267. W. P. Cochrane, R. Greenhalgh, and N. E. Looney, *J. Assoc. Off. Anal. Chem.*, **59**, 617 (1976).

268. J. Hurter, M. Manser, and B. Zimmerli, *J. Agric. Food Chem.*, **26**, 472 (1978).

269. H. A. Moye, presented at 4th International Congress of Pesticide Chemistry (IUPAC), Zurich, July, 1978.

270. R. A. Simonaitis and R. S. Cail, *J. Assoc. Off. Anal. Chem.*, **58**, 1032 (1976).

271. R. C. Holmstead and D. M. Soderland, *J. Assoc. Off. Anal. Chem.*, **66**, 685 (1975).

272. I. H. Williams, *Pestic. Sci.*, **7**, 336 (1976).

273. R. A. Chapman and H. S. Simmons, *J. Assoc. Off. Anal. Chem.*, **60**, 977 (1977).

274. Y. W. Lee, N. D. Westcott, and R. A. Reichle, *J. Assoc. Off. Anal. Chem.*, **61**, 869 (1978).

275. G. H. Fujie and O. H. Fullmer, *J. Agric. Food Chem.*, **26**, 395 (1978).

276. C. F. Keppel, *J. Assoc. Off. Anal. Chem.*, **54**, 528 (1971).

277. C. Bighi, *J. Chromatogr.*, **14**, 348 (1964).

278. W. H. Newsome, *J. Agric. Food Chem.*, **22**, 886 (1974).

279. IUPAC Special Report No. 1, Ethylenethiourea, Coordinator, R. Engst, *Pure Appl. Chem.*, **49**, 675 (1977).

280. S. Otto, W. Keller, and N. Drescher, *J. Environ. Sci. Health*, **B12**, 179 (1977).

281. R. R. Watts, R. W. Storherr, and J. H. Onley, *Bull. Environ. Contam. Toxicol.*, **12**, 224 (1974).

282. R. R. King, *J. Agric. Food Chem.*, **25**, 73 (1977).

283. J. H. Onley and G. Yip, *J. Assoc. Off. Anal. Chem.*, **54**, 165 (1971).

284. Z. Pecka, P. Baulu, and H. W. Newsome, *Pestic. Monit. J.*, **8**, 232 (1975).

285. R. G. Nash, *J. Assoc. Off. Anal. Chem.*, **58**, 567 (1975).

286. J. O. Watts, K. W. Boyer, A. Cortez, and E. R. Elkins, *J. Assoc. Off. Anal. Chem.*, **59**, 1226 (1976).

287. M. L. Schafer, U. Rhea, J. T. Peller, C. H. Hamilton, and J. Campbell, *J. Agric. Food Chem.*, **23**, 1079 (1975).

288. C. J. Capon and J. C. Smith, *Anal. Chem.*, **49**, 365 (1977).

289. W. O. Gauer, J. N. Seiber, and D. G. Crosby, *J. Agric. Food Chem.*, **22**, 252 (1974).

290. R. G. Lewis, *Res. Rev.*, **68**, 123 (1977).

291. Y. Shimoishi, *J. Chromatogr.*, **136**, 85 (1977).

292. C. F. Poole, N. J. Evans, and D. G. Wibberly, *J. Chromatogr.*, **136**, 73 (1977).

293. T. Stijve and E. Cardinale, *J. Chromatogr.*, **109**, 239 (1975).

294. Y. K. Chan, P. T. S. Wong, and P. D. Goulden, *Anal. Chem.*, **47**, 2279 (1975).

QUANTITATIVE IN SITU ANALYSIS OF PESTICIDES ON THIN LAYER CHROMATOGRAMS

JAMES D. MACNEIL

Laboratory and Scientific Services Division, Revenue Canada, Customs and Excise, Ottawa, Ontario, Canada

ROLAND W. FREI

Department of Analytical Chemistry, The Free University of Amsterdam, Amsterdam, The Netherlands

CONTENTS

3.1. INTRODUCTION

Thin layer chromatography (TLC) as used today in laboratories can probably best be attributed to Meinhard and Hall (1), who prepared mechanically stable adsorbent layers by the addition of a binder. Earlier research (2–7) had been with loose layers, which, lacking binders, were more difficult to handle. In the early 1950s, Kirchner and coworkers (8, 9) introduced "chromatostrips" and "chromatoplates." In 1954 TLC was first used for the determination of pesticide residues when Kirchner and Rice (10) analyzed for biphenyl in citrus fruits using silica gel chromatostrips for a cleanup separation. However, interest in TLC for pesticide analysis did not really develop until the 1960s. A review by Abbott and Thompson (11) in 1965 listed about 60 references to pesticide residues, most of which were published after 1961.

A growing use of TLC for pesticide analysis was reflected in more reviews published during succeeding years. The chromatography of triazines was reviewed by Fishbein (12), who also coauthored reviews of the TLC of carbamates (13), ureas and thioureas (14), and methylenedioxyphenol compounds (15). The TLC of organophosphorus compounds was reviewed by Watts (16), while Sherma has published a comprehensive review of the chromatography of fungicides (17). Methods reported, however, were largely qualitative or semiquantitative, as observed by Thornberg (18) in 1971. This situation has changed during the 1970s, as shown in a more recent review (19). This chapter will deal primarily with advances in quantitative TLC through in situ measurement of uv–vis absorbance, reflectance, and fluorescence.

3.2. ANALYSIS BY UV-VIS REFLECTANCE SPECTROSCOPY

Quantitative in situ spectroscopic analysis of pesticides separated on thin layer chromatograms was first reported in 1968 by Beroza et al. (20), Frei and Nomura (21), and Huber (22). In a previous study, Frei et al. (23) had shown that fluorescence quenching could be used to detect submicrogram quantities of a number of s-triazine herbicides separated on silica gel thin layers containing a fluorescent indicator. Subsequently, UV reflectance spectra of some s-triazine herbicides were measured in situ on silica gel thin layers (21). Chlorotriazines were shown to exhibit two absorption maxima, while others had only one. Reflectance spectra were found to be broader than corresponding transmission spectra and exhibited a bathochromic shift for these compounds. Measuring in situ, quantities of about 1 μg could be detected with an accuracy of about 10%. Improved accuracy was obtained by removing the developed spot of herbicide from the chromatogram along with the adsorbent to which it adhered, homogenizing this sample and then measuring the reflectance. With this approach an accuracy of within 3% was obtained for 10-μg samples.

Beroza et al. (20) described a fiber-optics scanner designed for in situ reflectance analysis of thin layer chromatograms and demonstrated its application for a number of chlorinated and thiophosphate pesticides. (A commercial version of this experimental instrument is now available.) Plots of measured peak area vs. concentration for 13 chlorinated insecticides were generally linear over a concentration range of 0.2–10.0 μg/spot. Variations observed between plates indicated a need to run standards on each plate.

This observation generally holds true for most in situ TLC methods. Similar results were found for the group of thiophosphate pesticides studied. Analysis of partially cleaned-up lettuce extracts fortified with the insecticide bensulide (0.05 ppm) and its oxygen analog (0.10 ppm) produced results of 0.04 ppm and 0.09 ppm, respectively.

The scanner described by Beroza et al. (20) was but one of the first of many instruments developed or modified for in situ spectroscopic analysis of compounds separated on thin layer chromatograms (24–26). These instruments provided an impetus for the development of new methodology and ended the need for the tedious removal of spots from chromatograms to make quantitative measurements in solution.

Further experiments with the fiber-optics scanner described by Beroza et al. (20) were reported by Getz (27). Organochlorine insecticides (p,p'-DDT, p,p'-DDE) separated by TLC were visualized by dipping the chromatogram in a solution of 1% silver nitrate in methanol containing 2% phenoxyethanol, followed by exposure to ultraviolet light. The resulting colored spots were measured in situ on the chromatogram. Broccoli extracts fortified with several organophosphorus insecticides, including methyl parathion, were also analyzed in a similar fashion, using color-forming reactions. Carrot extracts containing Landrin®, a carbamate insecticide, were analyzed by in situ reflectance measurement of the colored spots produced by reacting the chromatographed Landrin® with ninhydrin. These experimental results led Getz to suggest that quantitative TLC could supplement gas–liquid chromatography (GLC) methods. Getz and Hill (28) subsequently tested the sensitivity, selectivity, and reproducibility of a number of spray reagents used in quantitative TLC.

Kynast (29–32) studied an in situ reflectance method for the quantitative analysis of pesticide formulations under quality control conditions. Betenal®, a commercial herbicide, was analyzed for the active ingredient phenmedipham, as well as for a number of impurities (29). Phenmedipham was measured at 240 nm, while impurities were diazotized, coupled with β-naphthol and the resulting color measured at 495 nm. Additional work showed that impurities in formulations could be quantitatively detected at 0.1%, with a relative standard deviation of about ±5% (30). Improved accuracy, with a considerable saving in time, was achieved by developing a computer program which was used in routine quality control (31, 32).

Similar methodology was used by Kossman (33) to analyze for chlorphenamidine and formetanate in formulated products. TLC was preferred to GLC for these compounds due to their low volatility and thermal instability. Reflectance measurements were made at 270 nm, with a sensitivity of 0.5 μg/spot, equivalent to 0.02 ppm of pesticide in crop samples. Specificity was added by diazotization and coupling with β-naphthol, but resulted in decreased accuracy. Recoveries from crop material were about 80%, with relative standard deviations of 4–11% at concentrations of 0.2–1.3 ppm.

Reflectance measurements at 510 nm and transmittance measurements at 380 nm were used by Petrowitz and Wagner (34) for the in situ determination of isomers of 1,2,3,4,5,6-hexachlorocyclohexane separated on silica gel thin layers. Colored spots for measurement were produced by spraying the chromatogram with o-toluidine, then exposing it to uv–light (254 nm) for 10 min. A larger linear working range was observed for reflectance measurements than for transmittance measurements. Watanabe et al. (35) investigated the maximum uv absorbance of several pesticides, including fenthion and 2-isopropoxyphenyl methylcarbamate, and performed analyses of these compounds separated on thin layers by densitometry.

Electron donor–acceptor (EDA) complex formation has been surveyed as a means of detecting pesticides separated by TLC (36, 37) and the chromatography and detection of a number of carbamate and urea pesticides with electron-acceptor reagents has been reported (38). The electron-acceptor reagent 9-dicyanomethylene-2,4,7-trinitrofluorene (CNTNF) was used to form complexes with the carbamate insecticide Mobam® in a quantitative study (39). Reflectance analyses were performed at 490 nm on silica gel thin layers and at 500 nm on cellulose, with detection limits of about 1 μg/spot. Relative standard deviations reported were better than ±10%, except near the detection limit. Stability studies showed no loss in intensity on cellulose, but a gradual decrease on silica gel. A linear response of peak area vs. concentration was observed within a limited concentration range. Further investigations with EDA complexes (40) revealed linear response over a wider concentration range when a plot of reflectance vs the square root of concentration, an approximation of the Kubelka–Munk functions (41), was used.

EDA complex formation was used in a study of the photochemistry of methoxychlor to detect a number of the compounds formed (42). Methoxychlor standard was irradiated at 310 nm in heptane solution under a nitrogen atmosphere for 100 min. Thin films of methoxychlor were also irradiated in the laboratory under a sunlamp and fluorescent lamps having a spectral range similar to sunlight and outdoors under natural sunlight. In heptane solution, the degradation of methoxychlor and formation of methoxychlor olefin and DME, 1,1-dichloro-2,2-bis(p-methoxy-phenyl) ethane, was followed by removing aliquots at regular intervals. These were spotted on silica gel plates, chromatographed, sprayed with 2,4,7-trinitro-9-fluorenone (TNF) to form colored spots and then quantitated by measuring the reflectance at 430 nm for methoxychlor and DME complexes, and at 440 nm for methoxychlor olefin complex. Confirmation of these three compounds and identification of other compounds formed was by mass spectrometry of the EDA complexes or by GLC-mass spectrometry of bands recovered from preparative TLC. Compounds formed under the various photochemical conditions were compared.

The hydrolysis rates of the insecticides carbaryl and Mobam® and the herbicides propham and chlorpropham were also studied using a similar approach (43). Standards of these pesticides were left in aqueous solutions buffered at pH's 2, 7, 8, and 9 over a 5-week period. Aliquots were removed at regular intervals, extracted, dried, and the residue dissolved in acetone. Samples were chromatographed, complexed with CNTNF, and quantitated by in situ reflectance measurements. The two N-methyl carbamate insecticides were stable at acid pH, but degraded at an increasing rate as the basicity increased. No degradation of the N-phenyl carbamate herbicides was observed in the pH range studied. A final report investigated the effects of pesticide structure on complex formation and the suitability of some electron-donor reagents (44).

A study which may be of some interest to residue chemists is the report of a densitometric method for the determination of ecdysones (45). A number of these compounds, including α- and β-ecdysone, were separated on silica gel F_{254} thin layers and determined by densitometry. Solvent systems and chromatographic results were discussed.

Sherma (46) has used thin-layer densitometry to quantitate the fungicide benomyl and its metabolites, methyl 2-benzimidazole carbamate (MBC), and 2-aminobenzimidazole (2-AB). The compounds were chromatographed on fluorescent indicator plates and located by fluorescence quenching. Detection limits were about 0.5 μg, with a linear response for plots of peak area vs concentration from the detection limit up to 5 μg for benomyl and 9 μg for MBC and 2-AB. The method was viewed as a good alternative to GLC, high-performance liquid chromatography (HPLC), or spectrophotometry for residues of benomyl and its metabolites.

Onuska (47) reported the use of thin-layer densitometry to quantitate the fungicides ziram (zinc dimethyldithiocarbamate) and thiram (tetramethylthiuram disulfide), as well as some other N,N-dialkyldithiocarbamates in industrial wastewaters. These sulfur-containing compounds were complexed by adjusting 1-liter water samples to pH 10 and adding 50 ml of 1% copper dichloride in a 1:1 mixture of distilled water and glacial acetic acid. The complexed dithiocarbamates were extracted into chloroform and, following removal of water and volume reduction, aliquots were chromatographed on silica gel layers and quantitated by densitometric scanning. Detection limits were stable for at least 90 min subsequent to chromatography and reported coefficients of variation for analyses were 2–6%.

Sherma and Touchstone (48) reported the quantitation of chloramben residues in fortified tomato samples by in situ densitometry of the chromatographed herbicide. Prior to scanning, the chromatogram was sprayed with sodium nitrite and N-(l-naphthyl)ethylenediamine, resulting in a coupling reaction with chloramben to produce red spots. These were scanned at 505 nm and a linear plot of peak area vs concentration was obtained from the detection limit of 4 ng up to 80 ng.

Sherma et al. (49) have also studied the separation and quantitation of pesticide residues on precoated silica gel microscope slides. Results were compared with those obtained on conventional 20 × 20 plates. Pesticides included in the study were o,p'-DDT, carbaryl, phorate, and phorate metabolites. Carbaryl was reacted with p-nitrobenzenediazonium tetrafluoborate, while phorate and its metabolites were reacted with N-2,6-trichlorobenzoquinoneimine reagent, to produce visible spots after chromatography. Silver nitrate impregnated plates were used to produce a colored spot for o,p'-DDT. Quantitative densitometric measurements were found to have as good precision as those made on standard TLC plates. However, the short developing distance of 5 cm is a chromatographic restriction when separation poses a difficult problem. The use of the slides was shown to provide, for certain applications, a less expensive approach than the use of conventional TLC plates.

Sherma and Koropchak (50) have developed a quantitative procedure for the analysis of chlorophenoxy acid herbicides and their salts in water. Linear calibration plots were obtained by densitometry at concentrations of 100–1000 ng/spot for the compounds tested. A relative standard deviation of ±5.3% was calculated for 6 spots of MCPA on a single plate at a concentration of 500 ng/spot. Chromatography was on silica gel G plates impregnated with silver nitrate. Developed chromatograms were irradiated with ultraviolet light, causing the herbicides to appear as black spots on a light grey or tan background. With the addition of preliminary extraction and cleanup procedures, a sensitivity to 1 ppb of herbicide in water was achieved.

3.3. IN SITU FLUORESCENCE ANALYSES

Fluorescence measurements generally offer greater sensitivity and specificity than other quantitative TLC techniques, and it is in this area that much work in recent years has been concentrated. Native fluorescence, fluorogenic labeling, release of fluorescent chelates through complexation reactions, spray reagents sensitive to a polar environment, and conversion through heat or chemical treatment have all been used to develop quantitative in situ fluorescence–TLC methods (51).

3.3.1. Native Fluorescence

While a number of pesticides exhibit some native fluorescence, this has not usually been sufficient for quantitative TLC analysis. Some analytical methods were developed for measurements in solution, but most compounds required some pretreatment to produce fluorescence (52). However, there are some exceptions.

Norman et al. (53) used TLC and fluorometry for the determination of residues of the fungicide thiabendazole in and on citrus. Surface strips and extracts were chromatographed on alumina F_{254} thin-layer sheets following a preliminary cleanup by liquid partition. The chromatographed thiabendazole was located against the fluorescent background, and the area of the spot was removed from the chromatogram. Thiabendazole was eluted from the adsorbent with 25 ml methanol–0.1N HCl (99:1) and the fluorescence was measured in solution. Recoveries at fortification levels studied were 98%. Ebel and Herold (54) reported the direct determination of thiabendazole by in situ fluorometry on TLC. The plates were scanned at 355 nm, using a 310-nm excitation source, with the fluorometer interfaced to an integrator and a table-top computer. The computer corrected the curvilinear response observed between 0 and 400 ng to rectilinear, with an observed coefficient of variation for 20 spots on 5 plates at 50–200 ng of about 1.5%.

A routine method for the quantitative determination of thiabendazole residues by TLC and in situ fluorometry has also been reported by Ottender and Hezel (55). Results were compared with a previously reported reflectance method (56). Fluorescence was found to be less affected by coextractives than reflectance. Recoveries in excess of 90% were reported for fortified samples of oranges, grapefruit, lemons, and bananas. Standard deviation for the method was 2.67%.

A rapid TLC–in situ fluorometric method for analyzing residues of benomyl in surface washings from cherries has also been reported (57). The method was developed for quick assessment of the efficiency of application of the fungicide to cherries during packing operations to prevent storage rot. A linear relationship between peak area and concentration in the range 50–500 ng/spot was observed, with reproducibilities of ±10% or better. Less than 1 ppm of benomyl on cherries was detectable in commercially treated samples.

The acaricide–fungicide quinomethionate has been analyzed in water samples by in situ fluorometric measurements following TLC separation (58). Recoveries from fortified samples were better than 90% above 0.1 ppb, with coefficients of variation from 2.5 to 5.5% in the range 0.04–1.00 μg/spot. The method should be applicable to drinking water, lake and river water samples, and, with cleanup, to heavily polluted water samples. Preliminary results indicated the method should be applicable to fruit at the 1 ppb level. A method for the analysis of the organophosphate insecticide Maretin® by measurement of its native fluorescence has also been reported (59). Recoveries of 90% were achieved for water fortified at 0.1–2.0 ppb and of 90–100% for milk fortified at 10–50 ppb. In the range 0.02–1.0 μg/spot, coefficients of variation were 2.9–5.4%. Measurements were made at an excitation wavelength of 353 nm and an emission wavelength of 415 nm. The method provides an alternative to gas chromatography for this compound.

A method for the quantitative analysis of the plant growth regulator 3-indoleacetic acid has also been developed (60). Excitation and emission wavelengths were 293 nm and 356 nm, respectively, with a quantitative linear reponse in the working range 0.05–1.00 μg/spot. Coefficients of variation were from 7.3% at 1.0 μg to 11.3% at 0.05 μg. Fluorescence was enhanced by spraying the plates with 5% dimethyl sulfoxide in methanol, then drying the chromatogram for 1 hr prior to measurement to allow fluorescence to stabilize. Studies showed that only about 1% of the observed coefficients of variation was due to instrument error. The remaining error was due mainly to chromatographic factors and the difficulty in obtaining a uniform coating of the polar spray material causing fluorescence enhancement.

3.3.2. Fluorescence Through Release of Chelates

Frei and coworkers have studied a number of reactions involving the liberation of fluorescent chelates from spray reagents. An initial investigation (61) utilized a method of detection involving bromine treatment of organothiophosphorus insecticides, followed by use of a fluorogenic spray as described by Ragab (62). Further experiments (63) showed that, while a number of chelating agents could be used successfully, best results were obtained when the chromatograms were dried at 105° for 5 min, exposed to bromine vapor for 10 sec by placing the chromatographic plate in a chromatography chamber containing a solution of 10% Br_2 in CCl_4, then spraying with a reagent consisting of a 1:1 (v/v) mixture of manganese solution (0.1 g $MnCl_2 \cdot 4H_2O$ in 80% ethanol) and 0.05% (w/v) salicyl-2-aldehye-2-quinolylhydrazone (SAQH) in ethanol. Excitation and emission maxima were 365 nm and 460 nm, respectively. Relative standard deviations varied from ±10.2% at 0.1 μg/spot to ±2.4% at 1.0 μg/spot, with a linear response (peak area vs concentration) from 0.02 to 6.0 μg/spot. Recovery studies were carried out on water samples fortified with methyl azinphos and parathion.

This method was subsequently applied to the analysis of methyl azinphos in blueberries (64). Preliminary clean-up by liquid–liquid partition and column chromatography was required prior to TLC to remove interfering coextractives. Recoveries of 80–90% were obtained for samples fortified at the 0.5 ppm level, with a potential sensitivity of 0.05 ppm for the method. This compares favorably with GLC results which could be expected with a flame photometric detector but is less sensitive than the response obtained with electron-capture detection.

Sensitivity of this method is related to the number and oxidation state of sulfur atoms in the pesticide molecule, with thionate sulfur most actively promoting fluorescence. The measured fluorescence was that of free chelating agent, liberated by the presence of hydrobromic acid which was produced by the oxidation of sulfur-containing compounds by bromine. Groves (65) gave the following equation to illustrate the reaction which occurs:

$$\underset{\substack{\| \\ (C_2H_5O)_2-P-O-CH_2-S-C_2H_5}}{\overset{S}{}} + 5\ Br_2 + 6H_2O \rightarrow$$

$$\underset{\substack{\| \qquad\quad \| \\ (C_2H_5O)_2-P-O-CH_2-S-C_2H_5}}{\overset{O \qquad\quad O}{}} + H_2SO_4 + 10HBr \tag{1}$$

Oxidation of sulfur by bromine is thus the rate-determining step in this method of detection.

The bromination reaction has also been used with pH-sensitive spray reagents for fluorescence detection and quantitation of organothiophosphate insecticides (66–68). Initially, detection of a number of sulfur-containing pesticides and other organic compounds was attempted with the spray reagent 1,2-dichloro-4,5-dicyanobenzoquinone, which fluoresces in acid media, as well as some color-sensitive pH reagents (66). In a subsequent quantitative study (67), detection limits for organothiophosphate insecticides tested ranged from 50 ng for methyl azinphos and dimethoate to 10 ng for ethion. Detection limit for the sulfur-containing s-triazine herbicide prometryne was 100 ng. Relative standard deviations obtained for replicate dimethoate analyses ranged from 3.0 to 5.4% at 0.1–10 μg/spot. A final paper in this series discussed other pH-sensitive spray reagents which could be used for the detection of sulfur-containing pesticides (68).

A complexation reaction with release of a fluorescent chelating reagent has also been studied, using complexes of palladium chloride with the metallofluorescent indicators calcein and calcein blue to detect organothiophosphate insecticides (69). Palladium complexed preferentially with sulfur in these insecticides, releasing the fluorescent indicator. This complexation reaction of palladium chloride had previously been used to distinguish sulfur-containing organophosphates from those having no sulfur (70). Liberation of calcein was found to provide sensitivity of 10–50 ng/spot for phosphorodithioates and 50–100 ng/spot for phosphorothioates (69). Maxium stability for fluorescence measurement was obtained by storing chromatograms in a tank containing a saturated solution of calcium nitrate, thus providing a stable relative humidity of 51% for 24 hr prior to measurement. Coefficients of variation for dimethoate and methyl azinphos were 4–9% at concentrations of 50 ng/spot or greater, and about 15% near detection limits. Recovery studies were reported for lake water samples fortified with 2.0–20.0 ppb dimethoate.

Further studies were performed on lettuce extracts fortified with dimethoate and malathion (71). Dimethoate could be analyzed at the ppm level with no preliminary cleanup required. Recoveries from fortified samples were 85% or better. Interference from coextractives made a preliminary column cleanup according to Wessel (72) necessary for malathion.

This ligand-exchange process was also used for the detection and quantitative determination of other sulfur-containing compounds (73). Detection limits were in the 10–100 ng/spot range. A study of reproducibility for thioacetamide showed coefficients of variation from 5.9% at 100 ng/spot to 10% at 10 ng/spot. A linear relationship between peak area and concentration was observed in this concentration range. Reproducibility from plate-to-plate was sufficiently good that it was considered possible to plot a standard curve from accumulated data, which could be checked by running one or two standards on each chromatogram. In many cases, variations observed between chromatograms, especially when spray reagents are used, are so large that normal practice is to run sufficient standards to plot a calibration curve for each plate. With its excellent sensitivity for sulfur-containing compounds, this method could prove useful for rapid screening of samples for ethylene thiourea residues.

3.3.3. Polarity-Sensitive Spray Reagents

Compounds which fluoresce weakly in a nonpolar medium may fluoresce quite strongly in a polar environment. This phenomenon has been used for the detection of a variety of polar pesticides on thin layers with polarity-sensitive spray reagents (74–76). The reagents studied were 3-hydroxyflavones with an unsubstituted 5-position. When sprayed onto a chromatogram, they fluoresce in any area where a polar compound is located. As the pH of the chromatographic support is quite critical, best results were obtained on cellulose (75). However, organophosphorus insecticides could be detected on silica gel layers following bromination (77). Positive results were obtained with a variety of pesticides, including organophosphates, carbamates, s-triazines, and phenoxy acid herbicides (75). Some compounds were observed to produce quenching instead of fluorescence.

Further studies showed that best results were obtained when robinetin was used as a spray reagent, but flavenol and fisetin could also be used for quantitative work (76). Exposure to ultraviolet light reduced fluorescence dramatically. A linear response was obtained for plots of peak area vs concentration for 0.1–1.0 μg/spot, with a curvilinear response at higher concentrations. Below 0.1 μg/spot, background fluorescence interfered with quantitative work. Relative standard deviations varied from 3.1 to 12.2% in the concentration range studied. The nonspecific nature of the spray reagent was judged to be a limitation for its application to complex samples.

3.3.4. Conversion Reactions

In a study of the behavior of some naturally fluorescent pesticides on silica gel thin layers, Mallet and coworkers (78) observed that changes in the fluorescence spectrum and intensity could be produced by heating chromatograms at 200° for 45 min. These observations were attributed to chemical changes in the pesticides' structures caused by heating. Further research showed that selective heating (79) or spraying with strong acid or base, followed by heat treatment (80), produced fluorescence from a number of nonfluorescent or weakly fluorescent organophosphate insecticides. Optimum conditions for the detection of a number of pesticides were listed, with instrumental detection limits as low as 1 ng, though most were in the range 20–100 ng/spot. Using the combination of acidic or basic spray plus heating (80), fluorescence was produced from twelve of the 35 compounds tested.

These procedures were also tested on a variety of other pesticides (81). Fluorescence intensity was enhanced with alkali spray reagents. Acids produced spectral shifts, but also caused a decrease in fluorescence intensity. Again, sensitivities were in the submicrogram range for compounds studied, including quinomethionate, rotenone, and Warfarin®.

Quantitative studies were carried out for coumaphos and quinalphos in fortified lake and sewage water samples (82). Coumaphos was detected by heating the chromatogram at 200° for 20 min, while optimum conditions for quinalphos detection involved spraying the plate with $0.1N$ KOH (aqueous), then heating at 100° for 30 min. A linear relationship was observed for both compounds for plots of peak area vs concentration from 0.002–10 μg/spot, but poor reproducibility was observed from 0.002–0.01 μg/spot. Recoveries of both compounds exceeded 85% at fortification levels of 1.0 and 10.0 ppb in lake water samples and, for coumaphos, gave similar results even at the 0.01 ppb level. In sewage water, coumaphos recoveries ranged from 72% at 0.1 ppb to over 80% at higher concentrations, while quinalphos recoveries were poor (50–60%). Quantitative detection was not possible below 0.1 ppb in sewage water. Recovery problems in sewage water samples were attributed to emulsion formation during extraction.

A method has also been developed for the analysis of coumaphos and its analog in eggs using these procedures (83). A preliminary cleanup using liquid–liquid partition followed by chromatography on a Florisil® column was used to remove interfering coextractives. Two-dimensional chromatography was carried out on silica gel thin layers, with standards spotted on the unused part of the chromatogram after one-dimensional development for development in the second direction. Fluorescence was produced by heating chromatograms at 200° for 20 min. Recoveries were over 80% for both coumaphos and its oxygen analog at fortification levels of 0.02 ppm. Relative standard deviations for analyses were generally in the 12% range. The method was as sensitive as gas chromatography and offered some savings in time.

Carbaryl has been analyzed on TLC by conversion to naphtholate anion following chromatography (84). Spraying the chromatogram with $1N$ NaOH produces this fluorescent anion from both carbaryl and its major degradation product, 1-naphthol. Nanogram quantities of both compounds may be analyzed.

Francoeur and Mallet (85) have reported the conversion of captan and captafol to fluorescent species following chromatography on silica gel thin layers impregnated with silver nitrate. Fluorescence is produced by spraying the plate with $0.1M$ aqueous sodium chlorate and then heating it at 100° for 45 min. Analyses were conducted on samples of apples and potatoes fortified at a concentration of 0.2 ppm. Recoveries exceeded 90%, with a relative standard deviation of 7–8%. The detection limit reported was about 0.02 ppm.

3.3.5. Derivatization Reactions

Fluorogenic labeling is a sensitive and relatively specific technique that holds promise for the analysis of a number of pesticides, especially some of which are difficult to analyze by gas chromatography. This approach was first studied for N-methyl carbamate insecticides by Frei and Lawrence (86). Further study showed optimum reaction conditions for these compounds to be a prehydrolysis of 10–30 min, depending on the compound, followed by hydrolysis and coupling with the reagent dansyl chloride (1-dimethylaminonaphthalene-5-sulfonyl chloride) at 45° and pH 9 (87). A 4-fold excess of dansyl chloride over the carbamate concentration gave best results. Two fluorescent derivatives were produced for each compound, one from the coupled phenol moiety resulting from the hydrolysis and one from freed methyl amine, which also coupled with the reagent. As the phenol derivatives were characteristic of the compounds studied, these were chosen for analysis.

Fluorescence phenomena associated with dansyl derivatives of carbamate insecticides were investigated (88). A variety of polar spray reagents were tested to enhance fluorescence. It was shown that spraying the chromatogram with a solution of 20% triethanolamine in isopropanol stabilized fluorescence without changing the spectrum. Prolonged exposure to ultraviolet light reduced fluorescence, but heating chromatograms at 100° for 20 min had no effect.

Chromatographic separation of derivatized carbamates was also studied (89). Most derivatives could be separated on silica gel thin layers with benzene:acetone (98:2) or petroleum ether:triethylamine (3:1). Where a 2-dimensional development was needed, the benzene:acetone solvent system was used for the first dimension. Mobam® and carbaryl were not separated by either of these solvent systems, but could be resolved by carbon tetrachloride:methanol (99:1). Resolution of methiocarb and Landrin® could only be achieved when 0.2–0.5% silver oxide was incorporated in the silica gel layer. BUX® was not resolved into its two isomers, but appeared as a single spot. To prevent photochemical decomposition of the derivatives, chromatography should be carried out in the dark.

Quantitative studies revealed detection limits of 5 ng/spot (1 ng instrumentally with 2:1 signal-to-noise ratio) and a linear response (peak area vs. concentration) up to 300–400 ng/spot for most carbamate derivatives (90). Reproducibilities were near 5% in the range 5–500 ng/spot. Analysis of fortified water samples (tap, lake, and sea) for methiocarb, Landrin®. and aprocarb gave recoveries generally greater than 90% at concentrations of 2 ppb.

Further experiments were carried out with aminocarb and Zectran®, which in addition to phenol and methylamine dansyl derivatives, produce a dimethylamine derivative (91). Due to instability of the phenol derivatives, quantitative work was done with the amine derivatives. Results were comparable to those previously reported (90). Satisfactory recoveries were obtained from soil samples fortified with 200 ppb Zectran®. The methodology was also applied to the analysis of N-phenyl carbamate and phenylurea herbicides (92). Hydrolysis was carried out at 80° for 30–40 min, then the herbicide was extracted into hexane and a 10-μl aliquot was spotted on the chromatogram. A 4-μl aliquot of dansyl chloride solution was spotted over the herbicide, following which the chromatogram was covered with a clean glass plate and held in the dark at room temperature for 60 min prior to chromatography while the derivatization reaction occurred in situ. The chromatograms were developed with benzene:triethylamine:acetone (75:24:1) and sprayed with 20% triethanolamine in isopropanol or 20% paraffin oil in toluene to stabilize and enhance fluorescence. Reproducibilities and linear working ranges were similar to those reported for *N*-methyl carbamate insecticides (90). Recoveries near 100% were achieved for lake water samples fortified with 10 ppb swep and propham without any cleanup required.

Another reagent, 4-chloro-7-nitrobenzo-2,1,3-oxadiazole (NBD-Cl), was also used to form fluorescent derivatives of *N*-methyl and *N*,*N*-dimethyl carbamate insecticides (93). NBD-Cl is more specific in labeling reactions than dansyl chloride, as it reacts only with amino acids or primary amines, and does not form a fluoroescent hydrolysis product. Derivatives formed were more stable than dansyl derivatives when exposed to light. Relative standard deviations at concentrations 20–300 ng/spot were in the range 3–6% for NBD-methylamine and NBD-dimethylamine. However, as only amine derivatives were formed, the reaction could be used to analyze only total *N*-methyl carbamate or total *N*,*N*-dimethyl carbamate residues. It could not distinguish between individual compounds within each of these classes in a sample.

Analysis for carbamate and urea herbicides in foods has been carried out using dansyl labeling (94). Crops studied included beets, carrots, corn, oranges, parsnips, peas, potatoes, turnips, tomatoes, and strawberries. Samples were extracted with methanol, followed by a liquid partition step with chloroform and water. The organic phase was collected and evaporated to dryness, then hydrolyzed with 0.5 ml 2*N* KOH at 110° for 20–30 min, and finally extracted with hexane. A 5-μl aliquot was spotted on a silica gel plate along with 5-μl of dansyl chloride reagent as previously reported (92) and left in the dark for 30 min for reaction to occur. Chromatograms were developed in chloroform:hexane (3:1). Compounds were removed from the developed chromatogram and dissolved in acetone for quantitative analysis by fluorescence measurement in solution. Quantitative recoveries were obtained for linuron in peas at the 0.2 ppm level. Linuron and monuron were confirmed in extracts of beets, carrots, parsnips, potatoes, and turnips.

This approach was also followed in the analysis of residues of triazine herbicides in food crops (95). Extraction procedures were similar to those described for carbamate and urea herbicides (94), but hydrolysis was achieved by heating the dried extract with 1 ml 1*N* HCl at 150° for 16–18 hr. After cooling, 1 ml 1*N* NaOH, then 2 ml 5% borax solution, were added and the mixture was shaken. Dansyl chloride solution (1 ml) was then added and reacted at 45° for 15 min. Dansyl derivatives were then extracted into benzene and aliquots were chromatographed on silica gel layers using hexane:chloroform (3:1). As hydrolysis of different triazine herbicides produces different freed amines, the dansyl:amine derivatives detected are indicative of the triazine(s) present in the sample. Compounds were eluted from the plate following chromatography and the fluorescence was measured in solution. Recoveries for simazine spiked in corn samples at 0.2 ppm and carried through all steps of the analysis averaged 85%.

Sherma and Marzoni (96) reported the reaction of fluorescamine with anilines formed as pesticide degradation products. Quantitation was studied, with detection limits of about 4 ng/spot. Subsequently, Sherma and Touchstone (48) used this approach in analyzing for chloramben residues in tomato and lima bean samples fortified at the 0.1 ppm level. To produce fluorescence, the chromatogram was sprayed with a solution of fluorescamine in acetone (50 mg/100 ml) and then with triethylamine in methylene chloride (1:9 v/v). A linear calibration curve was obtained for concentrations of 20–500 ng/spot plotted vs peak area.

Derivatization techniques offer one of the most promising approaches to the development of rapid analytical methods of sufficient sensitivity to be of value to residue chemists. They should be particularly useful for some of the more polar compounds which are difficult to analyze by gas chromatography due to strong retention, thermal instability or poor detector response. Derivatization techniques and the reactions involved have been the subject of several reviews (51, 97).

3.4. OTHER TECHNIQUES

Researchers without access to thin-layer scanners may use other methods to obtain a semiquantitative estimate of concentrations. Huse and Adamovic (98), for example, have used spot area comparisons to estimate residue levels of ametryne and atrazine in soil. Extracts were chromatographed on alumina thin layers and spots of interest were eluted and chromatographed on silica gel thin layers containing a fluorescent indicator. Triazines were located by fluorescence quenching at 0.1 μg/spot, corresponding to 0.005 ppm in the soil sample. Areas of symmetrical spots were proportional to concentration for the range 0.5–2.0 μg/spot. Detection limits were comparable to those previously reported for fluorescence quenching (23).

Enzymic detection techniques have been reviewed by Mendoza (99, 100). Quantitation may be attempted by spot area measurement or by visual comparison with standards, with the latter method providing more accurate results (100). Greater accuracy is, however, achieved when pesticides are eluted from the plate following enzymic detection and analyzed by GC (101) or polarography (102). As far as we are aware, in situ scanning techniques have not yet been applied to enzymic detection methods.

3.5. RADIOTRACER STUDIES

The use of radiolabeled isotopes is of great value in many areas of pesticide research, including metabolism and degradation studies and investigations of extraction efficiencies. Work in these areas has been reviewed extensively (103–108), so discussion in this chapter will be limited. Analysis of isotope-labeled compounds separated on thin-layer chromatograms has been reviewed by Snyder (109, 110).

Frei and Duffy (111) compared counting efficiencies for ^{14}C-labeled ametryne in situ on thin-layer chromatograms and, following elution from the silica gel adsorbent, by liquid scintillation counting. The latter technique was found forty times more sensitive than the former. Extraction efficiencies in soil and water samples were also studied during this experiment. Improvements in scanners since this work was done have probably made them a more attractive alternative to liquid scintillation counting than these results would indicate.

3.6. CONCLUSIONS

Thin layer chromatography gained rather slow acceptance as a separation technique among pesticide chemists in the early 1960s (11). A parallel situation appears to have developed in the 1970s, but this time the slow acceptance is of quantitative TLC techniques. The increasing numbers of papers in the mid-1970s, however, may be indicative of greater interest in this method of analysis. This may be judged from the diverse applications of quantitative TLC discussed in a recent review (112). Zweig and Sherma have also discussed applications of TLC in pesticide analysis in *Analytical Methods for Pesticides, Plant Growth Regulators and Food Additives*, Vol. 7 (113). The use of new technology, such as the equipment now available which uses sintered glass ceramic sticks for TLC, combined with a flame ionization detector, should provide impetus for research on new applications. Okumura et al. (114) have reported the detection of sub-microgram quantities of compounds not suitable for gas chromatography using this approach. The speed, simplicity and relatively low cost of thin-layer chromatographic techniques combine to make this an attractive alternative to other methods of analysis for many applications.

REFERENCES

1. J. E. Meinhard and N. F. Hall, *Anal. Chem.*, **21**, 185 (1949).
2. N. A. Izmailov and M. S. Shraiber, *Farmatsiya*, **3**, 1 (1938); through *Chem. Abstr.*, **34**, 855 (1940).

3. M. O'L. Crowe, *Ind. Eng. Chem., Anal. Ed.*, **13**, 845 (1941).
4. T. I. Williams, *An Introduction to Chromatography*, Blackie, London and Glasgow (1946).
5. M. Mottier and M. Potterat, *Mitt. Lebensm. Hyg.*, **43**, 118 (1952).
6. M. Mottier and M. Potterat, *Anal. Chim. Acta*, **13**, 46 (1955).
7. M. Mottier, *Mitt. Lebensm. Hyg.*, **49**, 454 (1958).
8. J. G. Kirchner and G. J. Keller, *Anal. Chem.*, **23**, 420 (1951).
9. J. G. Kirchner and J. M. Miller, *J. Agric. Food Chem.*, **1**, 512 (1953).
10. J. G. Kirchner and R. G. Rice, *J. Agric. Food Chem.*, **2**, 1031 (1954).
11. D. C. Abbott and J. Thompson, *Residue Rev.*, **11**, 1 (1965).
12. L. Fishbein, *Chromatogr. Rev.*, **12**, 167 (1970).
13. L. Fishbein and W. L. Zielinski, Jr., *Chromatogr. Rev.*, **9**, 37 (1967).
14. L. Fishbein, H. L. Falk, and P. Kotin, *Chromatogr. Rev.*, **10**, 37 (1968).
15. L. Fishbein, H. L. Falk, and P. Kotin, *Chromatogr. Rev.*, **10**, 175 (1968).
16. R. R. Watts, *Residue Rev.*, **18**, 105 (1967).
17. J. Sherma, *J. Chromatogr.*, **113**, 97 (1975).
18. W. Thornberg, *Anal. Chem.*, **43**, 145R (1971).
19. J. D. MacNeil and R. W. Frei, *J. Chromatogr. Sci.*, **13**, 279 (1975).
20. M. Beroza, K. R. Hill, and K. H. Norris, *Anal. Chem.*, **40**, 1611 (1968).
21. R. W. Frei and N. S. Nomura, *Mikrochim. Acta*, **1968**, 565.
22. W. Huber, *J. Chromatogr.*, **33**, 378 (1968).
23. R. W. Frei, N. S. Nomura, and M. M. Frodyma, *Mikrochim. Acta*, **1967**, 1099.
24. M. S. Lefar and A. D. Lewis, *Anal. Chem.*, **42**, 79A (1970).
25. R. J. Hurtubise, P. F. Lott, and J. R. Dias, *J. Chromatogr. Sci.*, **11**, 476 (1973).
26. R. W. Frei and J. D. MacNeil, *Diffuse Reflectance Spectroscopy in Environmental Problem-solving*, CRC Press, Cleveland, Ohio (1973).
27. M. E. Getz, *Methods in Residue Analysis, Pesticide Chemistry*. Vol. 4 (A. S. Tahori, Ed.), Gordon and Breach, New York (1971).
28. M. E. Getz and K. R. Hill, Paper No. 54, American Chemical Society Division of Pesticides Chemistry, 163rd ACS National Meeting, Boston, Mass., 1972.
29. G. Kynast, *Z. Anal. Chem.*, **250**, 105 (1970).
30. G. Kynast, *Z. Anal. Chem.*, **251**, 161 (1970).
31. G. Kynast, *Chromatographia*, **3**, 425 (1970).
32. G. Kynast, *Z. Anal. Chem.*, **256**, 20 (1971).
33. K. Kossman, *Methods in Residue Analysis, Pesticide Chemistry*. Vol. 4 (A. S. Tahori, Ed.), Gordon and Breach, New York (1971).
34. H. J. Petrowitz and S. Wagner, *Chem. Ztg.*, **95**, 331 (1971).

35. S. Watanabe, H. Momo, and T. Kashiwa, *Noyaku Kensasho Hokuku*, **12**, 19 (1972).

36. O. Hutzinger, *Anal. Chem.*, **41**, 1662 (1969).

37. O. Hutzinger, W. D. Jamieson, J. D. MacNeil, and R. W. Frei, *J. Assoc. Off. Anal. Chem.*, **54**, 1100 (1971).

38. J. D. MacNeil, R. W. Frei, and O. Hutzinger, *Int. J. Environ. Anal. Chem.*, **1**, 205 (1972).

39. R. W. Frei, J. D. MacNeil, and O. Hutzinger, *Int. J. Environ. Anal. Chem.*, **2**, 1 (1972).

40. J. D. MacNeil, R. W. Frei, and O. Hutzinger, *Can. J. Chem.*, **51**, 500 (1973).

41. R. W. Frei, *CRC Crit. Rev. Anal. Chem.*, **2**, 179 (1971).

42. J. D. MacNeil, R. W. Frei, S. Safe, and O. Hutzinger, *J. Assoc. Off. Anal. Chem.*, **55**, 1270 (1972).

43. J. D. MacNeil, R. W. Frei, M. Frei-Haüsler, and O. Hutzinger, *Int. J. Environ. Anal. Chem.*, **2**, 323 (1973).

44. J. D. MacNeil, R. W. Frei, and O. Hutzinger, *Mikrochim. Acta*, **1973**, 641.

45. D. Sardini and J. Krepinski, *Farm. Ed. Prat.*, **29**, 723 (1974); through *Chem. Abstr.*, **82**, 133870y (1975).

46. J. Sherma, *J. Chromatogr.*, **104**, 476 (1975).

47. F. I. Onuska, *Anal. Lett.*, **7**, 327 (1974).

48. J. Sherma and J. C. Touchstone, *Chromatographia*, **8**, 261 (1975).

49. J. Sherma, K. E. Klopping, and M. E. Getz, *Am. Lab.*, **9**, 66 (1977).

50. J. Sherma and J. Koropchack, *Anal. Chim. Acta*, **91**, 259 (1977).

51. V. N. Mallet, P. E. Belliveau, and R. W. Frei, *Residue Rev.*, **59**, 51 (1975).

52. J. F. Lawrence and R. W. Frei, *J. Chromatogr.*, **98**, 253 (1974).

53. S. M. Norman, D. C. Fouce, and C. C. Craft, *J. Assoc. Off. Anal. Chem.*, **55**, 1239 (1972).

54. S. Ebel and G. Herold, *Dtsch. Debensm-Rundsch.*, **70**, 133 (1974).

55. H. Ottender and U. Hezel, *J. Chromatogr.*, **109**, 181 (1975).

56. H. Ottender, *Mittbl. GDCh. Lebensm-chem. Ger. Chem.*, **26**, 233 (1972).

57. J. D. MacNeil and M. Hikichi, *J. Chromatogr.*, **101**, 33 (1974).

58. V. N. Mallet, C. LeBel and D. R. Surette, *Analusis*, **2**, 643 (1974).

59. V. N. Mallet, J.-G. Zakrevsky, and G. L. Brun, *Bull. Soc. Chim. France*, **9—10**, 1755 (1974).

60. M. Haüsler, J. D. MacNeil, R. W. Frei, and O. Hutzinger, *Mikrochim. Acta*, **1973**, 43.

61. P. E. Belliveau, V. Mallet, and R. W. Frei, *J. Chromatogr.*, **48**, 478 (1970).

62. M. T. H. Ragab, *J. Assoc. Off. Anal. Chem.*, **50**, 1088 (1967).

63. R. W. Frei and V. Mallet, *Int. J. Environ. Anal. Chem.*, **1**, 99 (1971).

64. R. W. Frei, V. Mallet, and M. Thiebaud, *Int. J. Environ. Anal. Chem.*, **1**, 141 (1971).

65. K. Groves, *J. Agric. Food Chem.*, **6**, 30 (1958).

66. P. E. Belliveau and R. W. Frei, *Chromatographia*, **4**, 189 (1971).

67. R. W. Frei and P. E. Belliveau, *Chromatographia*, **5**, 296 (1972).

68. R. W. Frei and P. E. Belliveau, *Chromatographia*, **5**, 392 (1972).

69. T. F. Bidleman, B. Nowlan, and R. W. Frei, *Anal. Chim. Acta*, **60**, 13 (1972).

70. O. Antoine and J. Mees, *J. Chromatogr.*, **58**, 247 (1971).

71. J. D. MacNeil, B. L. MacLellan, and R. W. Frei, *J. Assoc. Off. Anal. Chem.*, **57**, 165 (1974).

72. J. R. Wessel, *J. Assoc. Off. Anal. Chem.*, **50**, 430 (1967).

73. R. W. Frei, B. L. MacLellan, and J. D. MacNeil, *Anal. Chim. Acta*, **66**, 139 (1973).

74. V. Mallet and R. W. Frei, *J. Chromatogr.*, **54**, 251 (1971).

75. V. Mallet and R. W. Frei, *J. Chromatogr.*, **56**, 69 (1971).

76. V. Mallet and R. W. Frei, *J. Chromatogr.*, **60**, 213 (1971).

77. R. W. Frei, V. Mallet and C. Pothier, *J. Chromatogr.*, **59**, 135 (1971).

78. V. Mallet, D. Surette, and G. Brun, *J. Chromatogr.*, **79**, 217 (1973).

79. G. L. Brun, D. Surette, and V. Mallet, *Int. J. Environ. Anal. Chem.*, **3**, 61 (1973).

80. G. L. Brun and V. Mallet, *J. Chromatogr.*, **80**, 117 (1973).

81. V. Mallet and D. P. Surette, *J. Chromatogr.*, **95**, 243 (1974).

82. V. Mallet and G. L. Brun, *Bull. Environ. Contam. Toxicol.*, **12**, 739 (1974).

83. J.-G. Zakrevsky and V. N. Mallet, *J. Assoc. Off. Anal. Chem.*, **58**, 554 (1975).

84. R. W. Frei, J. F. Lawrence, and P. E. Belliveau, *Z. Anal. Chem.*, **254**, 271 (1971).

85. Y. Francoeur and V. Mallet, *J. Assoc. Off. Anal. Chem.*, **60**, 1328 (1977).

86. R. W. Frei and J. Lawrence in *Proc. IUPAC Pesticide Congress, Tel Aviv, Israel, February, 1971*, pp. 193–202, Gordon and Breach, New York and London (1971).

87. R. W. Frei and J. F. Lawrence, *J. Chromatogr.*, **61**, 174 (1971).

88. J. F. Lawrence and R. W. Frei, *J. Chromatogr.*, **66**, 93 (1972).

89. J. F. Lawrence, D. S. Legay, and R. W. Frei, *J. Chromatogr.*, **66**, 295 (1972).

90. R. W. Frei and J. F. Lawrence, *J. Chromatogr.*, **67**, 87 (1972).

91. R. W. Frei and J. F. Lawrence, *J. Assoc. Off. Anal. Chem.*, **55**, 1259 (1972).

92. R. W. Frei, J. F. Lawrence, and D. S. Legay, *Analyst*, **98**, 9 (1973).

93. J. F. Lawrence and R. W. Frei, *Anal. Chem.*, **44**, 2046 (1972).

94. J. F. Lawrence and G. W. Laver, *J. Assoc. Off. Anal. Chem.*, **57**, 1022 (1974).

95. J. F. Lawrence and G. W. Laver, *J. Chromatogr.*, **100**, 175 (1974).

96. J. Sherma and G. Marzoni, *Int. Lab.*, **1974**, (Nov.–Dec.) 41.

97. J. F. Lawrence and R. W. Frei, *J. Chromatogr.*, **98**, 253 (1974).

98. M. Huse and V. M. Adamovic, *J. Chromatogr.*, **80**, 137 (1973).

99. C. E. Mendoza, *Residue Rev.*, **43**, 105 (1972).

100. C. E. Mendoza, *Residue Rev.*, **50**, 143 (1974).

101. C. E. Mendoza, *J. Chromatogr.*, **78**, 29 (1973).

102. J. Seifert and J. Davidek, *J. Chromatogr.*, **59**, 446 (1971).

103. W. Dedek, *Isotopenpraxis*, **6**, 204 (1970).

104. A. C. Houtman, *Chem. Weekbl.*, **67**, 14 (1971).

105. F. P. W. Winteringham, *Proc. Int. Congr. Pesticide Chem.* Vol. 6 (A. S. Tahori, Ed.), Gordon and Breach, New York, 1972, pp. 367–378.

106. J. R. Robinson, *Proc. Int. Congr. Pesticide Chem.*, Vol. 6 (A. S. Tahori, Ed.), Gordon and Breach, New York, 1972, pp. 389–403.

107. G. T. Brooker, *Proc. Int. Congr. Pesticide Chem.*, Vol. 6 (A. S. Tahori, Ed.), Gordon and Breach, 1972, pp. 405–418.

108. M. S. Chatrath and G. R. Sethi, *Indian Soc. Nucl. Tech. Agr. Biol. Newslett.*, **2**, 146 (1973).

109. F. Snyder, *Atomlight*, **1967**, 16.

110. F. Snyder, *Current Status of Liquid Scintillation Counting* (E. D. Bransome, Ed.), Grune and Stratton, New York, 1970, pp. 248–256.

111. R. W. Frei and J. R. Duffy, *Mikrochim. Acta*, **1969**, 480.

112. G. Zweig and J. Sherma, *Anal. Chem.*, **46**, 73R (1974).

113. G. Zweig and J. Sherma, *Analytical Methods for Pesticides, Plant Growth Regulators and Food Additives, Vol. VII, Thin Layer and Liquid Chromatography and Analyses of Pesticides of International Importance*, Academic Press, New York and London (1973).

114. T. Okumura, T. Kadono, and A. Isoo, *J. Chromatogr.*, **108**, 329 (1975).

HIGH PERFORMANCE
LIQUID CHROMATOGRAPHIC ANALYSIS
OF PESTICIDE RESIDUES

H. ANSON MOYE

Pesticide Research Laboratory, Food Science and Human Nutrition Department,
University of Florida, Gainesville

CONTENTS

4.1.   INTRODUCTION

Trace analysis of pesticide residues in foods, biological and environmental samples, since the mid-1950s, has been based primarily on a gas chromatographic (GC) determinative step, although other techniques, such as thin layer chromatography (TLC) (see Chapter 3) have also been employed. Most residue procedures officially recognized by various governmental and sanctioning organizations, such as the United States Food and Drug Administration, the United States Environmental Protection Agency, and the Association of Official Analytical Chemists, employ gas chromatography.

Use of high-performance liquid chromatography (HPLC) in residue determinations has been a very recent occurrence; the first to be reported in the literature was that of Henry et al. in 1971 (1). Abate insecticide was determined in pond water with a detection limit of 0.05 ppm. A year later Krzeminski et al. (2) used HPLC to isolate ^{14}C-labeled diphenamid. The following year Kirkland (3), and Kirkland et al. (4) employed HPLC to analyze residues of benomyl. These first appearances of HPLC procedures in the literature have preceded an ever increasing realization by residue chemists that the technique has real advantages in some instances over GC.

Many of the newer classes of pesticides at least in terms of popularity are inherently difficult to quantitate by GC due to their thermal instability, involatility, or lack of sensitivity when element specific detectors are used. Some examples of these are the carbamates, synthetic pyrethrins, and insect growth regulators. These problems can frequently be overcome by a judicious use of derivatization procedures; however, in so doing additional analysis time, expense and sometimes error are introduced, which make the approach unattractive to many chemists. Additionally, there are occasional problems of multiple derivatives and reagent peaks that confuse chromatograms and obscure analyte peaks. Derivatization is standard practice for acidic herbicides (see Chapter 6); these pesticides would seem to be prime candidates for HPLC analysis, since their anionic characteristics should lend them to highly selective ion-exchange chromatography. At this time, however, no published complete residue procedure exists.

Many of those applications that appear obvious undoubtedly have suffered difficulty in their associated method development due to lack of the availability of sensitive and selective detectors. This problem was recognized in 1975 (5) and since that time has not been fully addressed. The capability of performing highly efficient and rapid separations of pesticide standards at residue levels (ppm) has been available for a number of years for many pesticides; however, when the technique is applied to "real-world" samples there usually is difficulty with detector selectivity. Less frequently the molecular structure of the pesticide simply does not lend itself to being detected at nanogram levels. Special approaches in detector construction have obviated some of these difficulties, as shall be seen in subsequent sections of this chapter. Some residue chemists have been successful in incorporating the use of conventional HPLC detectors into procedures that employ sample extraction and cleanup approaches closely resembling those used in GC procedures. Careful selection of sample type and separation mode has been an important consideration in this, as we shall see later on.

To further promote the utilization of HPLC for pesticide residue analysis it is the intent of this chapter to consider briefly the types of columns and detectors available to the pesticide residue chemist, relate them to the problems faced in trace analysis of pesticides, review the literature of pesticide residue procedures to date, offer suggestions for future applications, and attempt to forecast what improvements in technology might be forthcoming that will be of significance. In addition, selected examples of analytical successes reported in the literature will be discussed in order to provide the reader with insight as to how HPLC might be incorporated into a residue procedure.

It is not the purpose of this chapter to provide the reader information relating to the fundamentals or theoretical aspects of HPLC as an anlytical technique. Several excellent books have been written on the subject to which the reader is referred (6–9). As well, numerous reviews have been written on the subject (10–18); for an in-depth understanding of HPLC and chromatography in general the reader is directed to the many fine theoretically based articles of J. H. Knox, C. Horvath, L. R. Snyder, J. J. Kirkland, R. P. W. Scott, B. L. Karger, and others which have appeared in the *Journal of Chromatography, Analytical Chemistry* and the *Journal of Chromatographic Science* over the past decade and a half.

Regulatory agencies in North America and Europe have lately become vitally interested in taking full advantage of HPLC and in incorporating it into their monitoring programs. In the United States the Food and Drug Administration held an HPLC workshop in Detroit on July 12 and 13, 1977, which featured several speakers generally recognized as leaders in the field of HPLC (19). From this workshop 11 recommendations were made that would commit the agency to train personnel, evaluate instrumentation, and standardize methods and equipment (columns, detectors, etc.). Of the eight papers dealing with the HPLC of pesticide residues that were presented at the Fourth International Congress of Pesticide Chemistry (IUPAC) in 1978, five were presented by Europeans.

One of these papers, which was subsequently published (20), provides detailed spectrophotometric characteristics of the 42 pesticides selected by the European Economic Community to be governed by specified residue levels; their order of elution on reverse phase HPLC was also given. In addition, a suitability factor (for HPLC with uv detection) was proposed, which combines the molar extinction coefficients with other factors to predict whether HPLC with uv detection would be suitable for the pesticide in question. This is the first step in evidence of a true "multiresidue" procedure approach utilizing HPLC combining both sensitivity and retention concepts. As we shall see in Section 4.4, usable residue procedures for many of these pesticides have already been developed; however, if detection is limited to only uv-vis absorbance of the intact pesticide there are more problems which appear than solutions when "real-world" residue samples are dealt with. To those working in the field it is obvious that much needs to be done in the way of detector development, extraction and cleanup integration, and separations mode selection.

4.2. SEPARATIONS

Some of the earliest examples of HPLC separations were of pesticides (21) in the form of standards. The literature is continuing to grow in this respect; indeed, an entire book could be dedicated to separations of mixed pesticide standards. However, the real challenge, in the opinion of this author, lies not in separating pesticides from themselves but rather in separating them from coextractives. Consequently, no definitive review of the literature will be given in this section concerning the chromatographic separation of pesticide mixtures. However, certain reports will be discussed in deference to their elegance or in order to stimulate further research into complete method development of which there is such need in certain cases. In addition, some generalizations will be presented, drawn from the experience of the author and others, which are aimed at assisting the prospective pesticide residue chemist in fully utilizing HPLC for trace analysis (pesticide residues).

Early HPLC columns were exceptionally inefficient compared to those commercially available now. A 1-m column packed with 37–50 μm diameter particles was considered to be a good column in 1972 if it exhibited 4000 theoretical plates; today, a 0.25-m column packed with 5 μm fully porous spheres can sometimes be expected to possess 10,000 theoretical plates. Concurrent with the smaller particles is an increase in required pumping pressure, sometimes approaching the 5000–6000 psi limit of many contemporary pumps. An excellent discussion of the interralationships and optimization of the parameters affecting the separations on small particle columns has been given by Snyder (22). This reference is part of a special issue of the *Journal of Chromatographic Science* (Sept. 1977) devoted to HPLC columns and contains other articles relating to the theoretical and especially the practical aspects of this subject. The guest editor for this issue, R. E. Majors, gave an excellent summary of column types and principles of separations (17).

Although the article by Knox (23) emphasizes the practical aspects of HPLC it also summarizes some limits imposed on column parameters as a result of theoretical considerations. For example, a good argument is made for keeping particle sizes below 10 μm for analytical applications; it is also argued that for preparative work all column dimensions, including particle diameter, should be increased proportionally. He also argues for analytical columns with a minimum of 5 mm bore and 100 mm in length when detector cells on the order of 8 μl are used in conjunction with 10–20 μl injections. At least one manufacturer is now supplying 100 mm columns with a bore of 8 mm; most others range from 3 to 4 mm in diameter.

⸱ The need for use of "zero dead volume" fittings and connecting tubes less than 100 mm × 0.25 mm was emphasized in order to avoid extra column band spreading when columns having N > 10,000 are employed.

Kirkland (19) has stated that approximately two thirds of all desired separations can be achieved by reverse phase (RP) chromatography. An excellent discussion of the theoretical aspects of this technique has been

presented by Horvath and Melander (24) using a model based on thermody-
namic aspects. Reverse phase is indeed a powerful separations tool
frequently giving unexpectedly high selectivities for similarly structured
compounds. For example, in the author's laboratory it has been
demonstrated (25) that the positional isomers, 4-OH and 5-OH car-
barylglucosides could be separated (Rs $\cong$ 2) on a μC_{18} column employing
dioxane:H_2O (10:90) (Figure 4.1). Separation of such water soluble
materials certainly brings to question the long believed postulate that RP
performs only for "hydrophobic" compounds.

A technique which greatly extends the use of RP columns, paired ion
chromatography (PIC) has also been reviewed recently (26). It is especially
valuable for strong acids ($pK_a < 2$) or strong bases ($pK_b > 8$), which would
normally be chromatographed by ion exchange. Reagents are currently
commercially available for PIC of both acids and bases.

In spite of the recognized versatility of RP, equal attention has thus far
been given to other modes of separation in the pesticide residue analytical
chemistry literature (see Table 4.1). The continuing popularity of adsorption
chromatography (AD), usually performed on silica, may be a carryover
from the analyst's chromatographic experience with large-diameter adsorption columns. More than likely it is probably due to certain advantages for
specific applications. For example, it is generally recognized that, since
selectivity is frequently lower in AD chromatography, it can be used to
separate two or more compounds possessing widely different structures
with shorter analysis time than for RP. In addition, when chromatographing
mixtures having one or more components that are insoluble in water a
deterioration in RP column performance is sometimes observed over a period
of time due to irreversible adsorption at the front of the column. By convert-

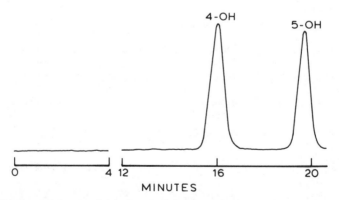

Figure 4.1 Chromatogram of 4-hydroxycarbarylglucoside and 5-hydroxycarbarylglucoside
(125 ng each). μC_{18} column, OPA-MERC postcolumn detection (Ref. 25).

TABLE 4.1. Sample and Chromatographic Conditions for the

Ref.	Pesticide	Sample Type	Extraction	Column Type
1	Abate	Pond water	CHCl$_3$	NP,[b] Zipax-BOP
2	Diphenamid, metabolites	Soybean plants	Benzene	RP,[c] Zipax-HCP
3	Benomyl, MBC, 2-AB	Cow's milk	Hexane	IE,[d] Zipax-SCX
		Urine	Hexane	IE, Zipax-SCX
		Feces	Hexane	IE, Zipax-SCX
		Tissues	Ethyl acetate	IE, Zipax-SCX
4	Benomyl, MBC, 2-AB	Soil	CHCl$_3$	IE, Zipax-SCX
		Plant tissue	CHCl$_3$	IE, Zipax-SCX
68	TH-6040 (Dimilin)	Bovine manure	CH$_2$Cl$_2$	RP, μC$_{18}$
69	Biphenyl	Citrus	Heptane	AD,[e] SI-60
49	TH-6040 (Dimilin)	Milk	Ethyl acetate	RP, Permphase–ODS
70	Chlorotoluron	Soil	MeOH	AD, SI-60
50	Benomyl, 2-AB	Plant tissues	Ethyl acetate	NP, Hitachi 3010
	Thiabendazole	Plant tissues	Ethyl acetate	NP, Hitachi 3010
52	DCNA, benzoyl-propethyl	Corn	Acetone[f]	AD, SI-60
	Terbacil	Corn, potato	Acetone[f]	AD, SI-60
	Linuron, propanil	Cabbage, wheat	Acetone[f]	AD, SI-60
71	Warfarin, metabolites	Blood	CHCl$_3$	RP, μC$_{18}$
72	TH-6040 (Dimilin)	Field water	CH$_2$Cl$_2$	RP, Micropak-CH
73	TH-6040 (Dimilin)	Soil, plants	CH$_3$CN	RP, Micropak-CH
		Water	CH$_2$Cl$_2$	RP, Micropak-CH
74	Parathion, paraoxon	Soil dust	Acetone:H$_2$O (9:1)	RP, Bondapak-C$_{18}$
		Foliar dust	H$_2$O:detergent	NP, Carbowax 400
53	Carbaryl	Potato	Acetone[f]	AD, SI-60
		Wheat	Acetone[f]	AD, SI-60
	Zectran	Corn	Acetone[f]	AD, SI-60
	Aminocarb	Corn	Acetone[f]	AD, SI-60
	Propoxur	Cabbage	Acetone[f]	AD, SI-60
	Swep	Potato	Acetone[f]	AD, SI-60
	Mobam	Corn	Acetone[f]	AD, SI-60
54	Carbofuran, metabolites	Turnips	Acetone[f]	AD, SI-60
75	Parathion, paraoxon		(See Ref. 74 for description of sample and	
60	Karbutilate, metabolites	Water	Ethyl acetate	AD, μPorasil
		Soil	MeOH:H$_2$O (2:1)	AD, μPorasil
		Grass	MeOH:H$_2$O (2:1)	AD, μPorasil

162

HPLC Analysis of Pesticide Residues

Parameters

Particle Size (μm)	Length (cm) × I.D. (mm)	Mobile Phase	Detector[a]
10	100 × 2.1	Heptane	uv, 254 nm
10	100 × 2.1	H$_2$O:MeOH (4:1)	uv, 254 nm
10	100 × 2.1	pH 5.15 acetate buffer	uv, 280 nm
10	100 × 2.1	pH 5.15 acetate buffer	uv, 280 nm
10	100 × 2.1	pH 5.15 acetate buffer	uv, 280 nm
10	100 × 2.1	pH 5.15 acetate buffer	uv, 280 nm
10	100 × 2.1	0.025N tetramethylammonium	uv, 254 nm
10	100 × 2.1	nitrate + 0.025N HNO$_3$	uv, 254 nm
10	30 × 6	CH$_3$CN:H$_2$O (57:43)	uv, 254 nm
20	90 × 2	Heptane	uv, 254 nm
37	100 × 2.1	MeOH:H$_2$O (1:1)	uv, 254 nm
10	20 × 4	Hexane:2-propanol (85:15)	uv, 240 nm
22	50 × 2.1	0.1% (v/v) acetic acid:MeOH	Fluorometry
22	50 × 2.1	5.0% (v/v) acetic acid:MeOH	Fluorometry
5	25 × 2.8	2% (v/v) 2-propanol:isooctane	uv, 254 nm
5	25 × 2.8	20% (v/v) 2-propanol:isooctane	uv, 254 nm
5	25 × 2.8	10% (v/v) 2-propanol:isooctane	uv, 254 nm
10	30 × 4	1.5% (v/v) acetic acid (adjust to pH 4.7 wt NH$_4$OH):CH$_3$CN (69:31)	uv, 313 nm
10	25 × 2	MeOH:H$_2$O (60:40)	uv, 254 nm
10	25 × 8	CH$_3$CN:H$_2$O (70:30)	uv, 254 nm
10	25 × 8	CH$_3$CN:H$_2$O (70:30)	uv, 254 nm
37–50	50 × 2.2	Program, 10% CH$_3$CN:H$_2$O to 100% CH$_3$CN	uv, 272 nm
37–50	50 × 2.2	1.5% Dioxane in hexane (parathion) or 22.5% dioxane in hexane (paraoxon)	uv, 265 nm
5	25 × 2.8	5% (v/v) 2-Propanol in isooctane	uv, 254 nm
5	25 × 2.8	5% (v/v) 2-Propanol in isooctane	uv, 254 nm
5	25 × 2.8	5% (v/v) 2-Propanol in isooctane	uv, 254 nm
5	25 × 2.8	5% (v/v) 2-Propanol in isooctane	uv, 254 nm
5	25 × 2.8	5% (v/v) 2-Propanol in isooctane	uv, 254 nm
5	25 × 2.8	5% (v/v) 2-Propanol in isooctane	uv, 254 nm
5	25 × 2.8	5% (v/v) 2-Propanol in isooctane	uv, 254 nm
5	25 × 2.8	5% (v/v) 2-Propanol in isooctane	uv, 254 nm
	HPLC conditions except for detector)		Auto Analyzer® colorimetric
10	30 × 6	3–7% Ethanol (v/v) in ethylene chloride	uv, 254 nm
10	30 × 6	3–7% Ethanol (v/v) in ethylene chloride	uv, 254 nm
10	30 × 6	3–7% Ethanol (v/v) in ethylene chloride	uv, 254 nm

Table 4.1

Ref.	Pesticide	Sample Type	Extraction	Column Type
55	Carbofuran, metabolites	Vegetables	Acetone^f	AD, SI-60
76	TH-6040 (Dimilin)	Crop, tissue, soil, eggs	CH$_3$CN	RP, μC$_{18}$
				AD, μPorasil
77	HCH	Wool fat	hexane	RP, μC$_{18}$
78	Rotenone, metabolites	Animal chow, tissues	benzene aceto-nitrile	RP, μC$_{18}$
66	2-Phenylphenol	Orange rind	CH$_2$Cl$_2$	RP, C$_{18}$ Corasil
57	Carbofuran, metabolites	Vegetables	Acetone^f	AD, SI-60
56	Monuron, diuron linuron	Corn	Acetone^f	AD, SI-60
61	Imazalil	Citrus	Ethyl acetate	AD, SI-60
79	Biphenyl	Citrus	Distillation from acid into hexane	AD, SI-60
	2-Phenylphenol	Citrus	Distillation from acid into hexane	AD, SI-60
	Thiabendazole	Citrus	CHCl$_3$	AD, SI-60
	Benomyl, 2-AB	Citrus	CHCl$_3$	AD, SI-60
62	β-Naphthoxyacetic acid, β-naphthol	Strawberries	Acetone-o-phos-phoric acid	RP, Partisil-ODS
58	Carbaryl	Potato, corn	Acetone^f	AD, SI-60
80	Dioxane herbicide	Soil, soybeans	MeOH:H$_2$O	RP, μC$_{18}$
59	Methomyl, oxamyl	Fruits, vegetables	Ethyl acetate	RP, μC$_{18}$
81	Permethrin	Cotton leaves	Heptane	AD, SiSA
82	Pirimiphos methyl, metabolites	Blood plasma	Deproteinized w/MeOH	RP, SAS-Hypersil
		Urine	Freeze-dried; MeOH	RP, SAS-Hypersil
83	Pesticide phenolic metabolites	Urine	Direct	RP, μC$_{18}$

Parameters

Particle Size (μm)	Length (cm) × I.D. (mm)	Mobile Phase	Detector[a]
10	25 × 2.2	3–5% 2-Propanol (v/v) in isooctane	uv, 254 and 280 nm
10	30 × 4	$CH_3CN:H_2O$ (60:40) or MeOH:H_2O (75:25)	uv, 254 nm
10	30 × 4	Isooctane:2-propanol (93:7) or CH_2Cl_2:MeOH (500:1)	uv, 254 nm
10	30 × 4	MeOH	uv, 254 nm
10	30 × 4	MeOH:H_2O (75:25)	uv, 295 nm
37–50	50 × 1.8	EtOH:H_2O (40:60)	Amperometric (oxidative)
5	25 × 2.2	3% Acetone in isooctane	Dansylation, fluorometry
5	25 × 2.8	Isooctane:2-propanol (90:10)	uv, 254 nm
10	25 × 3.2	CH_3CN:pH 7.5 phosphate (47:53)	uv, 202 nm
5		Isooctane	uv, 254 nm
5		1% EtOH in isooctane	uv, 254 nm
5		1% EtOH + 0.2% morpholine in $CHCl_3$	uv, 288 nm
5		1% EtOH + 0.2% morpholine in $CHCl_3$	uv, 288 nm
10	25 × 4.6	MeOH:H_2O (48:52)	uv, 223 nm
5	25 × 2.2	4% 2-propanol in isooctane	uv, 254 nm; fluorometry
10	30 × 4	CH_3CN:H_2O (47:53)	Derivatization; uv, 336 nm
10	30 × 4	CH_3CN:phosphate buffer (11:89)	uv, 240 nm
5	25 × 4.6	0.19% CH_3CN in isooctane	uv, 220 nm
5	10 × 4.6	MeOH:pH 4.5 phosphate	uv, 235 nm
5	10 × 4.6	MeOH:pH 4.5 phosphate	uv, 235 nm
10	30 × 4	MeOH:H_2O (3:2, 2:1)	Postcolumn prep; uv

Table 4.1

Column

Ref.	Pesticide	Sample Type	Extraction	Type
84	Hydroxy-s-triazines	Plant leaf	MeOH	AD, LiChrospher
85	Pentachlorophenol	Marine biota	CH_3CN	NP, μCN
86	Paraquat	Marijuana	$6N$ HCl	Pic,^{g} μC_{18}
87	Carbaryl	Forest foliage	$CHCl_3$	RP, C_{18} Corasil
		Soil	Acetone:H_2O (1:1)	RP, C_{18} Corasil
		Water	CH_2Cl_2	RP, C_{18} Corasil
63	NAA	Apples	$CHCl_3:H_2SO_4$	IE, LiChrosorb NH_2 RP, μC_{18}
64	NAA	Citrus fruit	$CH_2Cl_2:H_2SO_4$	RP, μCN
		Citrus by-products, water-soluble	$CH_2Cl_2 \cdot H_2SO_4$	RP, μCN
		Molasses	Ethyl ether: H_2SO_4	RP, μCN
		Oil	Ethyl ether	RP, μCN
65	Naphthaleneacetamide	Apples	$CHCl_3$	RP, RP-8
88	Ethoxyquin	Apples	Hexane	RP, Spherisorb ODS
89	Rotenoids (deguelin, rotenone, tephrosin, rotenalone)	Plants	$CHCl_3:MeOH$ (9:1)	RP, HC SIL-X-1
		Plants	$CHCl_3:MeOH$ (9:1)	RP, HC SIL-X-1
67	Glyphosate, amino-methylphosphonic acid	Vegetables, fruit	H_2O	IE, Aminex-A27
90	Monuron, diuron linuron, chlorbromuron, chlortoluron, monolinuron, chloroxuron, metobromuron	Grain, soil	MeOH	RP, Spherisorb ODS
		Water	CH_2Cl_2	RP, Spherisorb ODS

^{a} uv, ultraviolet absorbance.
^{b} NP, normal phase.
^{c} RP, reverse phase.
^{d} IE, ion exchange.

Parameters

Particle Size (μm)	Length (cm) × I.D. (mm)	Mobile Phase	Detector[a]
10	— × 3	$CHCl_3$: MeOH: H_2O: H_3PO_4	uv, 235 nm
10	30 × 4	2.5% 2-Propanol in isooctane	uv, 254 nm
10	30 × 4	CH_3CN: H_2O (27:73) w/10^{-3} M sodium octane sulfonate	uv, 254 nm
37–50	61 × 2	CH_3CN: H_2O (40:60)	uv, 254 nm
37–50	61 × 2	CH_3CN: H_2O (40:60)	uv, 254 nm
37–50	61 × 2	CH_3CN: H_2O (40:60)	uv, 254 nm
10	30 × 4	CH_3CN: H_2O, pH 3.2 (20:80)	uv, 220 nm; or fluorometry
10	30 × 4	CH_3CN: H_2O, pH 5.2 (20:80)	uv, 220 nm; or fluorometry
10	30 × 4	0.1M phosphate, pH 7	Fluorometry
10	30 × 4	0.1M phosphate, pH 7	Fluorometry
10	30 × 4	0.1M phosphate, pH 7	Fluorometry
10	30 × 4	0.1M phosphate, pH 7	Fluorometry
10	25 × 4.2	CH_3CN: H_2O (30:70, pH 3.5)	Fluorometry
10	25 × 4.6	MeOH: H_2O (80:20)	Fluorometry
—	25 × 2.6	Gradient, 35% CH_3CN in H_2O to 100% CH_3CN	uv, 254 and 244 nm
10	30 × 4		
13.5	25 × 4	0.1M H_3PO_4	Postcolumn fluorogenic labeling
5	30 × 4.6	MeOH: H_2O (60:40) w/0.6% NH_4OH	uv, 240 nm
5	30 × 4.6	MeOH: H_2O (60:40) w/0.6% NH_4OH	uv, 240 nm

[e] AD, adsorption.
[f] Reference 51.
[g] PIC, paired ion chromatography.

ing over to an AD column chances are good that the adsorbing components would elute early with the nonpolar mobile phase. Also, column permeability is usually higher in AD systems, with a concurrent reduction in head pressure resulting in some easily achieved pumping requirements (less expensive pumps). Since most RP packings are chemically bonded, beginning with silica, the silica packings (AD) are inherently less expensive and can be discarded more easily. Another advantage of AD columns is that easily obtainable TLC plates may be used for rapid solvent screening. These aspects and others regarding AD HPLC are discussed by Saunders (27).

The extreme versatility of RP HPLC was demonstrated early by Seiber (28) who, using a 1 m × 2.1 mm column packed with Vydac reverse phase eluted with MeOH:H$_2$O mixtures, succeeded in separating mixtures of chlorinated hydrocarbons, polychlorinated biphenyls, organophosphates and carbamates. A 254-nm absorbance detector was used; limits of detection of about 10 ng were observed for the strong absorbers, such as p,p'-DDE and dyrene. Thruston (21) observed somewhat better efficiency and resolution when he attempted (unsuccessfully) to separate 23 carbamates on a similarly sized column packed with Permaphase® ODS. An attractive feature of RP is that is can be easily and reproducibly solvent programmed; this became apparent in the author's laboratory early on (25) when two carbamoyl oximes (methomyl and aldicarb) were resolved from five carbamates (aminocarb, propoxur, carbofuran, carbaryl, and methiocarb) employing a μC$_{18}$ column with a gradient of 15 to 40% dioxane:H$_2$O over 20 min (Figure 4.2). Drinkwine, Bristol and Fleeker (29) were able to completely separate (Rs > 1.5) 2,4-D herbicide from seven of its metabolites in less than 20 min (Figure 4.3). They employed a μC$_{18}$ column and a linear gradient of 100% pH 7 0.001M phosphate buffer to 100% MeOH for 30 min. A somewhat similar reverse phase column marketed by Perkin Elmer, a PEC$_{18}$ Sil-X-11 was used to study the reverse phase characteristics of a number of herbicides of various structural types (30). Various MeOH:H$_2$O percentages were used ranging from 2.5 to 50%; the technique was particularly suitable for the nonelectron-capturing urea isoproturon, the uracils and the triazine cyanatryn, which tended to degrade under gas chromatographic conditions.

A detailed study was undertaken by Sparacino and Hines (31) to evaluate both AD and RP as separation techniques for carbamate pesticides. They also investigated columns which can be operated in the ion exchange, normal phase and mixed modes. Of all systems studied the RP mode using a μC$_{18}$ column and gradient elution of 20–60% CH$_3$CN in H$_2$O offered the best separation, even though a run of 77 min was required (Figure 4.4). Of the 24-component mix only Eurex, Vernolate, and Pebulate could not be at least partially resolved. Some interesting observations regarding uv absorp-

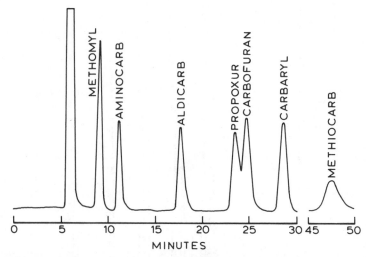

Figure 4.2 Chromatogram of two oximes (methomyl and aldicarb) and five carbamates (125 ng each). μC_{18} column, 15% to 40% dioxane/H_2O, 20 min, 1 ml/min, OPA-MERC postcolumn detection (Ref. 25).

tion characteristics will be discussed in the following section on detectors. Many of these same compounds were included in a brief report by Aten and Bourke (32), who also studied a number of herbicides. They were unable to separate carbaryl from one of its metabolites, 1-naphthol by RP but noted that this could be accomplished by AD on silica as reported by Sparacino and Hines (31). In a comparison between a derivatization GLC procedure and two HPLC procedures Glad, Popoff and Theander (33) demonstrated the efficiency and speed of modern HPLC techniques. Linuron and seven metabolites were ethylated and chromatographed by GLC on a mixed phase (OV-1/OV-255) column at 185°C. They were also chromatographed on RP (μC_{18}) and AD (μPorasil) columns. While the GLC procedure produced double derivatives for one metabolite (3,4-dichloroaniline) and no usable resolution for two others during a 12-min chromatogram, an unambiguous chromatogram with baseline separation of all components was achieved by AD HPLC in less than 18 min (Figure 4.5); coelution of two metabolites occurred during a 10-min RP chromatogram.

The reader is referred to an excellent review of HPLC chromatographic and spectrophotometric data covering 166 pesticides and related materials compiled by Lawrence and Turton (34). Information on column packings, dimensions, mobile phases, elution volumes, and absorption wavelengths is given. The review differs from the data presented in Table 4.1 of this

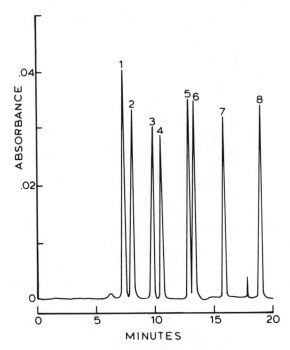

Figure 4.3 Separation of 2,4-D and hydroxylated metabolites. 1 = 3-hydroxy-2,4-D; 2 = 2-chloro-4-hydroxyphenoxyacetic acid; 3 = 4-hydroxy-2,3-D 4 = 4-hydroxy-2,5-D; 5 = 5-hydroxy-2,4-D; 6 = 4-chloro-2-hydroxyphenoxyacetic acid; 7 = 2,4-D; 8 = 6-hydroxy-2,4-D. μC_{18}, 100% pH 7 phosphate buffer to 100% MeOH, 30 min, 2.0 ml/min 280 nm uv (Ref. 29).

chapter in that most references are not for complete residue methods. However, it is extremely comprehensive and noteworthy.

This author concurs with Dr. Kirkland in the belief that most compounds, including pesticides, can be separated with good efficiencies and reasonable analysis times (<20 min) on a good RP column. These columns are generally long-lived when the manufacturer's precautions concerning minimum and maximum pH are observed as well as other techniques such as filtering the sample through 2–10 μm pore size filters before injection, and using "guard" columns. Since H_2O is the primary consitituent of the mobile phase economy of operation is also an advantage. If performance deteriorates over a period of time the column can frequently be "regenerated" using a sequence of solvent washes, progressing to a "nonpolar" solvent such as hexane, and back again. Most manufacturers provide a recommended set of solvents for doing this, which seem to function equivalently. An adsorption column (silica), a cation exchange column, and an anion exchange column complete the separations arsenal for the pesticide residue chemist with few exceptions.

4.3. DETECTORS

The earliest commercial HPLC detectors, and probably still the most widely used, employed low wattage germicidal lamps emitting the resonance line of Hg, 254 nm, as sources for light absorbance measurements on column eluents. Recent advances in sources and associated electronics have allowed manufacturers to design variable wavelength detectors which can be tuned to as low as 200 nm. While wavelength absorption maxima for many pesticides have yet to be determined Sparacino and Hines (31) recorded them for the 30 carbamates they studied. Except for carbaryl, methomyl, and Mobam they observed that absorption maxima lay in the 190–210 nm

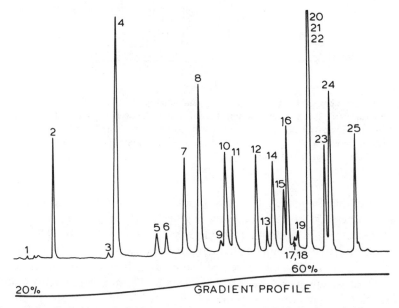

Figure 4.4 Chromatogram of 25 carbamate mix; μC_{18}, 20–60% MeCN:H$_2$O, concave 60 minute gradient, 1 ml/min, 220 nm uv. (1) solvent front, (2) methomyl, (3) aldicarb, (4) isolan, (5) baygon, (6) carbofuran, (7) moban, (8) carbaryl, (9) landrin, (10) propham, (11) banol, (12) mesurol, (13) zectran, (14) betanal, (15) chloropropham, (16) eptam, (17) bux, (18) captafol (19) barban, (20) eurex, (21) vernolate, (22) pebulate, (23) butylate, (24) avadex, (25) avadex BW (Ref. 31).

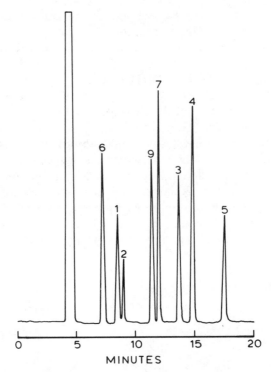

Figure 4.5 Chromatogram of linuron and its metabolites; μPorasil, 1-propanol: 25% NH₄OH (A) and hexane (B), 2–35% A over 30 min, 1.5 ml/min. (1) linuron, (2) 3,4-dichloroaniline, (3) 3-(3,4-dichlorophenyl)-1-methoxyurea, (4) 3-(3,4-dichlorophenyl)-1-methylurea, (5) 3,4-dichlorophenylurea, (6) methyl-*N*-(3,4-dichlorophenyl)-carbamate, (7) 3,4-dichloroacetanilide, (9) 3-(3,4-dichlorophenyl)-1-ethylurea (Ref. 33).

region and that frequently the extinction coefficients were 10^2 to 10^3 higher than at either 254 or 280 nm, wavelengths employed in most fixed wavelength detectors. In addition to being a superior modifier for RP separations acetonitrile was found to have the lowest background absorbance in the 195–210 nm range and hence resulted in lower limits of detection for the carbamates, as low as 1.1 ng for carbofuran at 205 nm, compared to limits of detection for the same compound of 59 ng at 254 nm and 4 ng at 280 nm. Clearly a well-designed variable wavelength detector has its advantages in terms of absolute sensitivity. The reader is referred to the excellent book by Scott (35), which thoroughly discusses commercially available variable wavelength detectors.

It has recently become apparent to most pesticide residue chemists that successes in the utilization of the variable wavelength detector have in great part been due to a realization of the need for *pure* chromatographic solvents, especially below 210 nm, where even trace quantities of impurities give rise to noise and elevated base lines. At least one supplier is offering HPLC grade water, as well, to insure purity for RP solvent systems.

The separations achievable by commercially available HPLC hardware are far superior to the results obtainable with commercially available detectors at this time; if HPLC has a weakness it is that it cannot take full advantage of the chromatography that can be achieved due to the lack of *specific* or *selective* detectors that are also sensitive. If variable wavelength absorbance detectors are greatly affected by mobile phase purity one can easily predict the problems which would, and do, arise when they are applied to trace analysis situations such as pesticide residues. Many exceptions exist, however, as we shall see in the subsequent section.

One alternative to the use of absorbance detectors for pesticide residues is the electroanalytical detector. Kissinger (36) has provided an excellent discussion of the principles and practical considerations concerning electrochemical detectors for HPLC. His early work in the field (37) provided the basis for production of a now commercially available detector (38). Liquid chromatography with electrochemical detection (LCEC) is possible if an analyte is capable of being oxidized or reduced at moderate potentials under the limitations imposed by the conditions of the separation (mobile phase solvents, buffers, column material, mobile phase impurities, etc.). LCEC performs best under reverse phase or ion-exchange conditions, and is less prone to problems from dissolved oxygen and trace metal ions when employed in the oxidative mode. Consequently, thus far most applications have been concerned with phenols and amines or similar compounds (36).

Many pesticides (carbamates, ureas, and anilides) can be degraded in the environment or in mammals to produced anilines. Some of these were studied by Lores, Bristol and Moseman (39) including aniline, 2-amino-4-chlorophenol, *p*-chloroaniline, *p*-bromoaniline, *m*-chloroaniline, *o*-chloroaniline, and 3,4-dichloroaniline. Using an RP column (15 cm Zorbax® C-18) and 80:20 CH_3CN:0.15M phosphate buffer at pH 2.1 they obtained a chromatogram for 3 of the anilines as seen in Figure 4.6; by changing to a 60:40 CH_3CN:buffer mix they were able to separate and detect the remaining 4 as seen in Figure 4.7. Limits of detection (based on 2 × noise) ranged from 0.23 ng for aniline to 0.38 ng for *p*-bromaniline. The detector oxidation potential was maintained at +1.1 V. Even though all 7 compounds could be easily separated in 12 min by RP using gradient elution with uv absorbance detection, it was not possible by LCEC due to the extremely slow equilibration of the detector; a 30-min waiting period was required between solvent changes. Using somewhat modified chromatographic conditions Lores and Moseman (40) were able to utilize the same

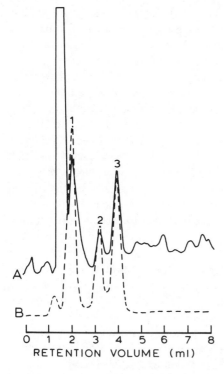

Figure 4.6 Chromatogram of anilines; Zorbax C_{18}, 80% pH 2.1 phosphate–20% CH_3CN, 1 ml/min; *A* (1) 0.52 ng aniline, (2) 0.32 ng 2-amino-5-chlorophenol, (3) 0.64 ng *p*-chloroaniline; *B* (1) 65 ng aniline, (2) 40 ng 2-amino-5-chlorophenol, (3) 80 ng *p*-chloroaniline (Ref. 39).

LCEC arrangement to detect a large number of chlorinated phenols as well as some phenols resulting from the degradation of the organophosphates and carbamates. Limits of detection of less than 1 ng were obtained in most cases. It is clear that with the availability of this detector and its inevitable modifications and improvements that many pesticide metabolites will be determined at residue levels when complete methods are devised.

One approach to supplementing commercially available HPLC detectors is that of adapting element specific GC detectors so that they may be interfaced to HPLC columns. Willmott and Dolphin (41) were the first to apply this to the detection of pesticides at trace levels. They passed the hexane mobile phase exiting a 10 cm × 2 mm column packed with 10 μm Spherisorb SIOW into a heated 40 cm × 0.25 mm stainless steel transfer tube to provide complete vaporization. From there the vapors were led into a Pye Unicam Series 104 [63]Ni electron capture detector also supplied with 30 ml/min of purge nitrogen. A 2m × 4.5 mm stainless steel tube at ambient temperature acted as a condenser. Sensitivity to aldrin was 100 times better than with a 254 nm UV absorbance detector allowing for detection of 40 pg; however, the column gave only partial separation of 5 chlorinated hydrocarbon pesticides. The arrangement used is shown in Figure 4.8.

The Coulson electrolytic conductivity detector has been used as an element specific detector in the GLC detemination of pesticides. In the reductive mode, without the aid of a pyrolytic catalyst, the detector can be used for the detection of halogenated compounds with a high degree of specificity. Dolan and Seiber (42) successfully adapted the Coulson detector as a monitor for HPLC. Introduction of the effluent (MeOH : H_2O) from the RP column was via a stainless steel ¼ in. to ⅛ in. Swagelok® reducing union, which was heated to 100–150°C by Nichrome resistance wire and through which the H_2 reaction gas was metered. Otherwise the detector was operated as specified by the manufacturer. A limit of detection of about 5 ng for lindane was observed. Relative standard deviation (3.4%) was nearly comparable to that observed for the uv absorbance detector (2.2%); the linear range was from 5 ng to 500 μg (10^5). Typical chromatograms for spiked and unspiked lettuce and water samples are shown in Figure 4.9. While not demonstrated, it was hypothesized that the detector could be similary used in the sulfur or nitrogen modes.

While the previously discussed approach relied on direct volatilization of the HPLC effluent, a technique investigated by Hill and coworkers employed a moving wire transport system (43–45), which is capable of removing the HPLC solvent before the solute is detected by the detector. The first reference, discussed in this section, is a published report on the interfacing of the Hall electrolytic conductivity detector to an HPLC column via the moving wire device. The second and third references (44, 45) describe an interfacing to the Melpar flame photometric detector (FPD) which was accomplished quite similarly. Figure 4.10 illustrates the equip-

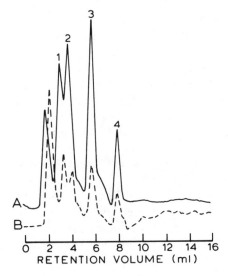

Figure 4.7 Chromatogram of anilines; Zorbax C_{18}, 60% pH 2.1 phosphate–40% CH_3CN, 2 ml/min; *A* (1) 15 ng p-bromoaniline, (2) 10.7 ng *m*-chloroaniline, (3) 18.7 ng *o*-chloroaniline, (4) 12.5 ng 3,4-dichloroaniline; *B* (1) 1.2 ng p-bromoaniline, (2) 0.86 ng *m*-chloroaniline, (3) 1.5 ng *o*-chloroaniline, (4) 1 ng 3,4-dichloroaniline (Ref. 39).

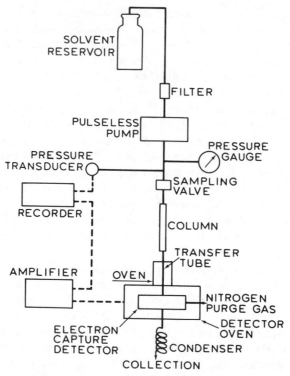

Figure 4.8   Diagram of HPLC coupled to GLC electron capture detector (Ref. 41).

ment. A supply spool of stainless steel wire feeds through a cleaner oven at a velocity of 0 to 18 cm/sec, through a coating block through which the effluent from the LC column is pumped. A portion of the effluent is retained by the wire, while a larger portion goes into waste; solvent from the column is evaporated away in a small evaporation oven under a steady stream of argon. The solute, now on the wire, is then fed through an "oxidizer" (pyrolyzer) oven (840°C), which is pumped out by the reduced pressure created by the entrainer which supplies H_2, reducing the fragments to NH_3 and halo acids. The halo acids are absorbed by a small strontium hydroxide scrubber at the exit of the reaction furnace. Ammonia, resulting from the organic nitrogen, is dissolved in the Hall cell solvent and measured conductimetrically. By comparison with known gas chromatographic responses of atrazine it was determined that there is an apparent transfer efficiency from injection onto the LC column to detector absorption of only 0.62%; undoubtedly much of the sample loss is at the coating block with other losses in the various oven sections of the instrument. Because of the losses a limit of detection of approximately 1 μg of atrazine was realized (2 × noise). By delaying wire movement until 4 min after injection the effects of lipid on the detector can be reduced, as can be seen in Figure 4.11 for a beef fat extract fortified at 10 ppm. This device, due to sensitivity and resolution restrictions, would obviously be best suited to screening applications where high levels of organo-nitrogen pesticides are suspected.

An approach for boosting HPLC sensitivity and selectivity for pesticide residue analysis is in the area of postcolumn chemistry. Rather than relying on the spectral or elemental characteristics of a pesticide this concept draws on the chemical reactivity of the pesticide toward various reagents. Such a technique has been called continuous flow analysis (CFA), flow injection analysis (FIA), dynamic fluorogenic labeling (DFL, when a fluorophore is formed), or most simply postcolumn derivatization. Detection of the analyte has thus far been limited to colorimetry or fluorometry, although, as

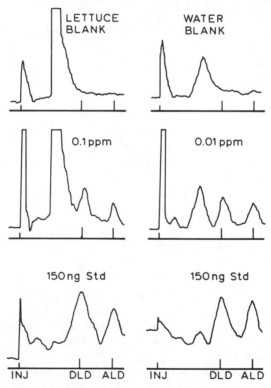

Figure 4.9 Chromatograms of lettuce and water extracts using HPLC/Coulson detection, showing responses to dieldrin (DLD) and aldrin (ALD) (Ref. 42).

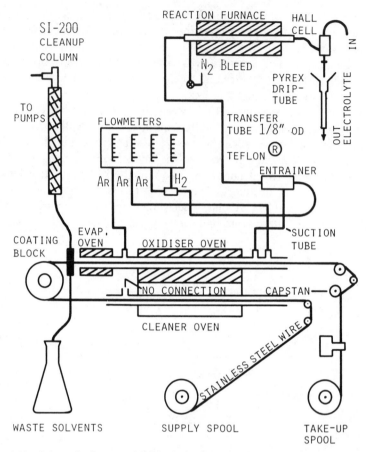

Figure 4.10 Schematic diagram of LC-Hall electrolytic conductivity detector assembly (Ref. 43).

pointed out by Kissinger (36), electrochemical reactions could be employed to modify the functionality of an analyte.

Ramsteiner and Hörmann (46) were the first to adapt a postcolumn derivation device to HPLC for the determination of pesticides. A Technicon Auto-Analyzer® peristaltic pump and derivatization cartridge were used to mix the effluent from a 50 cm × 3 mm ETH column (H₂O) with cholinesterase enzyme. The inhibited enzyme was prevented from hydrolyzing the substrate (acetylcholine), which consequently prevented a color-forming reaction with dithionitrobenzoic acid. Absorption was monitored at 420 nm. Limit of detection for CGA/18809, an experimental organophosphate, was found to be 20 ng (3 × noise); the ETH column could easily separate this compound from plum leaf extract.

Shortly thereafter Moye and Wade (47) used a somewhat similar approach but were able to observe considerable improvements in limits of detection, mainly due to the use of housefly heads as a source of cholinesterase and to the use of a fluorogenic substrate. A number of columns were studied for the separation of 4 carbamates and 2 organophosphates, with a 30 cm × 3 mm Permaphase ODS column, eluted with pH 7 phosphate buffer, being found to be superior. The substrate, N-methylindoxyl acetate produced the highly fluorescent N-methylindoxyl upon hydrolysis by the acetylcholinesterase; when pesticide eluted from the column it inhibited the enzyme and a proportionate decrease in fluorescence was observed. Limits of detection (2 × noise) ranged from 0.2 ng for carbofuran to 800 ng for dyfonate.

Additional work in the author's laboratory (25) has eliminated the need for employing difficult to obtain enzymes in the postcolumn derivatization of carbamate pesticides. The entire instrumental arrangement (Figure 4.12) was designed around the postcolumn chemistry of N-methylcarbamates, which, on alkaline hydrolysis, produce methylamine, which subsequently, and almost instantaneously, reacts with the primary amine specific reagent o-phthalaldehyde:2-mercaptoethanol (OPA-MERC) to produce a highly fluorescent isoindole. Whereas in the two previously described enzymic detectors postcolumn band spreading of the analyte was prevented by air segmentation, the detector described here utilized narrow bore Teflon® tubing

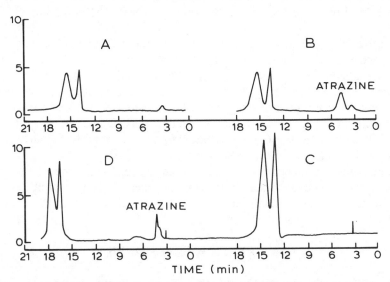

Figure 4.11 Chromatograms of a beef fat extract before and after fortification with atrazine at 10 μg/g. (*A*) Control extract; (*B*) fortified extract; (*C*) control extract, wire turn-on delayed 4 min; (*D*) fortified extract, wire turn-on delayed 4 min (Ref. 43).

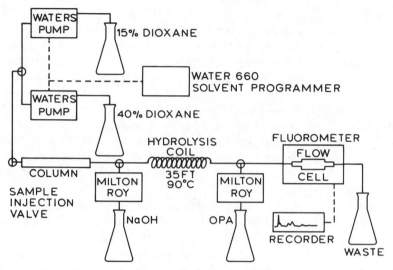

Figure 4.12 Modular dynamic fluorogenic labeling HPLC for the analysis of carbamoyl oxime and carbamate pesticides (Ref. 25).

(0.5 mm). A typical chromatogram in Figure 4.2 has already been presented, using gradient elution and a μC_{18} column; analytical curves were linear over 3 decades and 7 serial injections of methomyl (Lannate) produced an RSD of only 2.7%. Limit of detection was determined to be 0.1 ng for methomyl (2 × noise).

Krause (48) studied this arrangement in detail although some modifications were made such as using stainless steel instead of Teflon for the hydrolysis coil and pressurized reservoirs instead of piston pumps for reagent delivery. He found that the RSD for serial injections of standards was only 0.9% and retention times only 0.3%; also reported was a relatively low variability in sensitivity and retention over periods of several days (3.5% and 1.0%, respectively). Optimum parameters were found to be fairly close to those determined in the author's laboratory.

Actual residue determinations have been made using some of the previously described detectors and modifications thereof. These and others will be discussed in more detail in Section 4.4.

4.4. PESTICIDE RESIDUES

It has been the philosophy of the author (as many others) that if an HPLC analytical determinative step, in this case an HPLC separation–detector combination, offers no real advantages over those existing, then it does not significantly contribute to the field of pesticide analytical methodology and hence should not be utilized in the development of a complete analytical method. For that reason some of the detectors discussed in the preceding section have not been pursued by their developers and employed in complete methods. On the other hand, it has also been the author's philosophy (as also many others) that some well-described detectors stand clearly on their on merit and thus have been commended to the care of others for actual demonstration in applications.

This section will deal with the integration of HPLC detectors into complete pesticide residue methods. Descriptions of such methods published to date are summarized in Table 4.1; the reader is referred to the references listed for a more complete description of each method, since it is beyond the scope of this chapter to discuss each and every one. Instead, unique and also similar characteristics of representative methods will be discussed in order to provide the reader with insight so that he or she may be better equipped to fully utilize available HPLC equipment in analytical method development.

The first reported pesticide residue procedure employing HPLC was devised to determine the larvacide, Abate, in pond water (1); even though the nonspecific 254 nm uv absorbance detector was used and no sample cleanup was employed, the chromatograms were uncluttered and unambiguous (Figure 4.13). A fortuitous choice of sample (pond water) and the favorable partitioning characteristics between water and heptane undoubtedly contributed to the apparent success of the procedure. Subsequent work by Kirkland (3) and Kirkland et al. (4) with the fungicide benomyl and its metabolites gave hope for those who followed that some types of complex matrixes, such as cow milk, urine, feces, tissues, soil and plants could be similarly analyzed by HPLC with uv absorbance (here at 280 nm). Undoubtedly the high selectivity of the cation exchange analytical column greatly contributed to the success of the procedure; similar selectivity has been observed in the author's laboratory for anion exchange separations. Kirkland pointed out early the necessity of filtering out all particles from the injected sample by Millipore® filters (1.5 μm) in a Swinney® adapter in order to prevent column degeneration. A typical chromatogram of fortified cucumber is shown in Figure 4.14. An ethyl acetate extraction was employed for this residue determination which eliminated many interfering plant coextractives that were present when the acidic MeOH extraction procedure for soils was used.

Just as Kirkland demonstrated for benomyl, Corley et al. (49) were able to measure residues of TH 6040 (Dimilin) in milk following a simple partition with ethyl acetate, except they employed an RP separation and a 254

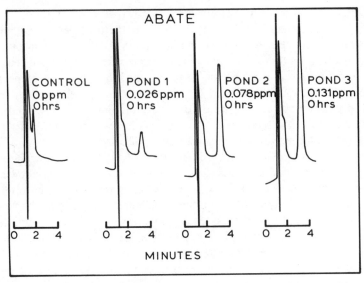

Figure 4.13 Chromatograms of Abate larvacide in salt marsh pond extracts; Zipax coated with 1.0 wt% β,β-oxydipropionitrile, heptane, 254 nm uv (Ref. 1).

nm uv detector. No interferences were observed, as seen in Figure 4.15, at the 1 ppm level; a limit of detection of 0.1 ppm was determined, adequate for the low toxicity insecticide.

One of the first applications of the fluorometer as a detector in a residue procedure was that of Maeda and Tsuji (50). Even through a normal phase column was used and an extraction procedure previously described by Kirkland (4) they were able to determine benomyl (as 2-AB) and thiabendazole (TBZ) in citrus tissues without any interferences appearing on the chromatogram whatsoever. Limits of detection for benomyl were 0.02 ppm and 0.001 ppm for TBZ. Recoveries for various plant tissues ranged from 90.5% for orange peel to 103% for cucumber.

It has long been recognized that many of the more water-soluble pesticides were not being extracted from various foods using the multi-residue procedure in the *Pesticide Analytical Manual* of the U.S. Food and Drug Administration. These include many of the carbamates, some of the organophosphates, most of the triazines and others. Luke et al. (51) in 1965 proposed an extraction procedure using acetone, which is quite similar to the *PAM* procedure in sample manipulation and glassware requirements. Since it recovers most of the water-soluble pesticides formerly lost to the *PAM* procedure, it would seem to be a good choice if it could be incorporated into HPLC procedures.

Beginning in 1976 Lawrence and coworkers published a series of papers (52–58) utilizing the acetone extraction procedure of Luke, Froberg and Masumoto (51) along with various modifications to the *PAM* Florisil cleanup procedure. Their attention focussed primarily on the carbamate insecticides and urea herbicides with chromatography performed exclusively in the adsorption (AD) mode on LiChrosorb SI-60 columns of 5-μm particle size. Major changes in the cleanup procedure of Luke et al. were in the solvents used for the Florisil cleanup; mixtures of CH_2Cl_2 in hexane and acetone in hexane were used instead of the ethyl ether–petroleum ether combination. Using 15% acetone–hexane on a 2% deactivated Florisil column he was able to quantitatively elute 10 carbamates and demonstrate detectabilities for several of them after addition to such foods as corn, cabbage, potato, and wheat. In one of the publications (55) the applicability of

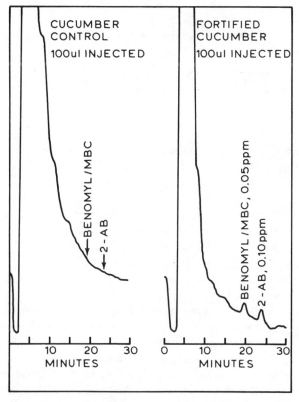

Figure 4.14 Chromatograms of cucumber extract; Zipax SCX, nitrate buffer, 0.5 ml/min, 254 nm uv (Ref. 4).

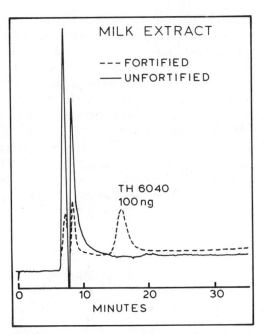

Figure 4.15 Chromatograms of milk extract unfortified and fortified (1 ppm) with TH6040; Permaphase-ODS, MeOH:H₂O (1:1), 254 nm uv (Ref. 49).

dual wavelength detection (254) nm, 280 nm) was demonstrated for the residue analysis of carbofuran, 3-ketocarbofuran and 3-hydroxycarbofuran. Figure 4.16 shows the detection of two of these in carrot fortified at 1.0 ppm and field-treated turnip. The selectivity for carbofuran at 280 nm is clearly evident here, allowing it to be quantitated, whereas at 254 nm a large peak appearing in check samples completely obscured the carbofuran.

Using a similar extraction and cleanup approach (56) Lawrence demonstrated that a number of urea herbicides could be quantitated in foods. Linuron and chlorbromuron could be eluted exclusively from the Florisil column with 15% acetone in hexane, whereas diuron, monuron, choroxuron, fenuron, and fluometuron eluted with 50% acetone in hexane. Cleanup was fairly efficient, as evidenced from the chromatogram obtained with the 254-nm absorbance detector (Figure 4.17). Dansylation as a precolumn derivatization was applied to carbaryl in potato and corn (58) with detection by uv absorbance and fluorometry. A comparison was made to direct uv detection before derivatization; cleaner chromatograms were obtained at the 0.1 ppm level without the derivatization step. These applications have clearly established HPLC as an important tool for the analysis of carbamates and urea herbicides, although the uv absorbance approach has not been demonstrated to be effective in a multiresidue scheme, applicable to all compounds within the class and to many commodities.

Two compounds frequently called carbamates, although they are carbamoyl oximes, are methomyl and oxamyl. Being highly water soluble they elute early from an RP column (see Figure 4.2). Thean et al. (59) used this characteristic to partition away potential interferences by first extracting vegetables with ethyl acetate, rotary evaporating and picking the residue up with water, which was then partitioned with hexane. After the water was saturated with NaCl the pesticides were extracted with $CHCl_3$; this was also rotary evaporated and the residue dissolved in mobile phase (11% CH_3CN

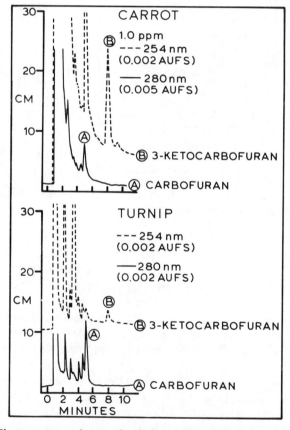

Figure 4.16 Chromatograms of carrot fortified at 1.0 ppm and field-treated turnip showing responses for carbofuran and 3-ketocarbofuran; LiChrosorb SI-60, 5% 2-propanol in isooctane, 1 ml/min (Ref. 55).

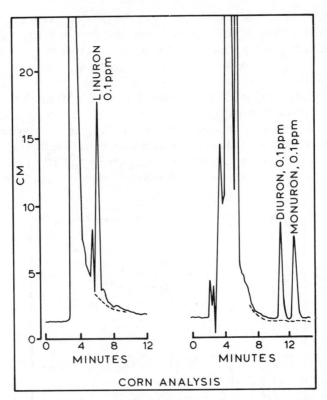

Figure 4.17 Chromatograms of corn fortified with linuron, diuron and monuron; LiChrosorb SI-60, 20% 2-propanol in isooctane, 1 ml/min, 254 nm uv (Rev. 56).

in pH 7 phosphate buffer). Chromatography was performed on a μC_{18} column with detection at 240 nm. A typical chromatogram is shown in Figure 4.18; recoveries for ten fruits and vegetables at 2 ppm ranged from 60 to 77% for oxamyl and 70 to 81% for methomyl.

Karbutilate, the active ingredient in Tandex® weed an bush killer, is both a substituted urea and a carbamate; its major metabolite in soil and grass is monomethyl karbutilate. Selim et al. (60) have reported an HPLC procedure for soil and grass which employs Florisil column chromatography for cleanup and determination by adsorption on a μPorasil silica column; detection was accomplished with the ubiquitous 254 nm uv absorbance detector. They found that Florisil cleanup could be achieved much more easily if the two chemicals were first hydrolyzed under alkaline conditions to the corresponding phenols, the urea portion of the molecule remaining intact, while the carbamate end was split. Producing the phenol did not improve absorption at 254 nm, which was low and limited detectability to 0.1 ppm for karbutilate and 0.2 ppm for monomethyl karbutilate; Figure 4.19 shows that early eluting interferences also limit the detectability considerably. Even so, recoveries from grass averaged 87% for karbutilate and 80% for the metabolite.

Many pesticides, even though they may possess highly conjugated or aromatic structures, do not absorb at 254 nm. Such was the case for the fungicide imazalil (1-[2-(2,4-dichlorophenyl)-2-(2-propenyloxy)ethyl]-1H-imidazole), which has an extinction coefficient at 202 nm approximately 1000 times than at 254 nm. Norman and Fouse (61), by using a variable wavelength detector at 202 nm were able to develop an extremely simple procedure for imazalil, employing an ethyl acetate extraction, partitioning into 0.05N H$_2$SO$_4$, adjusting to pH 8, back extracting into ethyl acetate and

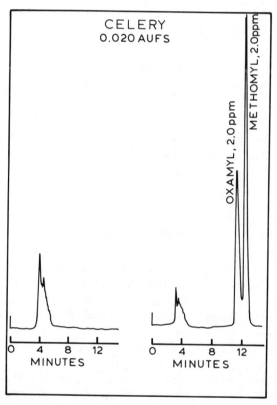

Figure 4.18 Chromatograms of celery fortified with oxamyl and methomyl; μC$_{18}$, 11% CH$_3$CN in pH 7 phosphate buffer, 1 ml/min, 240 nm uv (Ref. 59).

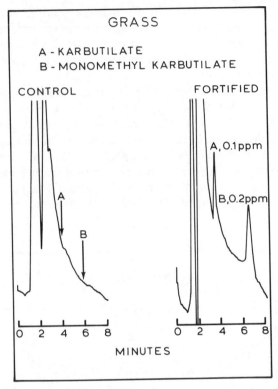

Figure 4.19 Chromatograms of grass fortified with karbutilate and monomethyl karbutilate; μPorasil, 3% ETOH/10% CH_3CN in ethylene chloride, 2.4 ml/min, 254 nm uv (Ref. 60).

chromatography on 10 μm LiChrosorb SI-60. The high degree of specificity is illustrated by the chromatogram in Figure 4.20, indicating an easily achievable limit of detection of about 0.1 ppm. Recoveries of 108% were measured for citrus rind surfaces sampled immediately after application.

Fluorometry of the underivatized pesticide has been employed by a few workers in the development of residue procedures (50, 62–65). We, in our laboratories, have found it extremely useful for the determination of naphthaleneacetic acid (NAA) in many types of citrus samples (64) even when using a conventional fluorometer equipped with a fairly large volume (300 μl) flowthrough cell. For citrus fruit, extraction by blending with CH_2Cl_2 in the presence of a small amount of concentrated H_2SO_4 was found to be satisfactory; the only cleanup required was a partition between ethyl ether and 0.2M K_2HPO_4. After acidification the NAA was extracted back into ethyl ether which was evaporated to dryness; water was added to dissolve the residue and 10 μl was injected onto the RP column, either a μCN or Permaphase ETH. Fluorescence was monitored at 340 nm with a 288 nm excitation setting. Few peaks were seen on the chromatogram other than NAA when extracts of whole fruit were prepared, as seen in Figure 4.21. A limit of detection of 0.008 ppm was possible (4 × noise); use of the fluorometer provided for a signal to noise ratio 200 times higher than that realized with a modern 254 nm uv detector without any measurable band spreading due to the larger flowthrough cell. Of particular interest was the fact that no difference in the chromatograms in terms of peak shape, height, or baseline noise, was observed whether a modern pulseless HPLC pump (Waters 6000) or a pulsatile piston pump (Milton Roy 196–0042–028) was used. Clearly fluorometry is less subject to pulsatile flow in the sample cell than is uv absorbance spectrophotometry.

Using an improved version of the amperometric detector evaluated by Lores et al. (39) Ott (66) was able to compare its performance with a 254

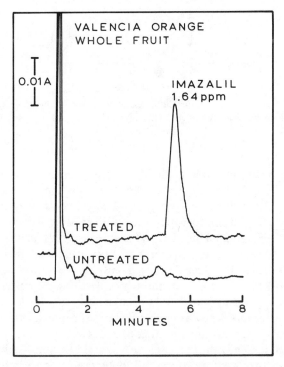

Figure 4.20 Chromatogram of untreated 'Valencia' orange and fruit found to contain 1.64 ppm of imazalil; LiChrosorb SI-60 reverse phase, 47% CH₃CN in pH 7.5 phosphate buffer, 1.3 ml/min, uv 202 nm (Ref. 61).

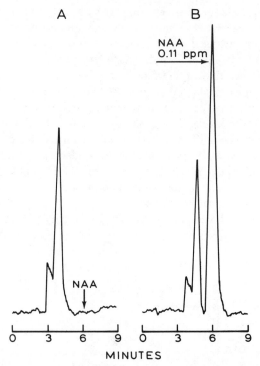

Figure 4.21 Chromatograms of untreated 'Pineapple' oranges (*A*) and those sprayed with 250 ppm solution of naphthaleneacetic acid (*B*); μCN, pH 7 phosphate buffer, 1 ml/min, fluorometer (Ref. 64).

nm uv detector. Set at +0.8 V, the Ag/AgCl reference electrode was also operated in the oxidative mode for the determination of the fungicide 2-phenylphenol (2PP). Fortified orange rind was extracted by blending with CH_2Cl_2 as eluant. Chromatography was performed by RP on a Corasil C_{18} column with 40% ethanol in H_2O. The amperometric detector was more specific for 2PP than the uv detector, and gave linear analytical curves from 10 to 50 ng; however, the slope decreased daily, necessitating frequent renewing of the working electrode. While no recoveries were determined it was simple to quantitate 2 ppm of 2PP in orange rind using this detector. At this writing, to the authors' knowledge this remains the only demonstrated use of the amperometric detector for a pesticide residue.

Postcolumn derivatization techniques for HPLC of pesticide residues have not been explored to their full potential. Recent research in the author's laboratory has illustrated that postcolumn reactions can achieve a degree of specificity and sensitivity that allows for the development of fast yet relatively simple pesticide residue procedures. Those familiar with the *PAM* method for glyphosate (*N*-phosphonomethylglycine, GLYPH) and its metabolite (aminomethylphosphonic acid, AMPA) can attest that the two compounds appear to be ill-suited for gas chromatographic determination. We have adapted the postcolumn fluorogenic labeling arrangement for carbamates previously described (25) so that it can respond to both GLPH, a secondary amine and AMPA, a primary amine, and have incorporated it into a complete residue procedure for them.

As can be seen from Figure 4.22 a pump capable of delivering 1 ml/min of 0.1*M* H₃PO₄ supplies mobile phase, through a loop type sample injection valve, to a strong anion-exchange analytical column which is thermostated at 62°C for optimum efficiency. Immediately postcolumn a dilute solution of Ca(C1O)₂ is metered in at 0.3 ml/min which is delayed by a 11 m × 0.5 mm coil during which time GLYPH is converted to glycine, a primary amine. A solution of o-phthalaldehyde–mercaptoethanol is then metered in, also at 0.3 ml/min, to form a highly fluorescent isoindole, which is monitored by the HPLC fluorometer with filters selected for excitation at 360 nm and emission at 455 nm. Typical chromatograms showing recovery of GLYPH and AMPA from cucumber are seen in Figures 4.23 and 4.24. A simple water extraction and cation-exchange cleanup on Dowex 50W-X8 was used to produce recoveries from 61 to 82% for AMPA and 70–90% at a

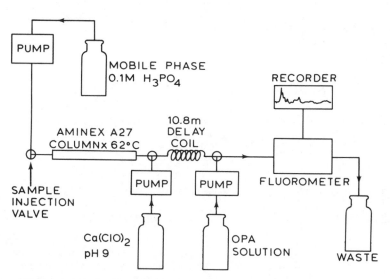

Figure 4.22 Post-column fluorogenic labeling HPLC for the analysis of glyphosate herbicide and its metabolite, aminomethylphosphonic acid (Ref. 67).

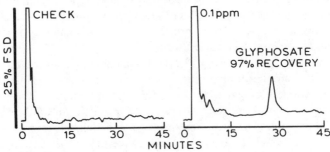

AMINEX A-27 COLUMN
0.1M H_3PO_4 1ml/min
20μl INJ   ATTEN 10X
CUCUMBER

CHECK   0.1ppm

GLYPHOSATE
97% RECOVERY

25% FSD

0   15   30   45 0   15   30   45
MINUTES

Figure 4.23 Chromatograms showing the analysis of untreated (check) cucumber and cucumber fortified at 0.1 ppm with glyphosate; HPLC as in Figure 4.22 (Ref. 67).

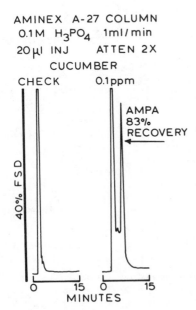

AMINEX A-27 COLUMN
0.1M H_3PO_4 1ml/min
20μl INJ   ATTEN 2X
CUCUMBER

CHECK   0.1ppm

AMPA
83%
RECOVERY

40% FSD

0   15 0   15
MINUTES

Figure 4.24 Chromatograms showing the analysis of untreated (check) cucumber and cucumber fortified at 0.1 ppm with aminomethylphosphonic acid; HPLC as in Figure 4.22 (Ref. 67).

192

fortification of 0.1 ppm in a variety of vegetables. Analysis time was considerably shortened compared to the *PAM* procedure (67).

4.5. CONCLUSIONS

It has clearly been demonstrated by many workers that HPLC has its place in pesticide residue analysis, its strength being that it can be used for those pesticides which for one or more reasons cannot be easily determined by gas chromatography. While many successes have been reported in the literature in the use of the 254 nm uv absorbance detector it has its limitations and is gradually being replaced by the variable wavelength detector; even so it is in the area of specific detectors that HPLC has its greatest weakness when applied to pesticide residue methodology. Efforts to remedy the situation by the development of new detectors, such as the electrochemical, and the adaptation of GLC detectors to HPLC have only been partially effective.

Undoubtedly refinements in electrode design will improve the performance of the electrochemical detector, and there may be some merit in continuing to explore devices for the removal of solvent (mobile phase) so that the highly developed element specific GLC detectors may be fully exploited. Acceptance by the residue chemist of techniques involving continuous flow analysis (CFA) that are already routinely used by the clinical chemist will undoubtedly improve detection capabilities. The potentials of such an approach have already been pointed out.

Most of the references listed in Table 4.1 decribe procedures in which sample cleanup was performed similarly or indentical to previously published techniques employing GLC as the determinative step. Since behavior of coextractives on an HPLC column is inherently different than on a GLC column, the residue chemist should consider reasonable alternatives to generally accepted cleanup techniques. Combining widely different modes of separation in the cleanup and determinative step would seem to offer the best chances of success in method development. The large variety of commercially available analytical columns at attractive prices will assist in this respect.

It would seem logical that to fully utilize HPLC for the analysis of pesticide residues in foods that a "multiresidue" approach should evolve, such as has been developed using GLC. The challenge appears great but can be met if a commitment is made by those governmental agencies charged with providing a safe and plentiful food supply to underwrite and coordinate it. Some bold rethinking of pesticide residue analytical chemistry will be necessitated if success is to be achieved in this manner.

REFERENCES

1. R. A. Henry, J. A. Schmit, and J. F. Dieckman, *Anal. Chem.*, **43**, 1053 (1971).
2. L. F. Krzeminski, B. L. Cox, and A. W. Neff, *Anal. Chem.*, **44**, 126 (1972).
3. J. J. Kirkland, *J. Agric. Food Chem.*, **21**, 171 (1973).
4. J. J. Kirkland, R. F. Holt, and H. L. Pease, *J. Agric. Food Chem.*, **21**, 368 (1973).
5. H. A. Moye, *J. Chromatogr. Sci.*, **13**, 268 (1975).
6. J. J. Kirkland (Ed.), *Modern Practice of Liquid Chromatography*, Wiley, New York, 1971.
7. L. R. Snyder and J. J. Kirkland, *Introduction to Modern Liquid Chromatography*, Wiley, New York, 1974.
8. J. N. Done, J. H. Knox, and L. Loheac, *Applications of High Speed Liquid Chromatography*, Wiley, London, 1974.
9. J. H. Knox (Ed.), *Practical High Performance Liquid Chromatography*, Simpson, Heyden and Son, London, 1976.
10. D. C. Locke, *J. Chromatogr. Sci.*, **12**, 433 (1974).
11. M. Martin, G. Blu, and G. Guiochon, *J. Chromatogr. Sci*, **12**, 438 (1974).
12. H. M. McNair and C. D. Chandler, *J. Chromatogr. Sci.*, **14**, 477 (1976).
13. Z. Deyl and J. Kopecky, *Bibliography of Liquid Column Chromatography*, *1971–73*, Elsevier, Amsterdam, 1976.
14. R. P. W. Scott, *Contemporary Liquid Chromatography*, Wiley-Interscience, New York, 1977.
15. C. Bollet, *Analusis*, **5**, 157 (1977).
16. R. E. Majors, *J. Assoc. Off. Anal. Chem.*, **60**, 186 (1977).
17. R. E. Majors, *J. Chromatogr. Sci.*, **15**, 334 (1977).
18. R. E. Majors, *Analusis*, **3**, 549 (1975).
19. H. L. Reynolds (Ed.), *FDA Bylines*, **9**, 31 (1978).
20. R. A. Hoodless, J. A. Sidwell, J. C. Skinner, and R. D. Treble, *J. Chromatogr.*, **166**, 279 (1978).
21. A. D. Thruston, Jr., EPA-R2-72-079 N.E.R.C., US EPA, Corvalis, Ore., 1972.
22. L. R. Snyder, *J. Chromatogr. Sci.*, **15**, 441 (1977).
23. J. H. Knox, *J. Chromatogr. Sci.*, **15**, 352 (1977).
24. C. Horvath and W. Melander, *J. Chromatogr. Sci.*, **15**, 393 (1977).
25. H. A. Moye, S. J. Scherer, and P. A. St. John, *Anal. Lett.*, **10**, 1049 (1977).
26. R. Gloor and E. L. Johnson, *J. Chromatogr. Sci.*, **15**, 413 (1977).
27. D. L. Saunders, *J. Chromatogr. Sci*, **15**, 372 (1977).
28. J. N. Seiber, *J. Chromatogr.*, **94**, 151 (1974).

29. A. D. Drinkwine, D. W. Bristol, and J. R. Fleeker, *J. Chromatogr.*, **174**, 264 (1979).

30. T. H. Byast, *J. Chromatogr.*, **134**, 216 (1977).

31. C. M. Sparacino and J. W. Hines, *J. Chromatogr. Sci*, **14**, 549 (1976).

32. C. F. Aten and J. B. Bourke, *J. Agric. Good Chem.*, **25**, 1428 (1977).

33. G. Glad, T. Popoff and O. Theander, *J. Chromatogr. Sci.*, **16**, 118 (1976).

34. J. F. Lawrence and D. Turton, *J. Chromatogr.*, **159**, 207 (1978).

35. R. P. W. Scott, *Liquid Chromatography Detectors.*, Elsevier, Amsterdam, 1977.

36. P. T. Kissinger, *Anal. Chem.*, **49**, 447A (1977).

37. P. T. Kissinger, C. F. Refshauge, R. Dreiling, L. Blank, R. Freeman, and R. N. Adams, *Anal. Lett.*, **6**, 465 (1973).

38. Bioanalytical Systems, Inc., P. O. Box 2206, West Lafayette, IN, 47906.

39. E. M. Lores, D. W. Bristol, and R. F. Moseman, *J. Chromatogr. Sci*, **16**, 358 (1978).

40. E. M. Lores and R. F. Moseman, paper no. 314, 30th Southern Regional Meeting, ACS, Nov. 8-10, 1978, Savannah, GA.

41. F. W. Willmott and R. J. Dolphin, *J. Chromatogr. Sci.*, **12**, 695 (1974).

42. J. W. Dolan and J. N. Seiber, *Anal. Chem.*, **49**, 326 (1977).

43. K. R. Hill, *J. Chromatogr. Sci.*, **17**, 395 (1979).

44. W. M. Jones and K. R. Hill, abstract no. 52, 166th National Meeting, ACS, Chicago, IL, Aug. 29-31, 1973.

45. K. R. Hill and W. M. Jones, abstract no. 009, Third International Congress of Pesticide Chemistry (IUPAC), Helsinki, Finland, July, 1974.

46. K. A. Ramsteiner and W. D. Hörmann, *J. Chromatogr.*, **104**, 438 (1975).

47. H. A. Moye and T. E. Wade, *Anal. Lett.*, **9**, 891 (1976).

48. R. T. Krause, *J. Chromatogr. Sci.*, **16**, 281 (1978).

49. C. C. Corley, R. W. Miller, and K. R. Hill, *J. Assoc. Off. Anal. Chem.*, **57**, 1269 (1974).

50. M. Maeda and A. Tsuji, *J. Chromatogr.*, **120**, 449 (1976).

51. M. A. Luke, J. E. Froberg, and H. T. Masumoto, *J. Assoc. Off. Anal. Chem.*, **58**, 1020 (1975).

52. J. F. Lawrence, *J. Chromatogr. Sci.*, **14**, 557 (1976).

53. J. F. Lawrence, *J. Agr. Food Chem.*, **25**, 211 (1977).

54. J. F. Lawrence, D. A. Lewis, and H. A. McLeod, *J. Chromatogr.*, **138**, 143 (1977).

55. J. F. Lawrence and R. Leduc, *J. Agric. Food Chem.*, **25**, 362 (1977).

56. J. F. Lawrence, *J. Assoc. Off. Anal. Chem.* **59**, 1066 (1976).

57. J. F. Lawrence and R. Leduc, *J. Chromatogr.*, **152**, 507 (1978).

58. J. F. Lawrence and R. Leduc, *J. Assoc. Off. Anal. Chem.*, **61**, 873 (1978).

59. J. E. Thean, W. G. Fong, D. R. Lorenz, and T. L. Stephens, *J. Assoc. Off. Anal. Chem.*, **61** 16 (1978).

60. S. Selim, R. F. Cook, and B. C. Leppert, *J. Agric. Food Chem.*, **25**, 567 (1977).

61. S. M. Norman and D. C. Fouse, *J. Assoc. Off. Anal. Chem*, **61**, 1469 (1978).

62. T. E. Archer and J. D. Stokes, *J. Agric. Food Chem.*, **26**, 452 (1978).

63. W. P. Cochrane and M. Lanouette, *J. Assoc. Off. Anal. Chem.*, **62**, 100 (1979).

64. H. A. Moye and T. A. Wheaton, *J. Agric. Food Chem.*, **27**, 291 (1979).

65. W. P. Cochrane, M. Lanouette, and R. Grant, *J. Assoc. Off. Anal. Chem.*, **63**, 145 (1980).

66. D. E. Ott, *J. Assoc. Off. Anal. Chem.*, **61**, 1456 (1978).

67. H. A. Moye and P. A. St.John, Ch. 6 "Pesticide Analytical Methodology," *ACS Monograph Series*, Washington, D.C., 1980.

68. D. D. Oehler and G. M. Holman, *J. Agric. Food Chem.*, **23**, 590 (1975).

69. S. K. Reeder, *J. Assoc. Off. Anal. Chem.*, **58**, 1014 (1975).

70. J. E. Smith and K. A. Lord, *J. Chromatogr.*, **107**, 407 (1975).

71. M. J. Fasco, L. J. Piper, and L. S. Kaminsky, *J. Chromatogr.*, **131**, 365 (1977).

72. C. H. Schaefer and E. F. Dupras, Jr., *J. Agric. Food Chem.*, **24**, 733 (1976).

73. C. H. Schaefer and E. F. Dupras, Jr., *J. Agric. Food Chem.*, **25**, 1026 (1977).

74. J. Kvalvåg, D. L. Elliott, Y. Iwata, and F. A. Gunther, *Bull. Environ. Contam. Toxicol.*, **17**, 253 (1977).

75. D. E. Ott, *Bull. Environ. Contam. Toxicol.*, **17**, 261 (1977).

76. S. J. DiPrima, R. D. Cannizzaro, J. C. Roger, and C. D. Ferrell, *J. Agric. Food Chem.*, **26**, 968 (1978).

77. S. L. Ali, *J. Chromatogr.*, **156**, 63 (1978).

78. M. C. Bowman, C. L. Holder, and L. I. Bone, *J. Assoc. Off. Anal. Chem.*, **61**, 1445 (1978).

79. J. E. Farrow, R. A. Hoodless, M. Sargent, and J. A. Sidwell, *Analyst*, **102**, 752 (1977).

80. S. Selim and R. F. Cook, *J. Agric. Food Chem.*, **26**, 106 (1978).

81. E. J. Kikta, Jr. and J. P. Shierling, *J. Chromatogr.*, **150**, 229 (1978).

82. C. J. Brealey and D. K. Lawrence, *J. Chromatogr.*, **168**, 461 (1979).

83. D. E. Ott, *J. Assoc. Off. Anal. Chem.*, **62**, 93 (1979).

84. K. A. Ramsteiner and W. D. Hörmann, *J. Agric. Food Chem.*, **27**, 934 (1979).

85. L. F. Faas and J. C. Moore, *J. Agric. Food Chem.*, **27**, 554 (1979).

86. L. Needham, D. Paschal, Z. J. Rollen, J. Liddle, and D. Bayse, *J. Chromatogr. Sci.*, **17**, 87 (1979).

87. G. R. Pieper, *Bull. Environ. Contam. Toxicol.*, **22,** 167 (1979).

88, G. F. Ernst and S. Y. Verveld-Röder, *J. Chromatogr.*, **174,** 269 (1979).

89. S. E. Moring and J. D. McChesney, *J. Assoc. Off. Anal. Chem.*, **62,** 774 (1979).

90. D. S. Farrington, R. G. Hopkins, and J. H. A. Ruzicka, *Analyst*, **102,** 377 (1977).

ANALYSIS OF CHLORINATED HYDROCARBONS

W. B. WHEELER and N. P. THOMPSON

*Pesticide Research Laboratory, Food Science and Human Nutrition Department,
University of Florida, Gainesville*

CONTENTS

5.1. INTRODUCTION

The advent of the chlorinated insecticides in the early 1940s caused a significant change in the field of residue chemistry. Until that time the analyst was largely concerned with inorganic chemicals such as heavy metal salts and sulfur. It was assumed that residues were of minor significance, since they could be washed from the products prior to consumption. The organochlorine compounds, however, exhibited the characteristics of persistence and penetration into foodstuffs. Thus, the residue chemist became a trace organic chemical analyst. This trend has continued to the present with the nearly exclusive use of synthetic organic chemicals for pest control.

In recent years the use of many chlorinated insecticides has been curtailed as a result of concerns for safety to human beings and to their environment. Although usage has been drastically reduced, the residue chemist still detects the chloroorganic insecticides in a number of substrates. In addition, a number of nonpesticidal chlorinated chemicals such as the polychlorinated biphenyls appear in residue samples. These environmental pollutants are of concern to regulatory agencies and are now measured along with pesticide residues.

This review will describe the techniques utilized for the analysis of organochlorine insecticides. A separate section briefly describes the analysis of chlorinated dioxins. The major emphasis will be upon newer chemical and instrumental approaches for the detection, identification, and quantification of the organochlorines.

5.2. SAMPLING, SAMPLE PREPARATION, AND STORAGE

This subject applies to any analytical determination and is not peculiar to the analysis of chloroorganics. It has been discussed in detail by Gunther (1) is discussed by Leng in Chapter 10 and, therefore, will not be dealt with extensively here.

A sample must be representative of the entire population, otherwise the analytical result will be inaccurate. In a regulatory situation incorrect results could lead, on the one hand, to unnecessary seizure of product caused by apparent above-tolerance residues and, on the other hand, the exposure of humans and animals to above-tolerance levels as a result of apparent below-tolerance residues.

Sample preparation and storage must be performed in such a way as to avoid any changes that will be reflected in the analytical results. Samples should not be contaminated during storage nor should storage result in loss or concentration of pesticide. Use of blank samples fortified with known quantities of the pesticide(s) in question and stored under the same conditions will help identify any effects that storage conditions might have on residue levels.

5.3. EXTRACTION

Historically, chlorinated insecticides have been extracted in one of three ways. If they were present on the surface of fruits, they may be removed by tumbling the fruits in a nonpolar solvent. While Soxhlet extraction is often employed to remove these insecticides from dry, finely ground plant materials, blending is the most commonly used method, particularly for the extraction of insecticides from fresh plant materials.

While the presence of internal residues of chlorinated hydrocarbon insecticides was well established (2–5), there had been little concern over whether the extraction methods quantitatively extracted these chemicals from the interior of plant materials. Extraction efficiencies of insecticidal residues were commonly measured by addition of known amounts of the chemical to an untreated sample of the crop often immediately prior to the extraction procedure followed by determination of the recovery. Despite the fact that such fortification procedures generally provide data concerning only the accuracy of the overall analytical technique and not extraction efficiency, there is no practical alternative to assessing extraction efficiency by these methods to measure the effectiveness of the extraction process.

Recognizing these limitations, Thornburg (6) suggested the use of a "weathered residue study" in an attempt to determine extraction efficiencies without sample fortification. Three subsamples were used: one was extracted using the routine method; a second was extracted three times; and the third was extracted using three times the normal volume of solvent. If all three methods yielded the same results, the routine procedure was considered trustworthy and usable. Although this procedure avoids sample fortification, it still gave no absolute measure of extraction efficiency, since a constant inefficiency might be present in all procedures.

Klein et al. (7) discussed the extraction of chlorinated insecticides. These investigators made two applications, 10 days apart, of radioactive methoxychlor [1,1,1-trichloro-2,2-bis(p-methoxyphenyl)ethane] to spinach plants. During analysis it was observed that extraction of an air-dried sample in a Soxhlet extractor using ethyl ether for 18 hr removed 89% of the label. The radioactive count of the material in the Soxhlet thimble remained the same, however, after 28 hr of additional refluxing. Changing the solvent to benzene and continuing the extraction reduced the measurable radioactivity by only 15% after 9 hr, and further refluxing in an isopropyl alcohol–benzene mixture for 5 hr reduced the count by only one third. The authors described this phenomenon as "fixation" and were able to achieve an absolute measure of extraction efficiency. This represents one of the few examples in the literature of such a direct measure of extraction efficiency.

The problem of obtaining complete extraction of solely internal chlorinated insecticides became apparent (8) when plants grown in substrates containing radioactive dieldrin were extracted by a commonly used method. Corn, orchard grass, and wheat were grown for 2 to 4 weeks in sand or in liquid culture to which ^{14}C-dieldrin had been added. The plant materials were extracted by blending in n-hexane:isopropyl alcohol (2:1 v/v), the solvent was decanted from the plant tissues and the blending was repeated using fresh solvent. Radioactivity measurements were then made on the plant material remaining from the blending to determine how well the dieldrin had been extracted. The presence of significant radioactivity (12–25% of the total quantity present) made it desirable to develop a method that would remove essentially all of the internal dieldrin. An additional extraction of plant materials using chloroform–methanol in a Soxhlet apparatus effectively removed 99% of the remaining labeled insecticide from these tissues (8, 9). This exhaustive extraction process consisting of blending followed by Soxhlet extraction has become a standard for comparison in a number of subsequent extraction efficiency studies. An example of the findings of Wheeler et al. (8) is illustrated in Figure 5.1.

Nash et al. (10), Nash and Beall (11, 12), and Wheeler et al. (13) have reported bound residues of chloroorganics in plants. Nash et al. (10–12) detected varying quantities of unextractable ^{14}C in soybeans and cotton after growing them in soil with ^{14}C-dieldrin, endrin, and heptachlor. Wheeler et al. (13) have found 20–25% bound ^{14}C in radish roots harvested 14 days after application of ^{14}C-dieldrin.

Bound residues as defined by these workers are residues unextractable using organic solvents. Such bound residues have not often been reported for chloroorganic compounds in plant materials. Bound residues may be the result of metabolized pesticide incorporated into naturally occurring bio-

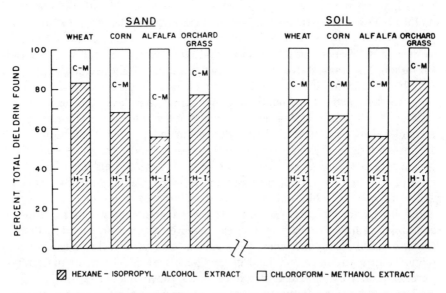

Fig. 5.1 Efficiency of dieldrin extraction by blending as applied to four plant species grown in sand and in soil. (Reprinted with permission from the *Journal of Agricultural and Food Chemistry*, **15**, 227–230, 1967. Copyright by the American Chemical Society.)

logical components or the incorporation of essentially intact pesticide into some biological constituent.

Methods currently used by most federal and state agencies responsible for monitoring pesticide residues on food and feed are contained in the *Pesticide Analytical Manual* (*PAM*) (14) prepared by the Food and Drug Administration. In general, the methods are classified as being applicable to fatty foods (>10% fat) (*PAM*, Vol. I, Sections 211.13–211.14) and nonfatty foods (*PAM*, Vol. I, Section 212). High fat content materials may be extracted with petroleum ether; or if the fat can be isolated, it may be dissolved directly in petroleum ether. In either case the petroleum ether solution is partitioned with acetonitrile, with the acetonitrile being back-partitioned with petroleum ether. The purposes of these manipulations are to extract the pesticides from the substrate and separate the fat from the pesticides. Extraction of nonfatty substrates is generally performed by blending with acetonitrile or acetonitrile–water mixtures. These methods apply across a broad spectrum of substrates and allow for the screening of multiple residues that may be present in foods or feed. Some of the specific aspects of various extraction parameters are discussed below.

Bertuzzi et al. (15) evaluated a variety of parameters applied to the extraction of dieldrin, DDTs, and methoxychlor from nonfatty samples containing less than 10% water. A number of conclusions were derived from this work:

1. Acetonitrile/water combinations were superior to a number of other solvents.
2. 35% H_2O/acetonitrile was the most efficient extracting solvent in the range of 0 to 40% water.
3. Hydration of the sample was critical to efficient extraction, but the manner in which hydration was accomplished was not important.
4. Differences in sample size, solvent volumes used for extraction (assuming sufficient was present to "wet" the sample), and blending times (2 min and 5 min) did not affect extraction efficiency.

Factors influencing the extraction of chloroorganic insecticides from soils have also been investigated. Chiba and Morley (16) evaluated a number of parameters which influenced the extraction efficiency of aldrin and dieldrin from three soil types. It was concluded that water added to soil prior to extraction allowed the most efficient extraction of both insecticides. Dimethylformamide was a highly efficient solvent and eliminated the need for adding water to the soil or for a relatively long contact time between soil and extraction solvent. These authors pointed out that fortification of soil and subsequent extraction is a poor method of determining extraction efficiency of weathered residues.

Saha et al. (17) also stated that fortification of air-dried soils with dieldrin should not be used for measuring recovery rates from field samples extracted in the air-dried state. It was pointed out that the addition of 20% water to soils immediately before extraction resulted in highly efficient extraction of dieldrin from four soil types (17) and of chlordane from a clay loam soil (18).

The extraction of soils using a number of methods to affect intimate mixing of soil and solvent was desribed by Johnson and Starr (19). These investigations compared Soxhlet extraction, roller extraction, blender extraction, Polytron® extraction, and one other ultrasonic extraction device. In general they concluded that the Polytron is the superior extraction device even for very short (30 sec) blending times. Further, acetone was the superior solvent and moist soils were more efficiently extracted than dry soils.

Extraction of chloroorganics from lipid material is often performed using the classic method of Mills (20), which utilizes liquid–liquid partitioning. Other methods include the coagulation of lipid materials (21) and the application of the fatty material to a solid support prior to extraction (22).

Extraction of pesticides from water represents a different problem. Extraction methods from water include partitioning with hexane or other organic solvents (23, 24), sorption by activated carbon (25), sorption by urethane foam plugs (26–28), sorption by Carbowax 4000 monostearate- and undecane-coated Chromosorb W (29), and sorption to cellulose triacetate membrane filters (30).

5.4. CLEANUP

Sample cleanup is designed to remove coextractives, which interfere with detection, quantification, and confirmation of the chloroorganic pesticide. Many of the commonly used techniques are described in the *PAM*, Vol. I, Sections 201 and 211. Procedures include liquid–liquid partitioning, sorption to and elution from various adsorbents, acidic and alkaline treatments and a distillation process. In many cases two or more of these methods are utilized sequentially.

The most frequently used liquid–liquid partitioning is the petroleum ether–acetonitrile procedure and is very well described in the *PAM*. Dimethyl sulfoxide has also been utilized (31) in place of acetonitrile for very high-fat content samples.

Sorption of extracts to Florisil and selective elution of pesticides is a frequently used cleanup technique. Elution of chlorinated insecticide is normally performed using 6% ethyl ether in petroleum ether followed by elution with a 15% solution. Mixtures of methylene chloride, acetonitrile, and hexane have also been used for elution. A recent report describes an ethyl ether prewash of Florisil columns, which results in superior cleanup and recovery of many more compounds in the 6% elution fraction and eliminates the need for the 15% elution (32).

Other sorption cleanup procedures include magnesium oxide–celite and silicic acid column chromatography.

Alkaline and acid hydrolyses designed to selectively remove interferences and allow the recovery of the insecticides of interest are sometimes used. The *PAM* (Vol. I, 211.15b) describes the use of a sulfuric acid impregnated Celite column to cleanup extracts containing DDT and a number of other chloroorganics. Alkaline hydrolysis (*PAM*, Vol. I, 211.15d) is also effective for the cleanup of a few chlorinated compounds (aldrin, dieldrin, endrin, and polychlorinated biphenyls) that are stable to the treatment. Sweep Co-Distillation has been shown to be satisfactory for chloroorganics particularly when applied to high fat samples (33).

An automated gel permeation chromatography system was described by Johnson et al. (34). This system, using a Bio-Beads SX-3 column and a toluene–ethyl acetate mobile phase, was reported to be effective, rapid, and resulted in excellent recoveries of chlorinated insecticides and PCBs.

5.5. DETECTION, MEASUREMENT, AND CONFIRMATION

These subjects will be discussed together because each technique provides information useful for the detection, measurement, or identity confirmation.

5.5.1. Gas–Liquid Chromatography

Gas–liquid chromatography (GLC) is the most often used technique for the separation, detection, and measurement of organochlorine pesticides. Prior to the advent of the election capture (EC) detector, GLC had only limited utility because of the nonspecific detectors and resulting complex chromatograms. The EC detector (35) provided greatly increased sensitivity to chloroorganics and a certain degree of specificity. Burke and Guiffrida (36) applied EC-GLC to the analysis of multiple residues of chloroorganics at a sensitivity level of 0.01 ppm.

The development of microcoulometric gas–liquid chromatography (MC-GLC) was reported in 1960 (37). The MC-detector was halogen-specific when operated in one mode and would detect sulfur in another mode. Thus the MC-detector offered high specificity to organochlorine but relatively low sensitivity.

The electrolytic conductivity detector was also developed by Coulson (38). It is capable of detecting organochlorine compounds based on the conductivity in water of combustion products (HCl). Another conductivity detector, similar in concept, was described by Hall (39). Hall reported that his modifications resulted in increased sensitivity and fewer operational problems.

The analysis of polychlorinated biphenyls (PCBs) is complicated by several factors, including separation from chlorinated insecticides, quantitative determinations resulting from the large number of possible isomers and the variable degradation of these isomers in samples from the environment. Some recent work has been accomplished toward solving these difficulties, although at this time it appears that continued research is needed.

Cook et al. (40) have reported separation and analysis of PCBs, polychlorinated naphthalenes (PCNs) and organochlorine pesticides by carbon skeleton gas chromatography where samples are catalytically converted to parent hydrocarbons. Detection was by a flame ionization detector (FID) following passage through a coupled catalytic converter containing 3% Pd (305°) or 5% Pt (180°) and a 5% SE-30 column. Absolute values for the concentrations of the polychlorinated mixtures were not obtained, but it was considered that concentrations in terms of equivalents of naphthalene or biphenyl are more equivalent to the original character prior to weathering or biological activity.

Sawyer (41) studied PCBs by GLC and concluded that quantitation by comparison with single peaks of individual Aroclors was equal to or better than total PCB methods using the Hall electrolytic conductivity detector in the chlorine mode and single ion monitoring chemical ionization mass spectrometry.

A system of PCB quantitation based on weights of individual PCB isomers was proposed by Webb and McCall (42), since the EC detector responds differently to each. They included a flowchart of PCB quantitation which allows measurement of several Aroclors from the same sample.

Luckas et al. (43) determined chlorinated pesticides and PCBs by derivatization gas chromatography using two injections of each sample, one yielding a basic chromatogram and a second, a derivatized chromatogram. They reported that adequate separation and quantitation was achieved. For the derivatized samples a microreactor containing MgO was attached to the injector of the GC so that direct injection of the sample was possible. DDT and metabolites were all converted to the corresponding olefins and estimations of PCBs could be made by comparing chromatograms. This method did not provide for DDE-PCB separation.

Sediment and oysters were analysed for chlorinated pesticides and PCBs by Teichman et al. (44) by GLC after adsorption chromatography on alumina and charcoal columns. Fractional hexane eluates of the alumina column followed by benzene and acetone–diethyl ether elution of the charcoal column provided good separation and recoveries. They stated that it may be possible to estimate the amount of PCBs in a sample using this procedure but, in many instances, their conclusion was that presence or absence of PCBs is more important.

5.5.2. Thin Layer Chromatography

Thin layer chromatography (TLC) is an excellent method for confirming the identity of chloroorganics already detected by GLC, but it is difficult to use quantitatively. Both silica gel and aluminum oxide adsorbants have been utilized, although aluminum oxide has been shown to be superior and is now routinely used for these analyses. Cleaned-up extracts are spotted, the plates are developed (usually with n-heptane) and the chromatogram components are visualized using $AgNO_3$ (either sprayed after development or incorporated into the adsorbant prior to development) as the chromogenic reagent (14).

5.5.3. p Values

The p value, determined by distributing a pesticide between equal volumes of two immiscible solvents, is defined as the fraction of the total solute remaining in the upper phase. Bowman and Beroza (45) originated this now classic approach for identifying or confirming 131 pesticides, among them the chloroorganics. It is simple, reliable, and offers the sensitivity of EC-GLC for the determination of concentrations. In fact, if an unknown component is being investigated, only relative amounts rather than absolute quantities need be known to calculate p values. By combining GLC relative retention times (retention compared to aldrin assigned a value of 1.0) and p values, the residue analyst can rapidly eliminate many possible unknowns and/or confirm identities. (See Chapter 1.)

5.5.4. Infrared Spectrophotometry

The applications of infrared (ir) spectrophotometry to pesticide analysis were reviewed by Blinn and Gunther (46) who emphasized that while ir can provide invaluable structural information for the identification or confirmation of pesticide residues, it was essential to prepare pure isolates of the compound under analysis.

Payne and Cox (47) reported the microinfrared analysis of several chlorinated pesticides isolated from sludge, soil, fish tissues, and aqueous industrial effluent. Samples were purified using Florisil column chromatography followed by TLC. At least 10–50 μg of the pesticides were required for satisfactory ir analysis.

Chen (48) reported the utilization of microtechniques for the ir analysis of methoxychlor. An ir spectrophotometer equipped with a beam condenser, scale expander, and auxiliary recorder was utilized and satisfactory spectra could be obtained with as little as 1 μg of material.

The requirements of relatively large sample size and rigorously pure compounds frequently preclude the use of ir for the analysis of chloroorganic residues. A relatively new approach, however, has been reported by Wall and Mantz (49). The method utilized a Fourier transform ir spectrophotometer interfaced to a computer for data collection and reduction, and samples were introduced into the spectrophotometer using a gas chromatograph. An eight-component mixture of mirex and mirex-related compounds was separated and an ir spectrum was obtained for each component after it eluted from the GC column. The separation and ir spectra are illustrated in Figure 5.2. This technique, requiring only in some cases as little as 10 ng of material, is sensitive and offers the potential of applying ir to residues of chloroorganics.

5.5.5. Polarography

The technique of polarographic analysis of organochlorine residues is not often utilized. Gajan and Link (50) described the polarography of DDT and its analogs present in formulations, standards, and as residues. Peak potentials vs a silver wire reference were reported as well as the ability to determine various isomer ratios. Since polarography required relatively large quantities of samples and other available methods could provide information necessary for residue analysis, its utility has not been exploited.

5.5.6. Mass Spectrometry

Mass spectrometry (MS) is a particularly powerful and sensitive tool in the identification of chlorinated insecticides. When coupled with a gas chromatograph (GC-MS), which provides a means of component separation, structural information for the confirmation and identification of chloroorganic insecticides can be obtained.

Mumma and Kantner (51) and Kantner and Mumma (52) were among the first to apply electron impact (EI) ionization MS to standards of chlorinated insecticides and to actual residues detected in plants. Samples were trapped from the effluent of a GC column and introduced as solid materials into the ionization source of their magnetic sector instrument. Molecular weights, information regarding numbers of chlorine atoms in the molecule, and fragmentation patterns were obtained.

The quadrapole mass spectrometer, a device that obtains mass separations by passing ions through radiofrequency and direct current fields applied to the quadrapoles, has become a very popular type of instrumentation.

Another mode of sample ionization is chemical ionization (CI). A chemical reagent (e.g., methane) is ionized in the ion source of the mass spectrometer and it in turn causes the ionization of the sample. CI spectra are generally simpler since less fragmentation of the molecule occurs; molecular ions per se are not observed, although "pseudo" molecular ions and other adducts are observed. These result, for example, from the addition of H^+, CH_5^+, or $C_2H_5^+$ (when methane is the reagent gas) to the molecular ion. A variety of reagent gases have been used in CI-MS.

Carlson et al. (53) utilized this technique to identify mirex and mirex degradation products. These spectra were characterized by fragments representing the molecular ion minus one chlorine atom.

The techniques described above involve the detection of positive ionic species. Negative ion detection has also been realized over the last few years. Rankin (54) reported the application of EI negative ion MS to a number of pesticides. As applied to the organochlorines, aliphatic chlorines were easily lost, leading to the formation of highly intense chlorine ions and also groups at m/e 70–75. The presence of negative chlorine ions in the mass spectrometer further resulted in an intense (M + Cl) ion as was seen for p,p'-DDT. This methodology using electron impact, negative ion generation, and detection has not seen much development in recent years. The use of negative ion (NI) CI techniques have, however, been developed. Deinzer et al. (55) used this technique to identify impurities formed in the manufacture of pentachlorophenol. A major class of impurities are the hydroxypolychlorinated diphenyl ethers. Figures 5.3, 5.4, and 5.5 illustrate the utility of NI-CI in detecting impurities and determining the members of chlorines on each phenyl ring. This information combined with the knowledge of the molecular ion allowed the unambiguous identification of several of the hydroxy ethers.

Field ionization (FI) MS has also found considerable usage. It, like CI-MS, shows relatively little fragmentation and an increased abundance of molecular ions. Application of FI-MS to four cyclodiene chlorinated insecticides has shown its great utility (56).

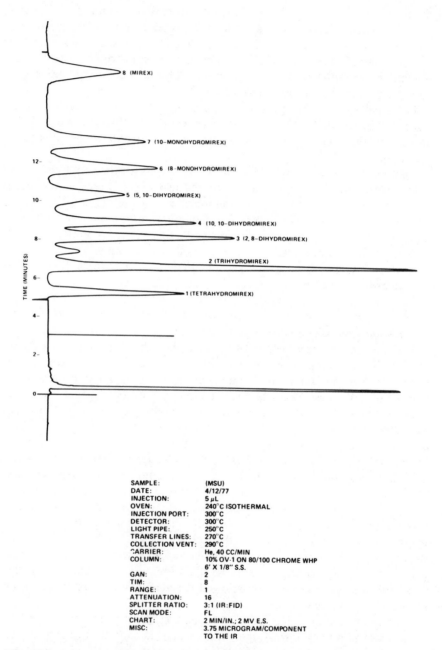

Figure 5.2 Mixture of mirex derivatives; 3.75 μg per component on-the-fly to the ir accessory. Society for Applied Spectroscopy.)

210

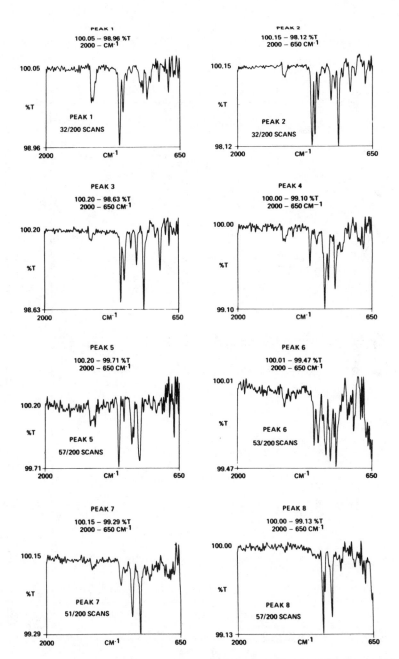

PEAK 1
100.05 – 98.96 %T
2000 – CM⁻¹

PEAK 1
32/200 SCANS

PEAK 2
100.15 – 98.12 %T
2000 – 650 CM⁻¹

PEAK 2
32/200 SCANS

PEAK 3
100.20 – 98.63 %T
2000 – 650 CM⁻¹

PEAK 4
100.00 – 99.10 %T
2000 – 650 CM⁻¹

PEAK 5
100.20 – 99.71 %T
2000 – 650 CM⁻¹

PEAK 5
57/200 SCANS

PEAK 6
100.01 – 99.47 %T
2000 – 650 CM⁻¹

PEAK 6
53/200 SCANS

PEAK 7
100.15 – 99.29 %T
2000 – 650 CM⁻¹

PEAK 7
51/200 SCANS

PEAK 8
100.00 – 99.13 %T
2000 – 650 CM⁻¹

PEAK 8
57/200 SCANS

(Reprinted with permission from *Applied Spectroscopy*, **31**, 552–560, 1977. Copyright by the

211

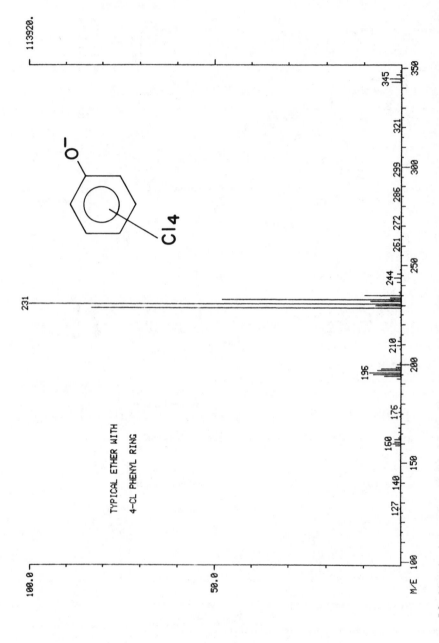

Figure 5.3 Negative ion methane chemical ionization mass spectrum of the tetrachlorophenate fragment of hydroxypolychlorodiphenyl ether. (Reprinted with permission from *Biomedical Mass Spectrometry*, **6**, 301–303. Copyright by Heyden and Son, Ltd.)

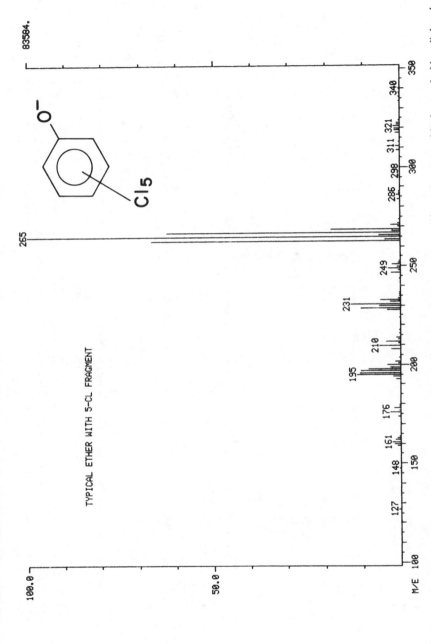

Figure 5.4 Negative ion, methane chemical ionization mass spectrum of the pentachlorophenate fragment of hydroxypolychlorodiphenyl ether. (Reprinted with permission from *Biomedical Mass Spectrometry*, **6**, 301–303. Copyright by Heyden and Son, Ltd.)

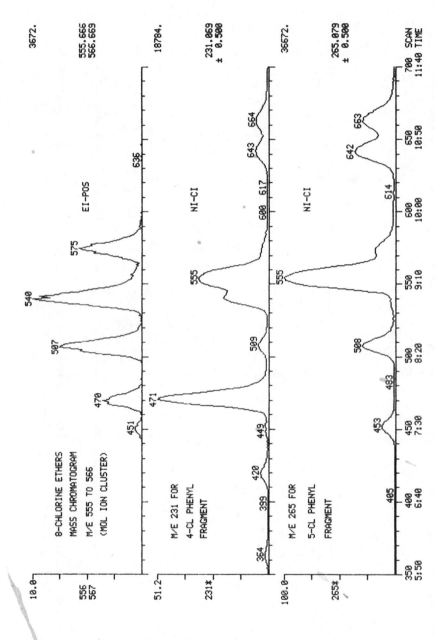

Figure 5.5 Mass chromatograms of hydroxyoctachlorodiphenyl ether, the tetrachloro- and the pentachlorophenate fragments (*Top*: Electron impact (positive) mass chromatogram of hydroxyoctachlorodiphenyl ether. *Middle*: Negative ion, methane chemical ionization mass chromatogram of the tetrachlorophenate fragment. *Bottom*: Negative ion, methane chemical ionization mass chromatogram of the pentachlorophenate fragment. (Reprinted through the courtesy of M. Deinjer, D. Griffin, T. Miller, and R. Skinner.)

214

In its present state of development, MS is highly sensitive requiring nanogram or smaller quantitites for satisfactory spectra. The isotopic distribution of the chlorine atoms provide spectra which are characteristic of a given number of chlorine atoms in a molecule. Thus this technique gives extremely valuable structural data for the unequivocal confirmation of organochlorine residues.

5.5.7. Derivatization

Chemical and photolytic derivatization have become prominent over the past several years as a method for the confirmation of identity of residues detected by other methods. The *PAM*, Vol. I, Sections 651 and 652 describe and summarize the literature regarding the utilization of these techniques. Cochrane (57) as quoted by the *PAM* lists criteria which successful derivatization confirmatory tests should possess:

1. Sensitivity at the residue level
2. Absence of interference from the sample or from other pesticides
3. Short GLC retention time of the derivative
4. Reaction should give a good yield of a specific derivative
5. Test must be facile and rapid

Chemical derivatization has been applied to aldrin (58–63), technical chlordane (*PAM*, Section 651.1, 64, 65) *cis*-chlordane (59, 66, 67), *trans*-chlordane (59, 66, 67), *p,p'*-DDE (*PAM*, Sections 251.2, and 64), *o,p'*-DDT (*PAM*, Sections 651.1 and 67, 65), *p,p'*-DDT (*PAM*, Sections 651.1, *PAM*, Sections 251.2 and 64, 65, 67–69), *p,p'*-dicofol (*PAM*, Section 651.1), dieldrin (70, 66, 71–74 61, 63, 75, 64, 76), endrin (70, 66, 71, 77, 73, 78, 61, 79, 62, 64, 76), heptachlor (70, 66, 71–74, 61, 63, 75, 64, 76), heptachlor epoxide (59, 66, 67, 64, 76), hexachlorobenzene (84–89), isobenzan (65, and *PAM*, Section 651.1), *p,p'*-methoxychlor (*PAM*, Sections 651.1 and 67, 64, 65), octachlor epoxide (*PAM*, Section 651.1), perthane (*PAM*, Sections 651.1 and 67, 65), polychlorinated biphenyl (40, 91, 69), strobane (*PAM*, Sections 651.1 and 65), *o,p'*-TDE (*PAM*, Sections 651.1 and 67), *p,p'*-TDE (*PAM*, Sections 651.1 and 251.2 and 67, 64, 65), tecnazene (*PAM*, Section 651.1) and toxaphene (*PAM*, Sections 651.1 and 92, 64, 65).

Photolytic derivatization of a number of chlorinated hydrocarbons may also be found in FDA's *PAM*, Section 652.

5.6. DIOXINS

The discovery of the toxicity of the chlorinated dibenzo-*p*-dioxins (CDD), particularly the 2,3,7,8-tetrachloro isomer (TCDD), has stimulated methodology research into their separation and quantitation.

Pfeiffer (93) reported early liquid chromatographic work by R. H. Stehl of Dow Chemical Co. using Porasil 60 with permanently bonded octadecane and brominated octadecane in the reverse phase mode. Sensitivity was 1 μg as limit of detection for CDD in pentachlorophenol (PCP). Pfeiffer separated nonphenolic impurities by ion exchange followed by reverse phase HPLC on ODS permaphase with 14% water in methanol. Limit of detection of hexadioxin was 0.2 ppm under these conditions and 0.1 ppm for octadioxin. Due to impurities interfering in the hexadioxin region, quantitation was unsuccessful in crude samples of pentachlorophenol.

Pfeiffer et al. (94) were able to determine CDD in pentachlorophenol (PCP) using high performance liquid chromatography (HPLC). They dissolved samples in caustic, extracted this with hexane, which was then passed through silica gel that retained chlorophenols, chlorophenoxyphenols (CPPs), or other polar extracted impurities. Hexane was then passed through basic alumina, which retains CDDs, chlorodibenzofurans (CDBFs), and chlorodiphenylethers (CDPEs), while chlorobenzenes or chlorobiphenyls are eluted with the hexane. 25% Carbon tetrachloride in hexane eluted CDPEs. The CDD fraction was eluted with 50% methylene chloride in hexane, which was evaporated to dryness and redissolved in chloroform and analyzed by LC with ODS/Zorbax at 245 nm (8 times more sensitive than 254 nm) using methanol as the mobile phase. This method correlated well when samples were checked by the GC-MS procedure of Blaser et al. (95) who used a DOWEX 21 K anion exchange resin eluted with 1:1 (v/v), benzene:methanol followed by GC with a 3% SE-30 column and mass spectrometry. Blaser et al. also verified that chlorinated phenols condensed to form dioxins in the gas chromatograph; however, the ion-exchange column removed chlorinated phenols so that this was not a problem in the method.

Crummett and Stehl (96) used both GC/MS detection after ion-exchange or silica gel cleanup and SE-30 or OV-17 GC columns. An LC method was also presented using ODS or C_{18} as columns as well as a laboratory prepared column which allowed determination of TCDD in the ppb range. Pohland et al. (97) reported a method for analysis and confirmation of CDDs based on their cation radicals using electron spin resonance (ESR) and absorption spectra. They synthesized a series of CDDs and showed that various isomers could be distinguished by absorption maxima following derivatization with trifluoromethanesulfonic acid (TFMS); 2,3,7,8-TCDD had a maximum at 845 nm. Octachlorodibenzodioxin (OCDD) had a maximum at 855 nm, but its rate of formation in TFMS was much slower (60 min) than the TCDD derivative (10 min). Higher chlorinated derivatives required addition of an oxidizing agent, potassium nitrate. They also illustrated with various spectra that ESR can be useful for confirmation.

Mieure et al. (98) isolated CDD components from PCP by a two-step column chromatographic procedure. A deactivated alumina column eluted with benzene removed phenols which was followed by an aluminum oxide column which retained the CDDs and CDFs; when eluted with 2% methylene chloride in hexane, interferences were removed. CDDs and CDFs were eluted with 50% methylene chloride in hexane. The three fractions separate these two compounds from technical grade PCP. Porter and Burke (99) had separated 2,3-, 2,3,7-, and 2,3,7,8-chloro-*p*-dioxins from PCB on an aluminum oxide column with 1% methylene chloride and then 20% methylene chloride in hexane using standard solutions.

Firestone (100) reported research on photodechlorination of dioxins for confirmation in oils and fats. Extracts irradiated at 254 nm confirmed presence of OCDD in tallow, oleic acid, and gelatin at 0.2–100 ppb.

Bradlaw et al. (101) have reported an enzyme induction assay for TCDD that paralleled results by GLC and stated the method might be useful as a screening tool in foods. Bradlaw et al. (102) stated that the enzyme induction method is capable of detecting 25 pg of TCDD.

Lengthy periods of alkaline digestion or elevated temperatures will drastically reduce hexachlorodibenzodioxin (HCDD) and OCDD concentrations according to Camparski et al. (103) in studies of recovery from bovine milk. To avoid loss, milk was made alkaline with a small amount of KOH. Recoveries averaged 80–100% at 0.05 ppb with limits of detection 0.025 ppb HCDD, and 0.050 ppb OCDD. In milk samples surveyed no dioxins were found. Mahle et al. (104) also reported no TCDD residues in milk following 2,4,5-trichlorophenoxyacetic acid (2,4,5-T) treatment of pastures. O'Keefe et al. (105) showed that neutral cleanup was equivalent to base–acid extraction in studies of TCDD in bovine fat and milk. Recoveries in the ppt range were about 80% from milk. In a rat given ^{14}C-TCDD intraperitoneally recovery was 70% by counting radioactivity. In these studies they modified the solvent system of Baughman and Messelson (106) and used an alumina minicolumn to reduce microgram quantities of DDE and PCB so that picogram quantities of TCDD could be measured. Ward and Matsumura (107) used ^{14}C-labeled TCDD to monitor disappearance in sediment and lake water under laboratory conditions. TCDD was found to be labile to photochemical degradation as shown by others but resistant to microbial breakdown.

Baughman and Messelson (106) in a study of samples from Viet Nam described a cleanup method which separates TCDD from DDE and PCBs. Following saponification with alcoholic KOH, extraction with hexane, and shaking with H_2SO_4, samples are eluted from an alumina column with carbon tetrachloride–hexane which removes most DDE and PCB. Chlorinated dioxins are then eluted with dichloromethane–hexane. Samples were analyzed by GC-MS using a 5% SE-30 GC column and a TC detector with signals at m/e 320 and 322 measured. Limit of detection of the method was 1 ppt with recoveries of 35%.

Camoni et al. (108) extracted soils with methanol and then methylene chloride which was evaporated to dryness. The residue was dissolved in light petroleum (bp 40–60°) and H_2SO_4 was added for cleanup. The cleaned up petroleum was added to a multilayer column of sodium sulfate, silica gel and sodium hydrogen carbonate and eluted with light petroleum. Further cleanup was performed on an aluminum oxide column which was eluted with light petroleum/methylene chloride and methylene chloride. The TCDD-containing methylene chloride fraction was analyzed by mass fragmentography.

In a study of decontamination of Seveso, Italy, soils, Bertoni et al. (109) used GC analyzes with an 1.5% OV-17/1.95% QF-1 column after activated alumina cleanup and elution with hexane:carbon tetrachloride, 1:1, and hexane:methylene chloride, 1:1 (v/v). The hexane:methylene chloride was shaken with H_2SO_4 and the organic phase evaporated to dryness and picked up in benzene.

REFERENCES

1. F. A. Gunther, *Adv. Pest Control Res.*, **5**, 206–215 (1962).

2. R. L. King, N. A. Clark, and R. W. Heimken, *J. Agric. Food Chem.*, **11**, 63 (1963).

3. E. P. Lichtenstein, K. R. Schulz, R. F. Skrentny, and P. A. Stitt, *J. Econ. Entomol.*, **58**, 742 (1965).

4. T. K. Wood, E. J. Armbrust, G. G. Gyrisco, W. H. Gutenmann, and D. J. Lisk, *J. Econ. Entomol.* **59**, 131 (1966).

5. W. B. Wheeler, D. E. H. Frear, R. O. Mumma, R. H. Hamilton, and R. C. Cotner, *J. Agric. Food Chem.*, **15**, 231 (1967).

6. W. W. Thornburg, *J. Assoc. Off. Anal. Chem.*, **48**, 1023 (1965).

7. A. K. Klein, E. P. Laug, and J. D. Sheehan, *J. Assoc. Off. Anal. Chem.*, **42**, 539 (1959).

8. W. B. Wheeler, D. E. H. Frear, R. O. Mumma, R. H. Hamilton, and R. C. Cotner, *J. Agric. Food Chem.*, **15**, 227 (1967).

9. R. O. Mumma, W. B. Wheeler, D. E. H. Frear, and R. H. Hamilton, *Science*, **152**, 530 (1966).

10. R. G. Nash, M. L. Beall, Jr., and E. A. Woolson, *Agron. J.*, **62**, 369 (1970).

11. R. G. Nash and M. L. Beall, Jr., *Agron. J.*, **62**, 365 (1970).

12. R. G. Nash and M. L. Beall, Jr., *J. Assoc. Off. Anal. Chem.*, **53**, 1058 (1970).

13. W. B. Wheeler, N. P. Thompson, R. L. Edelstein, R. C. Littell, and R. T. Krause, Abstracts of 93rd Annual Meeting of Assoc. of Off. Anal. Chemists, No. 97 (1979).

14. B. M. McMahon and L. D. Sawyer (Eds.), *Pesticide Analytical Manual*, Vol. I, U.S. Department of Health, Education and Welfare, Food and Drug Administration, Washington, D.C. (1979).

15. P. F. Bertuzzi, L. Kamps, and C. I. Miles, *J. Assoc. Off. Anal. Chem.*, **50,** 623 (1967).

16. M. Chiba and H. V. Morley, *J. Agric. Food Chem.*, **16,** 916 (1968).

17. J. G. Saha, B. Bhavaraju, and Y. W. Lee, *J. Agric. Food Chem.*, **17,** 874 (1969).

18. J. G. Saha, *J. Assoc. Off. Anal. Chem.*, **54** 170 (1971).

19. R. E. Johnson and R. I. Starr, *J. Agric. Food Chem.*, **20** 48 (1972).

20. P. A. Mills, *J. Assoc. Off. Anal. Chem.*, **42,** 734 (1959).

21. J. H. Onley and P. F. Bertuzzi, *J. Assoc. Off. Anal. Chem.*, **49,** 370 (1969).

22. W. M. Rogers, *J. Assoc. Off. Anal. Chem.*, **55,** 1053 (1972).

23. B. Berck, *Anal. Chem.*, **25,** 1253 (1953).

24. L. Kahn and C. H. Wayman, *Anal. Chem.*, **36,** 1340 (1964).

25. A. A. Rosen and E. M. Middleton, *Anal. Chem.*, **31,** 1729 (1950).

26. H. D. Gesser, A. Chow, F. C. Davis, J. F. Uthe, and J. Reinke, *Anal. Lett.*, **4,** 883 (1971).

27. J. F. Uthe, J. Reinke, and H. O'Brodovich, *Environ. Lett.*, **6,** 103 (1974).

28. J. F. Uthe, J. Reinke, and H. Gesser, *Environ. Lett.*, **3,** 117 (1972).

29. B. Ahling and S. Jensen, *Anal. Chem.*, **42,** 1483 (1970).

30. D. A. Kurtz, *Bull. Environ. Contam. Toxicol.*, **17** 391 (1977).

31. M. Eidelman, *J. Assoc. Off. Anal. Chem.*, **45,** 672 (1962).

32. R. M. Stimac, *J. Assoc. Off. Anal. Chem.*, **62,** 85 (1979).

33. B. Malone and J. A. Burke, *J. Assoc. Off. Anal. Chem.*, **52,** 790 (1969).

34. L. D. Johnson, R. H. Waltz, J. P. Ussary, and F. E. Kaiser, *J. Assoc. Off. Anal. Chem.*, **59,** 174 (1976).

35. J. E. Lovelock and S. R. Lipsky, *J. Am. Chem. Soc.*, **82,** 431 (1960).

36. J. A. Burke and L. Guiffrida, *J. Assoc. Off. Anal. Chem.*, **47,** 326 (1964).

37. D. M. Coulson, L. A. Cavanagh, J. E. DeVries, and B. Walther, *J. Agric. Food Chem.*, **8,** 399 (1960).

38. D. M. Coulson, *J. Gas Chromatogr.*, **3,** 134 (1965).

39. R. C. Hall, *J. Chromatogr. Sci.*, **12,** 152 (1974).

40. M. Cook, G. Nickless, A. M. Prescott, and D. J. Roberts, *J. Chromatogr.* **156,** 293 (1978).

41. L. D. Sawyer, *J. Assoc. Off. Anal. Chem.*, **61,** 272 (1978).

42. R. G. Webb and A. C. McCall, *J. Chromatogr. Sci.*, **11,** 366 (1973).

43. B. Luckas, H. Pscheidl and D. Haberland, *J. Chromatogr.*, **147,** 41 (1978).

44. J. Teichman, A. Bevenue, and J. W. Hylin, *J. Chromatogr.*, **151** 155 (1978).

45. M. C. Bowman and M. Beroza, *J. Assoc. Off. Anal. Chem.*, **48,** 943 (1965).

46. R. C. Blinn and F. A. Gunther, *Residue Rev.*, **2,** 109 (1963).

47. W. R. Payne, Jr. and W. S. Cox, *J. Assoc. Off. Anal. Chem.*, **49,** 989 (1966).

48. J-Y. T. Chen, *J. Assoc. Off. Anal. Chem.*, **48,** 380 (1965).

49. P. L. Wall and A. W. Mantz, *Anal. Spectros.*, **31**, (1977).

50. R. J. Gajan and J. Link, *J. Assoc. Off. Anal. Chem.*, **47**, 1119 (1964).

51. R. O. Mumma and T. R. Kantner, *J. Econ. Entomol.*, **59**, 491 (1966).

52. T. R. Kantner and R. O. Mumma, *Residue Rev.*, **16**, 138 (1966).

53. D. A. Carlson, N. K. Konya, W. B. Wheeler, G. P. Marshall, and R. G. Zaylski, *Science*, **194**, 939 (1976).

54. P. C. Rankin, *J. Assoc. Off. Anal. Chem.*, **54**, 1340 (1971).

55. M. Deinzer, D. Griffin, T. Miller, and R. Skinner, *Biomed. Mass Spectros.*, **6**, 301 (1979).

56. S. Safe and O. Hutzinger, *Mass Spectrometry of Pesticides and Pollutants*, CRC Press, Cleveland, Ohio, 1973, pp. 13–32.

57. W. P. Cochrane, *J. Chromatogr. Sci.*, **13**, 246 (1978).

58. M. Osadchuk and E. B. Wanless, *J. Assoc. Off. Anal. Chem.*, **51**, 1264 (1968).

59. A. S. Y. Chau and W. P. Cochrane, *J. Assoc. Off. Anal. Chem.*, **52**, 1092 (1969).

60. K. Noren, *Analyst*, **93**, 39 (1968).

61. J. H. Hammence, P. S. Hall, and D. J. Caverly, *Analyst*, **90**, 649 (1965).

62. W. W. Sans, *J. Agric. Food Chem.*, **15**, 192 (1967).

63. J. R. Duffy and N. Wong, *J. Agric. Food Chem.*, **12**, 46 (1964).

64. J. P. Minyard and E. R. Jackson, *J. Agric. Food Chem.*, **13**, 50 (1965).

65. G. A. Miller and C. E. Wells, *J. Assoc. Off. Anal. Chem.*, **52**, 548 (1969).

66. A. S. Y. Chau and W. P. Cochrane, *J. Assoc. Off. Anal. Chem.*, **52**, 1220 (1969).

67. A. S. Y. Chau and M. Lanouette, *J. Assoc. Off. Anal. Chem.*, **55**, 1058 (1972).

68. A. S. Y. Chau and W. P. Cochrane, *Bull. Environ. Contam. Toxicol.*, **5**, 133 (1970).

69. R. I. Asai, F. A. Gunther, W. E. Westlake, and Y. Iwata, *J. Agric. Food Chem.*, **19**, 396 (1971).

70. H. B. Pionke, G. Chesters, and D. E. Armstrong, *Analyst*, **94**, 900 (1969).

71. W. W. Wiencke and J. A. Burke, *J. Assoc. Off. Anal. Chem.*, **52**, 1277 (1969).

72. E. A. Baker and E. J. Skerrett, *Analyst*, **85**, 184 (1960).

73. D. W. Woodham, C. D. Loftis, and C. W. Collier, *J. Agric. Food Chem.*, **20**, 163 (1972).

74. R. B. Maybury and W. P. Cochrane, *J. Assoc. Off. Anal. Chem.*, **56**, 36 (1973).

75. E. J. Skerrett and E. A. Baker, *Analyst*, **84**, 376 (1959).

76. C. J. Musial, M. E. Peach, and D. A. Stiles, *Bull. Environ. Contam. Toxicol.*, **16**, 98 (1976).

77. A. S. Y. Chan and W. P. Cochrane, *J. Assoc. Off. Anal. Chem.*, **54**, 1124 (1971).

78. A. S. Y. Chau, *Bull. Environ. Contam. Toxicol.*, **8**, 169 (1972).

79. A. S. Y. Chau, *J. Assoc. Off. Anal. Chem.*, **57**, 585 (1974).

80. W. P. Cochrane and A. S. Y. Chau, *J. Assoc. Off. Anal. Chem.*, **51**, 1267 (1968).

81. W. P. Cochrane and A. S. Y. Chau, *Bull. Environ. Contam. Toxicol.*, **5**, 251 (1970).

82. T. Stijve and E. Cardinale, *Mitt. Geb. Lebensmittel. Hyg.*, **63**, 308 (1972).

83. A. S. Y. Chau and K. Terry, *J. Assoc. Off. Anal. Chem.*, **57**, 394 (1974).

84. B. E. Baker, *Bull. Environ. Contam. Toxicol.*, **10** 279 (1973).

85. M. V. H. Holdrinet, *J. Assoc. Off. Anal. Chem.*, **57**, 580 (1974).

86. H. L. Crist, R. F. Moseman, and J. W. Noneman, *Bull. Environ. Contam. Toxicol.*, **14**, 273 (1975).

87. B. Zimmerli and B. Marek, *Mitt. Geb. Lebensmittel. Hyg.*, **63**, 273 (1972).

88. G. B. Collins, D. C. Holmes, and M. Wallen, *J. Chromatogr.*, **69**, 198 (1972).

89. I. S. Taylor and F. P. Keenan, *J. Assoc. Off. Anal. Chem.*, **53**, 1293 (1970).

90. J. A. Armour, *J. Assoc. Off. Anal. Chem.*, **56**, 987 (1973).

91. J. N. Huckins, J. E. Swanson, and D. L. Stalling, *J. Assoc. Off. Anal. Chem.*, **57**, 416 (1974).

92. T. E. Archer and D. G. Crosby, *Bull. Environ. Contam. Toxicol.*, **1**, 70 (1966).

93. C. D. Pfeiffer, T. J. Nestrick, and C. W. Kocher, *Anal. Chem.*, **50**, 800 (1978).

94. C. D. Pfeiffer, *J. Chromatogr. Sci.*, **14**, 386 (1976).

95. W. W. Blaser, R. A. Bredeweg, L. A. Shadoff, and R. H. Stehl, *Anal. Chem.*, **48**, 984 (1976).

96. W. B. Crummett and R. H. Stehl, *Environ. Health Pers.*, **5**, 15 (1973).

97. A. E. Pohland, G. C. Yang, and N. Brown, *Environ. Health Pers.*, **5**, 9 (1973).

98. J. P. Mieure, O. Hicks, R. G. Kaley, and P. R. Michael, *J. Chromatogr. Sci.*, **15**, 275 (1977).

99. M. Porter and J. A. Burke, *J. Assoc. Off. Anal. Chem.*, **54**, 1426 (1971).

100. D. Firestone. *J. Assoc. Off. Anal. Chem.*, **60**, 354 (1977).

101. J. A. Bradlaw, D. Sims, and D. Firestone, Annual Meeting Assoc. Off. Anal. Chemists. Abs. 147 (1976).

102. J. A. Bradlaw, L. H. Garthoff, N. E. Hurley, and D. Firestone, Annual Meeting Soc. of Toxicol., Atlanta (1976).

103. L. L. Lamparski, N. H. Mahle, and L. A. Shadoff. *J. Agric. Food Chem.*, **26**, 1113 (1978).

104. N. H. Mahle, H. S. Higgins, and M. E. Getzendoner, *Bull. Environ. Contam. Toxicol.*, **18**, 123 (1977).

105. P. W. O'Keefe, M. Meselson, and R. W. Baughman, *J. Assoc. Off. Anal. Chem.*, **61**, 621 (1978).

106. R. Baughman and M. Messelson. *Environ. Health Pers.*, **5**, 27 (1973).

107. C. T. Ward and F. Matsumura, *Arch. Environ. Contam. Toxicol*, **7**, 349 (1978).

108. I. Camoni, A. Di Muccio, D. Pontecorvo, and L. Vergori, *J. Chromatogr.* **153**, 233 (1978).

109. G. Bertoni, D. Brocco, V. DiPalo, A. Liberti, M. Passanzini, and F. Bruner, *Anal. Chem.*, **50**, 732 (1978).

CHAPTER

6

ANALYSIS FOR RESIDUES OF ACIDIC HERBICIDES

DAVID J. JENSEN and RONALD D. GLAS

Agricultural Products Department, Dow Chemical U.S.A.,
Midland, Michigan

CONTENTS

6.1. INTRODUCTION

This chapter has been written to provide practical, proven approaches to solving problems associated with analysis of various substrates for residues of acidic herbicides. No attempt has been made to provide a comprehensive survey of the literature, as this has been done quite ably by the National Research Council of Canada (1) and Cochrane and Purkayastha (2). Rather, our purpose is to provide specific information, based on our personal experience, which can be used in evaluating the validity of specific residue data and to aid in choosing or devising an analysis scheme for acidic herbicides.

Impetus for writing this chapter came from two sources: first, a desire to provide a sound basis for evaluating studies on residues of acidic herbicides; these include 2,4-dichlorophenoxyacetic acid (2,4-D), 2,4,5-trichlorophenoxyacetic acid (2,4,5-T), 2-(2,4,5-trichlorophenoxy)propionic acid (silvex or fenoprop), 2-methyl-4-chlorophenoxyacetic acid (MCPA), 4-amino-3,5,6-

trichloro-2-picolinic acid (picloram), [(3,5,6-trichloropyridinyl)oxy] acetic acid (triclopyr) 3,6-dichloropicolinic acid and 2-methoxy-3,6-dichlorobenzoic acid (dicamba); second, to provide analytical procedures that have been carefully validated and successfully used by experienced analytical chemists. This chapter has been divided into two major sections. The first is method design, development, and validation, and the second consists of a compilation of validated methods for the phenoxy herbicides: 2,4-D, 2,4,5-T, silvex, and MCPA.

6.2. STEPS IN DESIGNING OR CHOOSING METHODS

The first step is to decide which moiety is the subject of the analysis. Acidic herbicides are generally formulated as esters or amine salts which rapidly hydrolyze in soil (3), water (4), and plants (5) to the acidic form. In a few instances the amount of ester present is significant, as found in air samples in agricultural areas (6,7), or water samples immediately after treatment (4). However, more often the major species of analytical interest is the acid form and most methods of analysis reflect this.

A second step is to determine the desired lower limit of detection. The limit of detection required will vary with compound toxicity and phytotoxicity and with the purpose of the experiment. Monitoring of toxicologically significant residues of acidic herbicides requires only a moderate lower limit of detection (0.1–0.01 ppm) due to their generally low acute and chronic toxicities. However, monitoring residue levels that will cause phytotoxicity in susceptible species may require a lower limit of detection of 0.001 ppm or less. The Environmental Protection Agency considers the toxicity and potential exposure in establishing tolerances, or acceptable levels, of pesticide residues in foods and feeds (8). Selected tolerances are shown in Table 6.1. The lower limit of detection should generally include a $2\times$ safety factor beyond the tolerance or phytotoxic level. In addition to tolerances and phytotoxicity, other factors may be important in setting the lower limit of detection.

These factors include background due to substrate and expected levels of residues to be found. As the lower limit of detection of the analysis moves from ppm to ppb the background interferences become much more significant. The effort to improve the limit of detection from 10 ppm to 1 ppm is minor compared to the effort that may be needed to improve it from 0.1 to 0.01 ppm, due to many naturally occurring organic acids in plants, animals, and soil. Fortunately, the necessary lower limit of detection for the acidic herbicides is generally attainable at moderate cost.

TABLE 6.1. Selected Tolerances for the Acidic Herbicides in PPM Compiled from 40 CFR §180

Substrate	Compound			
	2,4-D	Picloram	Dicamba	2-Methyl-4-chloro-phenoxyacetic acid
Grass	1,000	80	40	300
Hay	300	—	40	20
Forage of small grains	20	1.0	—	20
Small grains	0.5	0.5	—	0.1
Meat				
Cattle, hogs, sheep	0.2	0.2	—	0.1
Poultry and eggs	0.05	0.05	—	—
Milk	0.05	0.05	0.05	0.1

In some cases the chosen lower limit of detection of a method has been far lower than necessary when there have been high application rates. For example, in the analysis of treated grass or leaves, the application rate and the interval between exposure and sampling are more important in choosing the required limit of detection than are the tolerance levels. Studies covering herbicide applications in many states over several years have shown that the expected residue of an acidic herbicide on grass immediately after spraying is approximately 100 ppm per pound of herbicide applied per acre (9). This residue dissipates with a half-life of about 2 weeks, so that after a month the residue would be expected to be 25 ppm/pound of herbicide applied per acre. Therefore, samples collected the day of application where 2 lb of acidic herbicide were used require a method with a lower limit of detection of only 20–40 ppm. Correspondingly, 1 month after application a method with a detection limit of 5–10 ppm would be desirable.

6.3. METHOD

Once the analyst has decided on the moiety to be determined and the necessary lower limit of detection, the next step is to choose or design a suitable method. The important parts of the method are discussed in the following sections.

6.3.1. Sampling

The typical size of an analytical sample has decreased over the last 15 years from 100 to 10 g or even less as conditions warrant. This trend forces the analyst to be ever mindful of the need to ensure that the sample is a reliable representation of the whole. Field sampling in particular is susceptible to nonrepresentative sampling. Great care must be taken to insure that the sub sample analyzed is indeed representative of the bulk sample. Often the only available means is to take replicate samples with the assumption that the average level found will be a good approximation of the actual level present. More details about sampling techniques have been discussed in Chapter 7 and by Lykken (10).

To prevent loss of any of the residue that might be present, samples should be frozen soon after collection and maintained frozen until analyzed. Grinding or chopping samples to a homogeneous mixture before taking an aliquot for analysis is important in order to obtain a representative sample. This process is normally very simple but varies with sample type. For example, wheat straw or corn grain are more difficult to reduce to a homogeneous mixture than is beef liver. A more detailed report on sample preparation has been given by Grier (11).

The size of the actual analytical sample should depend on the homogeneity of the prepared sub sample and the desired lower limit of detection. Often 1- to 5-g aliquots of prepared animal tissues are adequate samples while 10- to 50-g aliquots are more appropriate for less homogeneous materials such as soils. Paradoxically, published methods generally specify 10- to 50-g aliquots for heterogeneous substrates, but much larger aliquots (up to 1000 grams) for such homogeneous substrates as water. The analysis of such large aliquots of water coupled with concentration to very small volumes has given rise to analytical sensitivities unjustified by reason of their biological insignificance.

6.3.2. Extraction

After preparing a proper sample the acidic herbicide must be separated or extracted from the sample matrix. The most desirable extraction procedure is one that is efficient enough to remove all of the herbicide residue but is as selective as possible in order to minimize coextraction of interfering materials.

6.3.2.1. Water

Extraction of acidic herbicides from water is a very simple process involving acidification of the sample with mineral acid, saturation with sodium chloride, and extraction with ether. As was mentioned earlier, occasionally the amount of ester present is of interest. This extraction procedure will allow simultaneous extraction of both the acid and ester if present. An alternate approach is to hydrolyze any remaining ester by alkaline hydrolysis, acidification with mineral acid, saturation with chloride salts, and extraction with ether to determine total herbicide present.

6.3.2.2. Soil

It is more difficult to remove acidic herbicides from soil than from water. There have been many different extraction methods used for removing acidic herbicides from soil. Published methodologies range from mild washes with water (12), to solvent extraction with organic solvent (13), to acidification and extraction with ether (14), to digestion and extraction with hot alkali (15). In studies conducted in this laboratory with a number of incorporated radiolabeled acidic herbicides, including 2,4-D, silvex, 2,4,5-T, triclopyr, and 3,6-dichloropicolinic acid, it was discovered that acidification and extraction with diethyl ether was very effective in removing residues of the acidic herbicides, even those "aged" as long as a year at room temperature. The critical factors for efficient extraction with acid–ether mixtures were thorough mixing of the phases and a long extraction time. Sufficient water and acid must be added to the soil to ensure low viscosity so that vigorous shaking will cause thorough mixing with the ether. Shaking for an hour each with three aliquots of diethyl ether was needed to completely extract the acidic herbicide residues. The efficiency of this extraction procedure was verified by showing that residue levels found in field "aged" samples by the acid–ether extraction were the same as those obtained using hot or cold alkali extraction.

6.3.2.3. Plants

Extraction of acidic herbicides from the cellular matrix of plants has proven to be much more difficult than their extraction from soil or water. Residues of acidic herbicides in plants may exist in several forms, characterized as "free acid," "conjugated residue," and "bound residue." Properly used, these terms refer to the chemical form of the acidic herbicide residue. "Free acid" refers to acidic herbicide residue whose carboxylic acid function is unreacted with any other compound. The "conjugated residue" refers to the fraction of the herbicide residue whose acid function is linked with another compound, while the "bound residue" is that fraction of the conjugated residue which requires the use of boiling alkali solution for removal from the sample matrix.

Early methods for the analysis of acidic herbicides in plant material utilized acidification and extraction with ether (16) or acetone (17). These methods determine "free acid" residues. More recent methods, based on the extensive literature on conjugation of acidic herbicides in plant materials, have utilized acidic and enzymatic hydrolysis (18) or alkaline hydrolysis (19), to release as much as ten times more residue than the acidification and ether extraction procedures. These additional residues have been characterizied as "bound residues." Further testing performed in this laboratory has shown that extraction with dilute aqueous sodium hydroxide at room temperature releases the same amount of residue as extraction with boiling alkali. These data would suggest that no "bound residue" is present in plant materials since all of the residue could be extracted at room temperature with dilute alkali solutions. The efficiency of the dilute alkali extraction procedure was independently verified by repetitive extraction and by hydrolysis with concentrated aqueous alkali or acid prior to extraction.

Residue data are presented in Table 6.2 for repetitive extraction of field applied residues of 2,4-D, silvex and 2,4,5-T from sugarcane and 2,4-D from wheat at room temperature with aqueous $0.5N$ NaOH. One extraction removed 89–100% of the residue and two extractions removed 97%+ of the residues. In no experiment listed in Table 6.2 was there found more than 1% "bound residue." In Table 6.3 the efficiency of extraction of 2,4-D from green plants and straw with aqueous $0.2N$ NaOH and from grain with $0.1N$ NaOH is compared with data obtained after acid or alkaline hydrolysis for 30–45 min at 100° prior to extraction. The highest residue was found using room temperature extraction with dilute aqueous sodium hydroxide. In addition, hydrolysis released many other materials which made the cleanup and determination steps much more difficult.

The acid/ether extraction procedure was compared with dilute alkaline extraction and with alkaline hydrolysis followed by ether extraction for 2,4,5-T in rice straw (Table 6.4). Alkaline extraction and alkaline hydrolysis prior to ether extraction gave essentially the same results indicating no analytically significant "bound" residues. Both procedures yielded considerably higher residue levels than the ether extraction procedure. These data clearly show that bound residues are not a practical problem in the analysis of the phenoxy herbicides in plant material if dilute aqueous alkali is used for extraction.

TABLE 6.2.   Percent Recovery of Ingrown Residues of Phenoxy Herbicides on Repeated Extraction With Dilute Aqueous Alkali Followed by Hydrolysis of the Substrate

	Sugar Cane					
	2,4-D			2,4,5-T		Silvex
Level (ppm)	10	1	0.5	17	0.4	5
First extract	93	97	98	90	91	100
Second extract	5	3	2	7	0	—
Third extract	1	0	0	1	0	—
Fourth extract	0.6	0	0	0.6	0	—
Hydrolysis (1·0N NaOH)	1	0	0	1	0	0
Hydrolysis (6N H$_2$SO$_4$)	—	—	—	—	—	0

	Wheat (2,4-D)					
	Green Plants			Straw		Grain
Level, ppm	22	23	3	11	0.02	0.02
First extract	89	90.8	92.8	95.7	100	100
Second extract	10.7	8.7	4.4	4.3	0	0
Third extract	0.2	0.4	2.3	0	0	0
Fourth extract	0.1	0.1	0.5	—	—	—
Hydrolysis (1.0N NaOH)	—	0	—	0.7	0	0

TABLE 6.3.   Comparison of the Effectiveness of Alkaline Extraction with Hydrolysis for Removing 2,4-D from Wheat and Barley

	2,4-D Found (ppm)					
	Green Plants		Straw		Grain	
	Wheat	Barley	Wheat	Barley	Wheat	Barley
Alkaline extraction	3.1	3.1	2.5	2.1	0.02	0.02
Hydrolysis 0.5N H$_2$SO$_4$	1.9	1.8	2.2	2.0	0.00	0.00
Hydrolysis 1.0N KOH	3.0	2.7	2.4	2.2	0.00	0.00

6.3.2.4.   Animals

While extraction procedures have been applied to animal tissues including extraction with hot ethanol followed by enzymic hydrolysis (20) and acidifi-

TABLE 6.4.   Comparison of the Effectiveness of Ether Extraction Versus Alkaline Extraction or Alkaline Hydrolysis for Removing 2,4,5-T from Rice Straw

	2,4,5-T Found (ppm)		
	Ether Extraction	Aqueous Alkaline Extraction	Aqueous Alkaline Hydrolysis
Sample 1	0.2	12.6	12.6
Sample 2	—	3.3	3.0
Sample 3	—	2.6	2.7
Sample 4	—	1.4	1.2
Sample 5	1.0	12.5	10.1
Sample 6	—	5.4	3.8
Sample 7	—	12.8	7.8
Sample 8	0.01	3.9	1.1
Sample 9	—	6.0	3.2
Sample 10	<0.01	3.2	1.1

cation followed by extraction with 2-propanol (21), there is no universally accepted procedure. Methanol containing 1–2% ammonium hydroxide is a very effective extraction solvent for ingrown phenoxy herbicides. The use of ammoniacal methanol generally yields larger values for phenoxy acids than does the hydrolysis of the tissue in aqueous alkali or acid, as can be seen in Table 6.5. Losses upon hydrolysis may be caused by the extra amounts of coextracted materials interferring in later separation steps.

Only ruminant animals cleave a significant amount of phenoxy acid to the corresponding chlorophenol (22, 23). It has been found that hydrolysis with alkali is necessary to recover all of the chlorophenol present in kidney and liver tissue (23). Muscle and fat, however, give no significant additional residue after alkaline hydrolysis compared to ammoniacal methanol extraction (Table 6.6). The additional chlorophenol obtained from kidney was found to be a "conjugate" soluble in ammoniacal methanol, but the extract required hydrolysis before the conjugate could be measured. Hydrolysis of liver samples was also necessary to release the "bound" chlorophenol from that tissue. Additional chlorophenol released by hydrolysis has ranged from 1 to 10 times the level obtained using ammoniacal methanol extraction.

6.3.3.   Cleanup and Concentration

Cleanup is the process of sorting out the coextracted materials from the acidic herbicides and preparing the herbicide for derivatization. The amount

TABLE 6.5. Effect of Hydrolysis on Release of Phenoxy Acids from Animal Tissues

Tissue	2,4,5-T Found (ppm)			2,4-D Found (ppm)	
	No Hydrolysis	Hydrolysis[a] of Whole Tissue		No Hydrolysis	Hydrolysis of Whole Tissue Alkaline[b]
		Alkaline	Acid		
Muscle	0.18	0.16	0.11	<0.05	0.05
	2.7	1.4	1.6	<0.05	<0.05
Fat	0.40	0.48	0.58	0.08	<0.05
	3.8	4.8	4.3	<0.05	<0.05
Liver	0.35	0.38	0.34	0.3	0.2
	9.3	9.7	8.5	3.8	3.2
Kidney	4.0	3.0	3.5	0.2	0.4
	22.3	22.3	16.7	11.2	7.8

[a] Hydrolysis of whole sample with either $6N$ KOH or $6N$ H_2SO_4 for 2 hr at 80°C followed by 16 hr at 50°C.
[b] Hydrolysis of whole sample with $0.5N$ KOH for 2 hr at 80°C.

TABLE 6.6. The Effect of Hydrolysis on Release of Chlorophenols from Animal Tissues

Tissue	2,4,5-Trichlorophenol (ppm)			2,4-Dichlorophenol (ppm)	
	No Hydrolysis	Hydrolysis[a]		No Hydrolysis	Alkaline[b] Hydrolysis
		Alkaline	Acid		
Muscle	<0.05	<0.05	<0.05	<0.05	<0.05
	<0.05	<0.05	<0.05	<0.05	<0.05
Fat	<0.05	<0.05	<0.05	0.08	0.15
	0.11	<0.05	0.10	<0.05	<0.05
Liver	0.04	0.10	0.07	<0.05	0.23
	1.04	3.3	1.0	<0.05	0.10
	0.05	0.56[b]	—	<0.05	0.16
	0.10	1.8[b]	—	<0.05	0.19
	0.16	2.7[b]	—	0.25	0.30
Kidney	<0.05	<0.05	0.05	0.3	0.7
	1.82	5.9	5.5	1.8	3.1

[a] Hydrolysis of whole sample with either $6N$ KOH or $6N$ H_2SO_4 for 2 hr at 80°C followed by 16 hr at 50°C.
[b] Hydrolysis of whole sample with $0.5N$ KOH for 2 hr at 80°C.

of cleanup required depends on the substrate to be analyzed, the extraction procedure, the method of detection, and the desired sensitivity. Factors that favor little or no cleanup include simple substrate (water), mild extraction (acid–ether), specific detection (electrochemical), and ppm or higher concentrations of residues.

In the procedures given at the end of this chapter, the alkaline aqueous extract is acidified and extracted with diethyl ether or a mixture of hexane and diethyl ether. For some substrates, due to the large number of unwanted compounds extracted by ether, a less polar extractant would be desirable. It was found that two extractions with a 70/30 mixture of hexane/diethyl ether (v/v) is very effective for transferring phenoxyacetic acids into the organic phase while minimizing the extraction of interferring materials.

Usually the first cleanup of the ether or hexane/ether extract is treatment with dilute aqueous sodium bicarbonate solution. Much of the coextracted material is left in a highly colored organic layer while the herbicide is extracted into the slightly colored bicarbonate layer. After removal of the organic layer, the bicarbonate solution is acidified and slowly saturated with sodium chloride (to control carbon dioxide evolution). The acidic herbicide is then reextracted into diethyl ether which can be evaporated to a small volume for derivatization. Some investigators (24, 25) use an oxidation with saturated aqueous $KMnO_4$ solution after acidification of the bicarbonate solution but prior to the addition of ether. Sodium bisulfite is added to remove the excess permanganate before saturation with sodium chloride and extraction with ether. The ether extract is then evaporated to a small volume suitable for derivatization. This technique often removes a great many interferences but also has the potential of destroying the herbicide. Decomposition of 2,4-D has occurred using this technique but is probably only a problem in relatively "clean" extracts which do not use up the majority of the potassium permanganate present.

Potentially troublesome steps such as drying organic solvents with anhydrous sodium sulfate, evaporation to dryness, and phase separation using separatory funnels have been avoided in the methods given at the end of this chapter. These techniques can be replaced with simpler, more reliable procedures. Drying ether extracts with sodium sulfate can be eliminated by saturating the acidified aqueous phases with sodium chloride before extraction with ether. This technique also aids in the partitioning of the herbicide into the ether. Evaporation to dryness is only necessary when using chlorinated solvents, which must be removed before electron capture gas chromatography and should be avoided, particularly after derivation. In general, two commonly used techniques will eliminate the need for evaporation to dryness and will minimize losses of acidic herbicide residues when evaporating the solvents. The first, which works for the underivatized (free) acidic herbicides, is the addition of a solution of weak alkali to keep the herbicide in the nonvolatile salt form. A 1% solution of ammonium hydroxide in methanol is very effective, and is compatible with BF_3/MeOH esterification. Further, several milliliters of methanol can remain after evaporation without affecting the derivatization step. The second technique, suitable for both the acidic and esterified forms, is the addition of a high boiling point solvent in which the herbicide is soluble, such as benzene or isooctane, which acts as a "keeper" and prevents volatilization of the herbicide.

The use of screw cap vials with polyethylene-lined caps has several advantages over the use of separatory funnels for phase separations. First, they are readily and conveniently machine shaken. Second, emulsions, which often form during extractions, can be easily broken by centrifuging the screw cap vial. The vials will survive up to 6000 G if rubber stoppers are inserted into the centrifuge cup to provide a flat base for the vials. The desired phase (usually the upper) can be easily removed with a pipet, or if it is to be discarded, it can be removed using a disposable pipet connected by tubing to a suction flask and vacuum line. Third, screw cap vials have no stopcock to plug with excess sodium chloride or coextracted solids. In addition they are inexpensive enough to be considered disposable.

6.3.4. Derivatization

The acidic herbicides have very low vapor pressures and cannot be effectively gas chromatographed without first being derivatized. The formation of a number of ester derivatives; methyl (26), ethyl (27), chloroethyl (28), propyl (29), butyl (6), 2-butoxyethyl (30), and silyl (31) has been reported. Diazomethane has been the most consistent methylating reagent in terms of simplicity, lack of reagent blank and quantitative reaction (26). This reagent, however, does methylate any active hydrogen, so ethers are formed as well as esters. Additional cleanup is often required to remove these extra interfering compounds. Another method of methylation using borontrifluoride or borontrichloride in methanol gives fewer interfering compounds but is less convenient to use when further cleanup is required. For a few of the acidic herbicides such as 3,6-dichloropicolinic acid and dicamba, diazomethane must be used in order to achieve complete esterification (26).

Cleanup following methylation can be simply and efficiently performed using micro alumina columns. These columns consist of a disposable capillary pipet containing a glass wool plug and 0.5–1.0 g of deactivated acidic alumina (1–6% water). Before use, the column is prewashed with the same solvent in which the derivatized herbicide is dissolved. The sample is then added to the column in a total volume of 1–2 ml of solvent and the ester is eluted with the same solvent in a total volume of 3–5 ml. Generally ether or benzene are used and the amount of deactivation is adjusted accordingly, 1–2% deactivation for ether and 4 to 6% deactivation for benzene. The ether eluate is evaporated with isooctane or benzene keeper and then diluted to volume while the benzene eluate from the column is merely diluted to volume and then chromatographed.

6.3.5. Gas Chromatography

6.3.5.1. Columns

A great deal of the separation between substrate interferences and the acidic herbicide esters occurs in the gas chromatograph where a variety of column packings have been used. A standard column has been a 180 cm × 3 mm ID glass column containing 4% LAC-446 and 1% H_3PO_4 on Chromosorb W-HP (30). This column gives a reasonable separation of the esterified phenoxies (silvex and 2,4-D overlapped to some extent) and allows chromatography of the underivatized chlorophenols (see Figures 6.7 and 6.8). This column was used to obtain a large amount of animal residue data (24, 33). Other packings have given even better separation of the methyl esters of the acidic herbicides. A very useful column has been 3% OV-3 on Chromosorb W-HP. The relative retention times of the acidic herbicides chromatographed in a 3% OV-3 column are shown in Table 6.7 and a chromatogram is shown in Figure 6.1. Other columns that have been used include the mixed phase QF-1/OV-17 columns and 3% OV-17. An interesting new column is the Ultrabond PEGS which alters the normal order of elution of the acidic herbicides and may be useful for confirmation of acidic herbicide residues.

6.3.5.2. Detectors

The most commonly used detection systems for gas chromatography of acidic herbicides are: electron capture (EC), microconductivity, microcoulometry and mass spectrometry. Of these, the most widely used detector is EC because of its ease of operation and excellent sensitivity for most of these compounds. Although EC detection has been the principal detection system used, its response to compounds other than the acidic herbicides necessitates careful verification of all residues found.

Many laboratories prefer to use the less sensitive, but more specific electrochemical detectors. These detectors rely on conversion of organic chlorine in the gas chromatographic effluent to inorganic chloride. The

TABLE 6.7. **Relative Retention Times of Methyl Esters of Acidic Herbicides on 3% OV-3**

Common Name	Compound	Relative Retention Time (2,4-D - 1.00)
—	Methyl-3,6-dichloropicolinate	0.56
Dicamba	Methyl-2-methoxy-3,6-dichlorobenzoate	0.67
MCPA	Methyl-2-methyl-4-chlorophenoxyacetate	0.78
2,4,-DP	Methyl-2-(2,4-dichlorophenoxy) propionate	0.91
2,4-D	Methyl-2,4-dichlorophenoxyacetate	1.00
Fenac	Methyl-(2,3,6-trichlorophenyl) acetate	1.09
Triclopyr	Methyl-[(3,5,6-trichloropyridinyl)oxy] acetate	1.27
Silvex	Methyl-2-(2,4,5-trichlorophenoxy) propionate	1.51
2,4,5-T	Methyl-2,4,5-trichlorophenoxyacetate	1.72
Picloram	Methyl-4-amino,3,5,6-trichloropicolinate	3.34

amount of chloride in the effluent is titrated with silver in the microcoulometric detector (34, 35), or the change in conductivity is measured by the microconductivity detector. These detectors are limited in sensitivity to about 1 ng of phenoxy herbicide. In this laboratory the EC detector has been found to present fewer operating difficulties than the microconductivity detectors, but discussions with others indicate many very happy users of the electrochemical detectors. Instead of using electrochemical detectors to give specificity, this laboratory uses gas chromatography/mass spectrometry.

Gas chromatography/mass spectrometry (GC/MS) systems have become reliable and quite sensitive (7, 36). Less sample cleanup is required when using GC/MS because the specificity of the detector eliminates signals for most interferences. In this laboratory the GC/MS was applied to the analysis of residues of MCPA in soil. A method utilizing microcoulometric detection had shown major interferences for MCPA in some soils (Figure 6.2). Normally, such interferences would force extensive redevelopment of the method, but by changing over to GC/MS detection the method could be simplified to extraction, derivatization, and chromatography. The methyl ester of MCPA was measured in the quadrapole mass spectrometer by monitoring m/e's 214 and 216. The GC-electron impact ionization mass

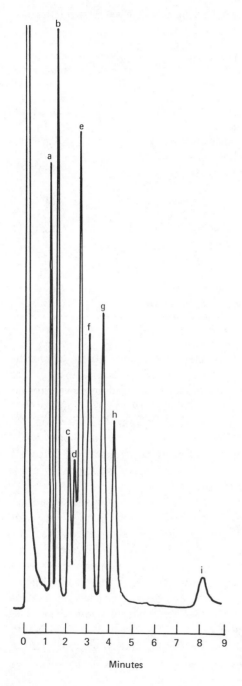

Figure 6.1 Chromatogram of the methyl esters of some acidic herbicides chromatographed on a 180 cm × 3 mm glass column packed with 3% OV-3 on 100/120 mesh Chromosorb W-HP at 175°C and 35 cc/minute. (*a*) 3,6-Dichloropicolinic acid. (*b*) Dicamba. (*c*) 2,4-DP. (*d*) 2,4-D. (*e*) Fenac. (*f*) Triclopyr. (*g*) Silvex. (*h*) 2,4,5-T. (*i*) Picloram.

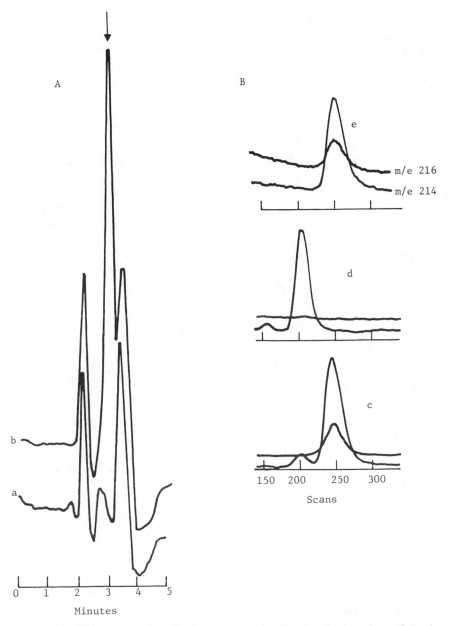

Figure 6.2 Chromatograms from the determination of MCPA in soil using microcoulometric and mass spectrometric detection. (*A*) Microcoulometric detection: (*a*) Control soil, (*b*) control soil fortified at 0.05 ppm MCPA. (*B*) Mass spectrometric detection (peaks normalized): (*c*) control soil fortified with 1 ppm MCPA, (*d*) control soil, (*e*) MCPA standard 1 ng.

237

spectrometer gave a rapid, interference-free assay for residues of MCPA in soil (Figure 6.2). Present "state-of-the-art" equipment gives a multiple ion monitoring capacity at less than 1 ng sensitivity with computer capabilities to standardize and get quantitative data while operating unattended with autosampling devices.

6.3.6. Validation

Proper validation of the analytical procedure and identification of the residue is critical to correct analysis. Proper validation is an expensive and sometimes difficult task but is necessary to avoid supplying incorrect or at least misleading data. A case in point is a study in which a number of pesticides including 2,4,5-T were determined in soil by a multiresidue method (35). The authors showed no recovery data, no chromatograms, and no confirmation of residues. No less than 4 steps in the procedure precluded the possibility of recovery of any 2,4,5-T residues present, yet the authors reported finding 2,4,5-T in 24% of the soil samples analyzed. Cases such as this one clearly show how important the validation process is to the analysis as a whole.

6.3.6.1. *Validation of the Method*

Validation of residue data requires several steps. First, the analytical procedure must be validated. Ideally, control samples (free of acidic herbicide residues) are fortified with known amounts of compound, the samples processed and the percent recovery determined. Acceptable recovery of the herbicides is 80% or better. The level of fortification should be close to the level being sought. Fortification only at 0.5 ppm for samples containing 0.05 ppm levels is not proper as the method is then unproven at the level of residue being determined. A small constant loss of herbicide equivalent to 0.1 ppm is not significant when determining levels of 100 ppm, 10 ppm or even 1 ppm, but it certainly has an effect for a 0.1 ppm determination where there would be no recovery of the herbicide. In addition, a blank should be run on the same sample without spiking and on the reagents without sample present to show that there are no significant interferences due to sample or reagents.

For a method to be properly validated one must also prove the completeness of the extraction procedure by reextraction or hydrolysis of samples with ingrown residues as was discussed in Section 6.3.3.

6.3.6.2. *Confirmation of Residues*

Careful verification of the analysis procedure is not the last step in a valid analysis. One also needs to confirm the identity and quantity of the measured residue to be sure that it has not been confused with another compound. Several procedures can be used to confirm that the peak seen is the acidic herbicide. First, one should demonstrate that the peak did not come from reagents by running through the method using water in place of the substrate. Next, one should chromatograph the extract on another GC column of quite different polarity so that retention times are changed (Section 6.3.5.1). Another approach has been to use p values (36, 37) obtained by shaking a portion of the extract with another solvent which will extract 30–60% of the ester and then chromatographing both solvents. A ratio, p value, is obtained which is somewhat characteristic of the herbicide and aids in confirmation of its identity.

Methyl esters can be transesterified to longer chain esters such as ethyl or butyl with BF_3/ethanol or BF_3/butanol. The corresponding change in GC retention times are characteristic for the particular acidic herbicide. Here again, control samples are necessary as chromatography of the longer chain esters may give an interference not encountered with the methyl esters. Each of these techniques is only a partial confirmation and simply increases the probability that the peak seen is indeed the herbicide.

The best confirmatory test available today is the use of GC/MS. The GC separates the compounds and the mass spectrometer can be tuned to see only several strong ions characteristic of each individual herbicide. Under these conditions the compound must elute from the gas chromatograph at the proper time and also have the proper masses and in the correct mass ratio for the number of chlorine atoms present. A list of suggested masses for monitoring some phenoxy and picolinic acids after electron impact ionization is presented in Table 6.8. An alternate method of analysis should be applied to positively confirm a compound. Such a method may be liquid or thin layer chromatography, and so on. However, at the sensitivity levels needed for residue analysis this is often impractical. Perhaps high performance liquid chromatography will advance enough to give an alternate

TABLE 6.8. Mass/Charge Ratios for GC-MS Determination of Several Phenoxy and Picolinic Acids Using Electron Impact Ionization

Compound	Mass/Charge Ratios
Methyl-4-amino-3,5,6-trichloropicolinate	196,198
Methyl-3,6-dichloropicolinate	146,148
Methyl-2,4-dichlorophenoxyacetate	234,236
Methyl-2,4,5-trichlorophenoxyacetate	268,270
Methyl-2-methyl-4-chlorophenoxyacetate	214,216

analysis to confirm acidic herbicide residues, but at present, HPLC is still limited by lack of sensitivity.

6.4. ANALYSIS PROCEDURES

Methods used in this laboratory for the determination of the phenoxy herbicides in various substrates are given in this section. These methods have been found to be effective for the analysis of acidic herbicides and serve as a convenient guide for the analysis of acidic herbicide residues in most substrates. Each of the methods is thoroughly validated. The extraction procedures have been proven by rigorous hydrolysis tests as described in Sections 6.3.3. and 6.3.6. Residues were quantitated by gas chromatography with electron capture detection, with the exception of MCPA and 2-chloro-4-methylphenol which were quantitated by gas chromatography with microcoulometric detection. Recoveries of acidic herbicides added to control samples are given in Table 6.9. For each method the table lists: the compounds added, the fortification levels, the number of separate recovery determinations performed, the mean recovery, and the standard deviation of the recovery data for each compound. Representative chromatograms obtained by each method are presented in Figures 6.3–6.8. The analyst is always charged with responsibility to confirm that these or any other analysis procedures work before using them to generate data.

6.4.1. Water Method

1. Add a well-mixed, representative 100.0-ml sample of the water to an 8-oz glass bottle.

2. Add 1 g KOH to the sample, cap, and shake for 10 min at approximately 135 exursions/min.

3. Add 40 ml diethyl ether, cap, and shake for 5 min. Carefully remove as much ether as possible without disturbing the aqueous layer. Discard the ether.

4. Add 4 ml 18N H_2SO_4, 35–40 g NaCl, and 100.0 ml diethyl ether. Cap and shake at about 270 excursions/min for 5 min.

5. After the phases separate, pipet a 25-ml aliquot of the ether into a 12-dram vial.

6. Add 1 ml of 5% NH_4OH in methanol and several small boiling chips, and evaporate the sample on a 70°C water bath until the liquid just wets the bottom (approximately 0.5 ml).

7. Add 2.0 ml BCl_3/MeOH reagent to the 12-dram vial, cap, and immerse the vial in the 70°C water bath to a depth of approximately 4 cm for 20 min.

8. Remove the cap and evaporate the methanol under a gentle stream of nitrogen in a 70°C water bath until the bottom is just wet (approximately 0.5 ml). Do not allow the sample to go dry.

9. Add 5 ml of 5% aqueous Na_2SO_4 solution and 5.0 ml of isooctane to the vial, cap, and shake (270 excursions/min) for 2 min.

10. Allow phases to separate. Inject 2 μl of isooctane solution into a 180 cm $\times$ 3 mm column packed with 3% OV-3 on 80/100 mesh Gas Chrom Q.

11. Determine the concentration in μg/ml of each of the esters in the final solution by comparing peak heights to those of the standard. Multiply μg/ml found by 200 to obtain gross μg/l for each compound.

12. Subtract apparent gross μg/l of each compound found in control samples from that found in the treated samples to obtain net μg/l.

13. Divide net μg/l of each compound by its average percent recovery and multiply by 100 to obtain corrected μg/l.

Recovery data are shown in Table 6.9 and representative chromatograms are shown in Figure 6.3.

6.4.2. Soil Method

1. Prepare the soil sample (without drying) by grinding in a Wiley Mill with dry ice. Weigh 10.0 g of soil into a 50-ml centrifuge bottle.

2. Add 10 ml deionized water, enough $6N$ H_2SO_4 to reduce the pH to <2 (generally 1–2 ml), and 20 ml diethyl ether. Shake for 1 hr (270 excursions/min). Centrifuge briefly (1–5 sec) at 2000 rpm (see Section 6.4.2.3, No. 1)

3. Pipet the ether layer into a 12-dram vial and add 10 ml of $0.1N$ sodium bicarbonate solution.

4. Shake the ether–bicarbonate mixture for 5 min (135 excursions/min), let settle, then discard the ether.

5. Repeat extraction of the soil with 2 additional 20-ml aliquots of ether, shaking for 1 hr each time, centrifuging, and partitioning the ether with the same bicarbonate solution each time (see Section 6.4.2.3, No. 2).

6. After all the ether is removed from the bicarbonate solution for the third time, add 1 ml $6N$ sulfuric acid, and sprinkle in 4–6 g sodium chloride. Add 10 ml ether, shake 5 min (135 excursions/min), and place the ether into a conical centrifuge tube.

7. Reextract the solution with 10 ml ether, shake 5 min and combine the ether with the first ether extract.

8. Evaporate the ether to approximately 2 ml on a 60° water bath under a gentle stream of N_2. Add 0.2 to 0.4 ml of diazomethane solution (18 mg/ml) in ether [prepared as per DeBoer (40)]. Evaporate as before to 1.0 ml.

TABLE 6.9. Analytical Method Validation: Recovery of Acidic Herbicides Added to Control Samples

Sample Type	Compound Added	Compound Added (ppm)	Number of Determination	Mean Recovery	Standard Deviation	Method
Water	2,4-D	0.001–10.0	24	89.1	3.7	4.1
	Silvex	0.001–10.0	24	84.3	6.2	4.1
	2,4,5-T	0.001–10.0	24	83.3	5.0	4.1
	MCPA	0.01–10.0	20	85.3	6.3	4.1
	Triclopyr	0.001–10.0	24	87.7	4.4	4.1
Soil	2,4-D	0.01–10.0	33	88.5	5.6	4.2
	Silvex	0.005–5.0	33	90.5	4.7	4.2
	2,4,5-T	0.005–5.0	33	91.6	6.2	4.2
	3,6-Dichloropicolinic acid	0.005–1.0	25	90.4	6.3	4.2
Grass	2,4-D	1.0–2000	41	95.3	5.5	4.3
	Silvex	1.0–2000	13	94.6	7.6	4.3
	2,4,5-T	1.0–2000	26	93.7	6.4	4.3
	Triclopyr	1.0–2000	15	97.9	4.9	4.3
	Picloram	1.0–400	33	83.0	6.4	4.3
Beef muscle	2,4-D	0.05–1.0	10	94.0	5.0	4.4.1
	2,4-Dichlorophenol	0.05–1.0	10	89.0	5.0	4.4.1
	2,4,5-T	0.05–2.0	9	79.0	11.0	4.4.1
	2,4,5-Trichlorophenol	0.05–0.5	9	76.0	9.0	4.4.1
	Silvex	0.05–2.0	12	87.0	6.0	4.4.1
	2,4,5-Trichlorophenol	0.05–2.0	12	81.0	12.0	4.4.1
	MCPA	0.05–2.0	11	91.0	11.0	4.4.2
	2-Chloro-4-methyl phenol	0.05–2.0	13	86.0	8.0	4.4.2
Beef fat	2,4-D	0.05–1.0	9	92.0	9.0	4.4.1
	2,4-Dichlorophenol	0.05–0.5	9	88.0	6.0	4.4.1

2,4,5-T	0.05–5.0	12	85.0	5.0	4.4.1
2,4,5-Trichlorophenol	0.05–1.0	13	87.0	5.0	4.4.1
Silvex	0.05–5.0	13	81.0	6.0	4.4.1
2,4,5-Trichlorophenol	0.05–5.0	13	87.0	4.0	4.4.1
MCPA	0.05–0.5	13	86.0	6.0	4.4.2
2-Chloro-4-methyl phenol	0.05–0.5	10	75.0	7.0	4.4.2
Beef liver					
2,4-D	0.05–1.0	9	92.0	9.0	4.4.1
2,4-Dichlorophenol	0.05–1.0	9	87.0	7.0	4.4.1
2,4,5-T	0.05–2.0	10	85.0	5.0	4.4.1
2,4,5-Triclorophenol	0.05–1.0	11	82.0	6.0	4.4.1
Silvex	0.05–2.5	8	92.0	6.0	4.4.1
2,4,5-Trichlorophenol	0.05–2.5	8	78.0	5.0	4.4.1
MCPA	0.05–2.0	14	81.0	10.0	4.4.2
2-Chloro-4-methyl phenol	0.05–2.0	14	79.0	6.0	4.4.2
Beef kidney					
2,4-D	0.05–1.0	18	96.0	7.0	4.4.1
2,4-Dichlorophenol	0.05–5.0	14	94.0	11.0	4.4.1
2,4,5-T	0.05–2.0	14	85.0	7.0	4.4.1
2,4,5-Trichlorophenol	0.05–5.0	15	85.0	6.0	4.4.1
Silvex	0.05–2.5	8	83.0	8.0	4.4.1
2,4,5-Trichlorophenol	0.05–2.5	8	83.0	3.0	4.4.1
MCPA	0.05–1.0	13	81.0	10.0	4.4.2
2-Chloro-4-methyl phenol	0.05–1.0	13	82.0	4.0	4.4.2
Sheep muscle					
2,4-D	0.05–0.1	4	99.0	7.0	4.4.1
2,4-Dichlorophenol	0.05–0.1	6	86.0	5.0	4.4.1
2,4,5-T	0.05–0.5	5	95.0	4.0	4.4.1
2,4,5-Trichlorophenol	0.05–0.5	5	94.0	8.0	4.4.1
Silvex	0.05–2.0	7	90.0	6.0	4.4.1
2,4,5-Trichlorophenol	0.05–2.0	7	80.0	10.0	4.4.1
MCPA	0.05–2.0	11	91.0	11.0	4.4.2
2-Chloro-4-methyl phenol	0.05–2.0	13	86.0	8.0	4.4.2

TABLE 6.9. (Continued)

Sample Type	Compound Added	Compound Added (ppm)	Number of Determination	Mean Recovery	Standard Deviation	Method
Sheep fat	2,4-D	0.05–0.1	4	94.0	5.0	4.4.1
	2,4-Dichlorophenol	0.05–0.1	5	86.0	6.0	4.4.1
	2,4,5-T	0.05–0.1	8	84.0	4.0	4.4.1
	2,4,5-Trichlorophenol	0.05–0.1	8	83.0	4.0	4.4.1
	Silvex	0.05–2.0	7	84.0	7.0	4.4.1
	2,4,5-Trichlorophenol	0.05–2.0	7	93.0	13.0	4.4.1
	MCPA	0.05–0.5	13	86.0	6.0	4.4.2
	2-Chloro-4-methyl phenol	0.05–0.5	10	75.0	7.0	4.4.2
Sheep liver	2,4-D	0.05–0.5	8	89.0	8.0	4.4.1
	2,4-Dichlorophenol	0.05–0.5	8	85.0	10.0	4.4.1
	2,4,5-T	0.1–5.0	4	89.0	8.0	4.4.1
	2,4,5-Trichlorophenol	0.05–0.25	4	88.0	9.0	4.4.1
	Silvex	0.05–2.5	8	90.0	8.0	4.4.1
	2,4,5-Trichlorophenol	0.05–2.5	8	72.0	5.0	4.4.1
	MCPA	0.05–2.0	14	81.0	10.0	4.4.2
	2-Chloro-4-methyl phenol	0.05–2.0	14	79.0	6.0	4.4.2
Sheep kidney	2,4-D	0.05–1.0	8	90.0	8.0	4.4.1
	2,4-Dichlorophenol	0.05–1.0	11	83.0	6.0	4.4.1
	2,4,5-T	1.0–5.0	5	102.0	6.0	4.4.1
	2,4,5-Trichlorophenol	0.05–1.0	5	84.0	4.0	4.4.1
	Silvex	0.05–2.5	8	86.0	5.0	4.4.1
	2,4,5-Trichlorophenol	0.05–2.5	8	78.0	4.0	4.4.1
	MCPA	0.05–1.0	13	81.0	10.0	4.4.2
	2-Chloro-4-methyl phenol	0.05–1.0	13	82.0	4.0	4.4.2
Beef muscle	2,4,5-T	0.025–0.5	10	82.0	7.0	4.4.4
	2,4,5-Trichlorophenol	0.025–0.2	10	90.0	5.0	4.4.4

Matrix	Compound					
Beef fat	2,4,5-T	0.025–0.5	11	86.0	8.0	4.4.4
	2,4,5-Trichlorophenol	0.025–0.2	8	84.0	5.0	4.4.4
Beef liver	2,4,-D	1.0–10.0	3	104.0	—	4.4.4
	2,4-Dichlorophenol	1.0–5.0	3	86.0	—	4.4.4
	2,4,5-T	0.025–5.0	13	84.0	7.0	4.4.4
	2,4,5-Trichlorophenol	0.025–0.2	10	93.0	5.0	4.4.4
	Silvex	0.1–2.0	3	79.0	—	4.4.4
	2,4,5-Trichlorophenol	0.1–2.0	2	83.0	—	4.4.4
	MCPA	0.1–0.2	6	85.0	8.0	4.4.4
	2-Chloro-4-methyl phenol	0.1–0.2	6	98.0	15.0	4.4.4
Beef kidney	2,4-D	0.05–5.0	3	74.0	—	4.4.4
	2,4,5-T	0.025–3.0	10	80.0	5.0	4.4.4
	2,4,5-Trichlorophenol	0.025–0.2	9	94.0	4.0	4.4.4
	Silvex	0.02–5.0	3	78.0	—	4.4.4
	2,4,5-Trichlorophenol	0.1–0.2	2	78.0	—	4.4.4
Sheep muscle	2,4,5-T	0.01–0.2	9	88.0	6.0	4.4.4
	2,4,5-Trichlorophenol	0.01–0.2	9	88.0	4.0	4.4.4
Sheep fat	2,4,5-T	0.025–2.0	10	83.0	8.0	4.4.4
	2,4,5-Trichlorophenol	0.025–0.2	9	90.0	7.0	4.4.4
Sheep liver	2,4,5-T	0.025–0.2	9	86.0	15.0	4.4.4
	2,4,5-Trichlorophenol	0.01–0.2	14	91.0	9.0	4.4.4
Sheep kidney	2,4,5-T	0.025–2.0	10	86.0	4.0	4.4.4
	2,4,5-Trichlorophenol	0.01–0.2	11	98.0	4.0	4.4.4
Chicken muscle and skin	2,4,-D	0.05–2.0	10	93.0	8.0	4.4.4
	2,4-Dichlorophenol	0.05–2.0	14	95.0	6.0	4.4.4
Chicken fat	2,4-D	0.05–0.5	10	86.0	10.0	4.4.4
	2,4-Dichlorophenol	0.05–0.5	12	94.0	8.0	4.4.4
Chicken liver	2,4-D	0.05–0.5	11	82.0	6.0	4.4.4
	2,4-Dichlorophenol	0.05–0.5	10	87.0	9.0	4.4.4
Chicken egg	2,4-D	0.05–0.5	12	85.0	11.0	4.4.4
	2,4-Dichlorophenol	0.05–0.5	7	90.0	13.0	4.4.4

245

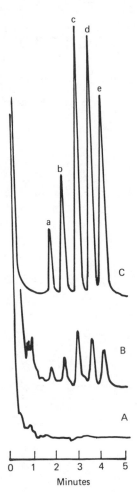

Figure 6.3 Typical chromatograms from the determination of acidic herbicides in water. (*A*) Control water. (*B*) Control water fortified at 10 μg/l MCPA and 1 μg/l each of 2,4-D, triclopyr, silvex, and 2,4,5-T. (*C*) Standards: 400 pg (*a*) MCPA and 40 pg each of (*b*) 2,4-D, (*c*) triclopyr, (*d*) silvex, and (*e*) 2,4,5-T. Equivalent to water concentrations of 40 μg/l and 4 μg/l, respectively.

9. Prepare a microcolumn by placing a loose glass wool plug in a 146 mm × 7 mm OD disposable pipet and adding 1 inch of 1.5% deactivated (with H$_2$O) acidic alumina. Wash the column with 1.5 ml diethyl ether (see Section 6.4.3.2, Nos. 3 and 4).

10. Place a 5-ml volumetric flask under the column. Add the 1.0 ml of methylated extract to the column. Allow the ether to run down to the top of the alumina, then rinse the centrifuge tube twice with 0.5 ml of ether, adding each rinse to the column soon after the previous addition reaches the top of the alumina. Complete the elution with 0.5 ml ether. Collect all of the eluant.

11. Add 1.0 ml benzene and a large carborundum boiling chip to the volumetric flask and evaporate on a steam bath until the ether is gone (15–25 min). Do not allow to go dry. Dilute to volume with hexane.

12. Inject 4 μl into a 180 cm $\times$ 2 mm column of 11% OV-17 + QF-1 on 80/100 mesh Gas Chrom Q.

6.4.2.1. Determination of Moisture Content of Soil

1. Transfer about 20 g of well-mixed soil into a tared glass container and weigh to the nearest 0.01 g.

2. Place sample in 110°C oven and allow to dry at least 3 hr.

3. Remove sample from oven, place in desiccator until cooled to room temperature, and then reweigh.

4. Calculate the percent moisture in the soil using the equation:

$$\% \text{ moisture} = \frac{\text{g of sample before drying} - \text{g of sample after drying}}{\text{g of sample before drying}} \times 100$$

6.4.2.2. Computation of Corrected Residues

1. Determine gross ppm phenoxy acids by comparing peak height to the appropriate standard.

2. Correct for any apparent phenoxy residue in the control samples (net ppm).

3. Calculate the corrected residues using the following equation:

$$\text{corrected residues} = \frac{\text{net ppm phenoxy acid}}{\% \text{ recovery}} \times (100 + \% \text{ moisture})$$

4. Recoveries are listed in Table 6.9 and representative chromatograms are shown in Figure 6.4.

6.4.2.3. Notes

1. For soils high in organic matter it has occassionally been necessary to centrifuge longer than 1–5 sec to effect complete separation of the aqueous and ether phases.

2. When repeating the extraction of the soil one should be certain that all the soil is freed from the side and bottom of the centrifuge tube. It has been found that tapping the tube rapidly against a large rubber stopper followed by vigorous manual shaking is quite effective in resuspending the soil.

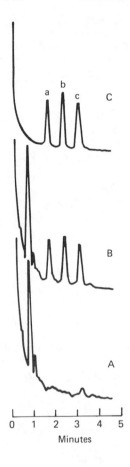

Figure 6.4 Typical chromatograms from the determination of acidic herbicides in soil. (*A*) Control soil, fargo clay (7.0% organic matter and 10% sand, 36% silt, and 54% clay). (*B*) Control soil fortified at 0.01 ppm 2,4-D and 0.005 ppm each of silvex and 2,4,5-T. (*C*) Standards: 80 pg (*a*) 2,4-D and 40 pg each of (*b*) silvex and (*c*) 2,4,5-T. Equivalent to soil concentrations of 0.01 ppm and 0.005 ppm, respectively.

3. The partially deactivated acidic alumina is prepared by adding the requisite amount of deionized water to the alumina, and rolling or gently shaking the alumina for an hour to promote equilibrium. The container should be kept tightly capped at all times. The elution pattern of each fresh batch of alumina should be checked to determine that methyl esters of the phenoxy acids elute quantitatively from the column within the first 2.5 ml of ether eluant.

4. The adsorption chromatography on the partially deactivated acidic alumina is a crucial step for both cleanup and recovery. Using a small tight glass wool plug will decrease recovery due to greater contact time. Recovery can be improved (with a concurrent loss of cleanup) by increasing the deactivation of the acidic alumina to 2%.

6.4.3. Grass Method (Plants)

1.   Weigh a representative, homogenized, 10.0 g sample of grass into an 8-oz bottle, and add 180 ml of 0.1N NaOH.

2.   Blend the sample with a Polytron homogenizer (Model PT 10-35 with a PT 20ST generator) for 30 sec at full speed. Rinse the blades with 20 ml of 0.1N NaOH contained in a 12-dram vial.

3.   Add the rinse to the sample bottle, and cap it. Shake for 30 min at 180 excursions/min.

4.   After shaking, allow to settle, then pipet a 10-ml aliquot of the caustic supernatant into a 12-dram vial and add 7 g NaCl, 1 ml 6N H_2SO_4, and 20 ml of 3/7 ether/hexane. Cap and shake for 5 min at 280 excursions/min.

5.   Centrifuge 1–2 min at 2000 rpm. Remove the ether/hexane layer by pipet and place in a 12-dram vial containing 15 ml of 0.1N NaHCO$_3$.

6.   Shake the ether/hexane–bicarbonate mixture for 5 min (180 excursions/min), centrifuge 1–2 min at 2000 rpm and discard the ether/hexane.

7.   Repeat extraction with an additional 20-ml portion of ether/hexane, partioning with the same bicarbonate solution and discarding the ether/hexane portion as in Step 6.

8.   Add 10 ml diethyl ether to the bicarbonate solution, shake 5 min (180 excursions/min), centrifuge 1–2 min at 2000 rpm and discard the ether.

9.   Add 4–6 g of NaCl to the bicarbonate solution. Slowly add 1 ml of 6N H_2SO_4 and let stand until CO_2 is no longer evolved. Add 10 ml of ether, shake 5 min (180 excursions/min), centrifuge 1–2 min at 2000 rpm, and place the ether in a 40-ml conical centrifuge tube.

10.   Repeat addition of 10 ml ether to the acidified bicarbonate solution, shake, centrifuge, and combine the ether with the first ether extract.

11.   Evaporate the combined ether extracts to approximately 2 ml on a 60° water bath under a gentle stream of nitrogen or dry air. Add 0.25–0.5 ml diazomethane solution (18 mg/ml) and swirl to mix. Evaporate to 1.0 ml on the water bath.

12.   Prepare a microcolumn by placing a loose glass wool plug in a 15-cm × 7 mm OD disposable pasteur pipet and adding 2.0–2.5 cm of 4% water-deactivated acidic alumina. Wash the column with 1.5 ml diethyl ether.

13.   Place a 10-ml volumetric flask under the column. Place the 1.0 ml of methylated extract on the column. Allow the ether to run down to the top of the alumina, then rise the centrifuge tube twice with 0.5 ml of ether, adding each rinse to the column just as the previous addition reaches the top of the column. Complete the elution with 1.0 ml ether. Collect all of the eluant.

14. Add 2.0 ml isooctane and a 10 mesh carborundum boiling chip to the volumetric flask and evaporate on a steam bath until the ground glass joint is dry (15–25 min). Dilute to volume with isooctane.

15. Inject 2 μl into a 180 cm $\times$ 4 mm column of 3% OV-3 on 80/100 mesh Gas Chrom Q.

16. Determine the concentration in μg/ml of the esters in the final solution by comparing peak height to that of the standard. Multiply μg/ml by 20 to give gross ppm.

17. Subtract apparent gross ppm found in control samples from that found in the treated samples to obtain net ppm.

18. Divide net ppm by average percent recovery and multiply by 100 to obtain corrected ppm.

Recovery data are presented in Table 6.9 and representative chromatograms are shown in Figure 6.5.

6.4.4. Animal Tissue Methods

6.4.4.1. *Animal Tissue Method A-1*

For the determination of phenoxy acids in all tissues and the related phenol in muscle and fat. Alkaline hydrolysis is required to determine chlorophenols in liver and kidney.

1. Cut 20.0 g of frozen homogenized tissue into small pieces and place in a 4-oz French square jar.

2. Add 6 g of Hyflo Super-Cel, 50 ml of methanol, and 1 ml of concentrated ammonium hydroxide. Attach the jar to a homogenizer and blend for 5 min at 50–70% of full speed, keeping the jar immersed in an ice bath. Rinse the spindle, collecting the rinsings in the jar, cap with a polyseal lined cap, and shake for 10 min. Filter the sample through a 60-ml fritted Buchner funnel containing a 1/8-inch pad of Hyflo Super-Cel filteraid, attach to a Fisher Filtrator and collect the filtrate in a 100-ml graduated mixing cylinder. Wash the jar three times with 10 ml of methanol, putting the washings through the filter. Continue washing the filter cake with methanol until a total of 100 ml of filtrate is collected. Avoid allowing the filter cake to go dry during filtration and washing. Stopper the receiver and mix.

3. Pipet 15 ml of the extract into a 12-dram vial, add a boiling chip, and boil down on a steam bath until foaming occurs (about 5 ml volume). Cool.

4. Add 15 ml of water (to give 20 ml of water phase), 0.5 ml of concentrated phosphoric acid, 2 g of NaCl and 12 ml of ether. Cap the vial and shake at the rate of 660 excursions per minute for 10 min, and separate the layers by centrifugation.

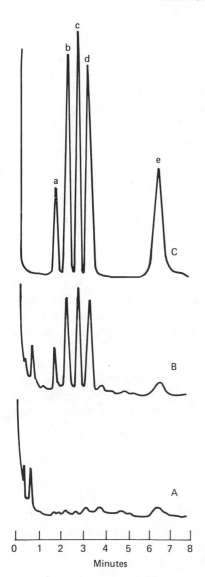

Figure 6.5 Typical chromatograms from the determination of acidic herbicides in grass. (*A*) Control grass. (*B*) Control grass fortified at 1.0 ppm each of 2,4-D, triclopyr, silvex, 2,4,5-T, and picloram. (*C*) Standards: 200 pg each of (*a*) 2,4-D, (*b*) triclopyr, (*c*) silvex, (*d*) 2,4,5-T, and (*e*) picloram. Equivalent to a grass concentration of 2.0 ppm each.

5. Prepare a 2-g column of alumina in a 1 × 13 cm glass column and wash it with 15 ml of ether.

6. Transfer the ether extract to the column and allow it to run through, collecting the effluent in a 12-dram vial. Reextract the water phase with two more 12-ml portions of ether adding each extract to the column. Wash the column with 10 ml of ether collecting all of the ether in the vial. Add a boil-

ing chip and concentrate the ether to 5–10 ml volume by heating on a steam bath.

7. Add 10 ml of hexane and 10 ml of 0.5N KOH to the ether concentrate and shake for 5 min (660 excursions/min), centrifuge 3 min at maximum speed, 2000 rpm, and remove the ether layer. To the water layer add 1.0 ml of concentrated phosphoric acid, 2 g NaCl and 10.0 ml of benzene and shake (660 excursions/min) for 5 min. Centrifuge and then inject 2 μl of this benzene solution into the gas chromatograph equipped with an electron-capture detector. Measure the height of the peak obtained for the phenol in terms of percent full-scale deflection and determine the weight of chlorophenol found (in nanograms) by reference to an appropriate standard curve.

8. Elute the alumina column with 20–25 ml of 0.25N sodium bicarbonate or ammonium hydroxide solution, collecting the eluate in a 12-dram vial. Add 2.0 ml of concentrated phosphoric acid, 4 g of NaCl, and 10.0 ml of ether.

9. Shake the mixture at the rate of 660 excursions/min for 5 min and centrifuge. Pipet 5 ml of the ether phase into a 10-ml volumetric flask, add 1 ml of diazomethane reagent prepared as per DeBoer (40), 2 ml of benzene, and a boiling chip. Remove the ether by heating on a steam bath. Cool and dilute to volume with benzene.

10. Chromatograph 2 μl of the benzene solution on a chromatograph equipped with an electron capture detector. Measure the height of the peak obtained for the ester in terms of percent full-scale deflection and determine the weight of acid found (in nanograms) by reference ·to the appropriate standard.

Recovery data are presented in Table 6.9 and representative chromatograms are shown in Figures 6.6 and 6.7.

6.4.4.2. Animal Method A-2

Animal Method A-2 is for MCPA and 4-chloro-2-methyl phenol in all tissues.

1. Cut 20.0 g of frozen homogenized tissue into small pieces and place in a 4-oz French square jar.

2. Add 6 g of Hyflo Super-Cel, 50 ml of methanol, and 1 ml of concentrated ammonium hydroxide. Attach the jar to a homogenizer and blend for 5 min at 50–70% of full speed, keeping the jar immersed in an ice bath. Rinse the spindle, collecting the rinsings in the jar, cap with a polyseal lined cap and shake for 10 min. Filter the sample through a 60-ml fritted Buchner funnel containing a ⅛-inch pad of Hyflo Super-Cel filteraid, attach

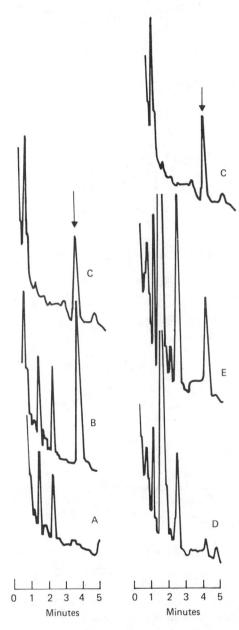

Figure 6.6 Typical chromatograms from the determination of 2,4,5-T in beef fat and liver by Method 4.4.1. (*a*) Control fat. (*b*) Control fat fortified with 0.1 ppm 2,4,5-T. (*c*) Standard: 10 pg 2,4,5-T. Equivalent to a tissue concentration of 0.05 ppm. (*d*) Control liver. (*e*) Control liver fortified with 0.05 ppm 2,4,5-T.

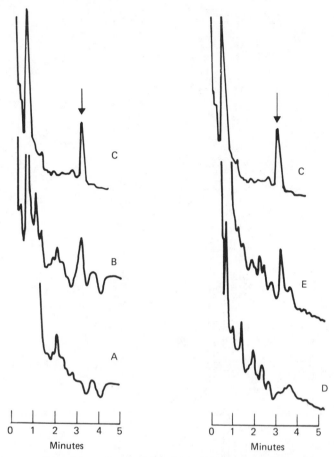

Figure 6.7 Typical chromatograms from the determination of 2,4,5-trichlorophenol in beef muscle and fat by Method 4.4.1. (*A*) Control muscle. (*B*) Control muscle fortified with 0.05 ppm 2,4,5-trichlorophenol. (*C*) Standard: 20 pg 2,4,5-trichlorophenol. Equivalent to a tissue concentration of 0.05 ppm. (*D*) Control fat. (*E*) Control fat fortified with 0.05 ppm 2,4,5-trichlorophenol.

to a Fisher Filtrator and collect the filtrate in a 100-ml graduated mixing cylinder. Wash the jar three times with 10 ml of methanol, putting the washings through the filter. Continue washing the filter cake with methanol until a total of 100 ml of filtrate is collected. Avoid allowing the filter cake to go dry during filtration and washing. Stopper the receiver and mix.

3. Transfer all of the extract into a 200-ml beaker containing 0.5 ml of 1*M* NaOH, rinsing the sides of the graduated cylinder with several milliliters of methanol to insure complete transfer.

4.   Add three boiling chips and place the beaker on a hot plate set at moderate heat to evaporate the contents to about 15 ml. (A good draft across the top of the beakers eliminates most of the foaming problems.)

5.   Allow the beaker to cool and transfer the concentrated extract to a 12-dram vial. Rinse the beaker with 5 ml of water and add the water to the vial.

6.   Rinse the sides of the beaker with 12 ml of diethyl ether and let stand to help dissolve residual material while adjusting the volume of aqueous liquid in the vial (step 5) to approximately 25 ml.

7.   Acidify the contents of the vial with 1 ml of H_3PO_4, add 4 g of NaCl, and pour in the ether from the beaker (step 6), rinsing the beaker with an additional 2–3 ml of ether.

8.   Cap the vial and shake for 10 min, then separate the layers by centrifugation.

9.   Prepare an acidic alumina column by adding 10 g of alumina as received into an 18×30 mm glass column with gentle tapping to settle the column.

10.   Draw off the ether layer (step 7) with a 20-ml pipet and transfer it to the alumina column. Allow the ether to run through and collect the eluate in a 100-ml beaker.

11.   Reextract the water phase with two more 12-ml portions of ether, shaking for 5 min each time, then centrifuging, and transferring each extract onto the column.

12.   Wash the column with 12 ml of ether and blow out all ether with air. Retain the column for determination of MCPA.

13.   Add 5 ml of hexane and two boiling chips to the beaker containing the combined eluates and place on a moderately heated steam bath.

14.   Evaporate the ether to a volume of approximately 3 ml. (The remaining solution may be clear or rather cloudy, and it may be colorless or quite yellow, with no apparent difference in the final result.)

15.   After cooling, transfer the contents of the beaker to a clean 12-dram vial, using about 10 ml of hexane to wash down the sides of the beaker. Add a few milliliters of hexane to bring the volume to approximately 15 ml.

16.   Add 10 ml of $1M$ NaOH and cap the vial.

17.   Shake the vial for 5 min and separate the layers by centrifugation.

18.   Using a disposable pipet connected to a vacuum source with rubber tubing, draw off and discard the hexane layer and any fatty layer which floats on top of the NaOH solution.

19.   Evaporate any remaining hexane from the surface of the aqueous phase by directing a jet of air down the neck of the vial for several minutes.

20.   Acidify the contents of the vial with 1 ml of 85% H_3PO_4 and add 4 g of NaCl and 2.0 ml of benzene.

21. Cap and shake 5 min and centrifuge.

22. Inject a 20-μl aliquot of supernatant liquid into the gas chromatograph.

23. Measure the height of the peak obtained for the 2-methyl-4-chlorophenol and determine the weight of residue found (in nanograms) by reference to the appropriate standard.

24. Wash the alumina column (step 9) with 20 ml of 0.1M NaHCO$_3$, collecting the eluate in a 12-g vial for discard.

25. Elute the MCPA from the alumina column with 65 ml of 0.1M NaHCO$_3$, added in amounts of 40 and 25 ml collecting in 100-ml centrifuge bottle. (Air pressure may be used during this step to prevent the accumulation of bubbles and to maintain a good flow rate.)

26. Remove the bottle and acidify with 1 ml of 85% H$_3$PO$_4$, then carefully add 10 g of NaCl to prevent excessive foaming, then swirl.

27. Add 30.0 ml of diethyl ether, cap the bottle and shake the bottle for 5 min and allow it to stand or centrifuge to separate the phases.

28. Using a pipet, transfer 25.0 ml of the ether phase into a 50-ml beaker.

29. Add about 0.5 ml of benzene and two boiling chips, then place the beaker on a covered steam bath and evaporate the contents down to a volume of approximately 0.5 ml.

30. Cool and transfer to a 2-ml volumetric flask, rinsing the beaker with several portions of benzene to nearly fill the flask.

31. Gently evaporate the benzene solution to a volume of about 0.5 ml by directing a stream of air down the neck of the flask.

32. Add about 0.75 ml of ethereal diazomethane and a boiling chip to the flask and place on a steam bath at moderate heat (watch for possible boilover at first) until the refluxing solution no longer reaches the ground-glass portion of the flask or until the volume of solution remaining in the flask is about 0.5 ml.

33. Allow the flask to cool to room temperature and dilute to the volume with benzene and inject a 20 μl aliquot of the benzene solution into the gas chromatograph.

34. Measure the height of the peak obtained for the methyl-2-methyl-4-chlorophenoxyacetate and determine the weight of residue found (in nanograms) by reference to the appropriate standard.

6.4.4.3. Notes

1. For liver and kidney increase the amount of filter aid in (b) to 8 g.
2. For fat substitute the following for step (b):
   (a) Add 50 ml of methanol and 1 ml of concentrated ammonium hydroxide.

   (b)   Place the open powder jar in a 60–65°C water bath for at least 15 min.

   (c)   Add about 8 g of filter aid and attach the jar to the homogenizer.

   (d)   Blend for 1 min at full speed, suspending the jar in a wide-mouth metal or plastic container, for protection of the operator.

   (e)   Rinse the spindle, collecting the rinsings in the jar, and place the uncovered jar in an ice bath for about 15 min.

   (f)   Filter the sample and continue with the procedure.

3. For fat, no air draft should be caused to flow across the top of the 250 ml beaker during evaporation (step 11 in Section 6.4.4.2).

Recovery data are presented in Table 6.9.

6.4.4.4. Animal Method B

Animal Method B is for phenoxy acids and related phenols in all tissues.

1. Cut 5.0 g of frozen homogenized tissue into small pieces in a 12-dram vial.

2. Add 20 ml of $0.5N$ NaOH, cap with a polyseal cap and shake for 5 min on a horizontal shaker. For fat, shake 10 min. Place the tube into the steam bath so that all liquid is below the top of the bath, but the rest of the tube extends above the opening of the steam bath. After 15 min, release the pressure in the tube and shake several times. For fat, shake for 10 minutes on a mechanical shaker at the rate of 660 excursions/min to get complete mixing of the fat and aqueous phases. Shake briefly by hand after 30 min and 45 min. Remove tube after 1 hr and allow to cool to room temperature.

3. Extract the alkaline hydrolysate with 20 ml of ether/hexane (3/7, v/v) for 5 min. Centrifuge and remove the upper organic layer. (Keeping the samples warmed allows the removal of most of the white lipid that solidifies when samples are allowed to cool to room temperature.)

4. Add 5 ml of $6N$ H_2SO_4 and 10 ml of ether/hexane (3/7, v/v). Cap with the polyseal lined cap and shake in a horizontal shaker (660 excursions/min) for 15 min.

5. Centrifuge the tube at 2000 rpm at 1 min. If a gel forms, open tube and add 0.5 ml of methanol, cap and recentrifuge for 1 min.

6. Transfer the ether/hexane layer to a clean 12-dram vial. Repeat the extraction (step 4) with another 10-ml portion of ether/hexane. Centrifuge, etc., (as in step 4) combining the two extracts. Discard the aqueous layer.

7. Add 10 ml of $0.5N$ NaOH to the combined ether/hexane extract, cap with a new polyseal cap and shake (660 excursions/min) for 5 min. Centrifuge and remove the organic layer.

8. Wash the aqueous layer with 10 ml of ether/hexane (3/7, v/v) and discard the ether/hexane layer.

9. Add 2 ml of concentrated phosphoric acid, 4 g of NaCl and 5.0 ml of ether/hexane (3/7, v/v). Cap the vial and shake at the rate of 660 excursions/min for 10 min on a horizontal shaker and separate the layers by centrifuging for 1 min at 2000 rpm.

10. Transfer the ether/hexane to a clean 12-dram vial and repeat the extraction with another 5.0 ml portion of ether/hexane combining the two extracts. Discard the aqueous layer.

11. Extract the phenoxy herbicide from the ether/hexane with 10 ml of 1% sodium bicarbonate shaking (660 excursions/min) for 5 min. Separate the layers by centrifugation at 2000 rpm for 1 min.

12. Inject 1 μl of the ether/hexane solution containing the chlorophenol into the gas chromatograph. Measure the height of the peak obtained for the chlorophenol in terms of percent full-scale deflection and determine the weight of the chlorophenol found (in nanograms) by reference to a standard curve.

13. Wash the bicarbonate solution with 10 ml of ether/hexane (3/7, v/v) by shaking (660 excursions/min) for 5 min followed by separating the layers by centrifugation. Discard all of the organic layer.

14. Add 1 ml of concentrated phosphoric acid to the bicarbonate solution, 4 g of NaCl and 5 ml of the ether/hexane (3/7, v/v). Cap with clean polyseal cap and shake (660 excursions/min) for 10 min on a horizontal shaker. Separate the layers by centrifugation at 2000 rpm for 1 min. Transfer the ether/hexane to a 10-ml volumetric flask. Add a boiling chip then place the flask on a steam bath and reduce the volume of solvent to about 1 ml. Remove and cool the flask to room temperature.

15. Repeat the extraction with another 5 ml portion of ether/hexane. Transfer this extract to the volumetric flask.

16. Add 0.1 ml MeOH and 0.5 ml of diazomethane reagent and 1 boiling chip to the extract in the flask and shake. Allow to stand 5 min and heat on a steam bath until colorless. Cool and dilute to volume with ether/hexane.

17. Chromatograph 1 μl of the ether/hexane solution. Measure the height of the peak obtained for the ester in terms of percent full-scale deflection and determine the weight of herbicide found (in nanograms) by reference to an appropriate standard curve.

6.4.4.5. Notes

1. With chicken liver recovery of added herbicide was poor unless step 4 in Section 6.4.4.4 was modified by hydrolyzing the tissue for 10 min on a

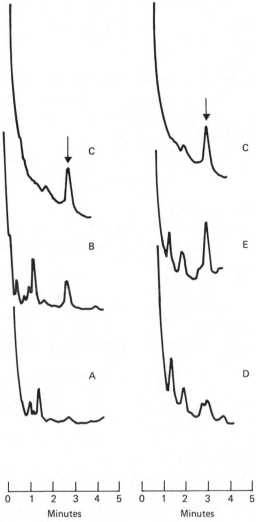

Figure 6.8 Typical chromatograms from the determination of 2,4,5-trichlorophenol in beef muscle and kidney by Method 4.4.3. (*A*) Control muscle. (*B*) Control muscle fortified with 0.025 ppm 2,4,5-trichlorophenol. (*C*) Standard: 12.5 pg 2,4,5-trichlorophenol. Equivalent to a tissue concentration of 0.05 ppm. (*D*) Control kidney. (*E*) Control kidney fortified with 0.05 ppm 2,4,5-trichlorophenol.

steam bath after adding H_2SO_4 but before cooling and extracting with ether/hexane.

2.   Recoveries are listed in Table 6.9 and typical chromatograms are shown in Figure 6.8.

6.4.4.6.   Method for Milk and Cream

An excellent method for the determination of the phenoxy acids and their chlorophenols is contained in reference 32 along with appropriate recovery and validation data for each compound.

REFERENCES

1.   *Phenoxy Herbicides; Their Effects on Environmental Quality*, National Research Council of Canada, #16075 (1978).

2.   W. P. Cochrane and R. Purkayastha, *Toxicol. Environ. Chem. Rev.*, **1**, 137 (1973).

3.   A. E. Smith, *Weed Res.*, **16**, 19 (1976).

4.   R. G. Zepp, N. L. Wolfe, T. A. Gordon, and G. L. Baughman, *Environ. Sci. Technol.*, **9**(13), 1144 (1975).

5.   A. S. Crafts, *Weeds*, **8**, 19 (1960).

6.   S. S. QueHee, R. G. Sutherland, and M. Vetter, *Environ. Sci. Technol.*, **9**, 62 (1975).

7.   R. Grover, L. A. Kerr, K. Wallace, K. Yoshida, and J. Maybank, *J. Environ. Sci. Health* B 11., (4), 331 (1976).

8.   Environmental Protection Agency Title 40 Code of Federal Regulations, Part 180, Subparts, .142, .227, .292, .339

9.   H. L. Morton, E. D. Robison, and R. E. Meyer, *Weeds*, **15**, 268 (1967).

10.   L. Lykken, *Residue Rev.*, **3**, 19–34 (1963).

11.   J. D. Grier, *J. Assoc. Offic. Anal. Chem.*, **49**(2), 291 (1966).

12.   A. J. Pik and G. W. Hodgson, *J. Assoc. Offic. Anal. Chem.*, **59**, 264 (1976).

13.   R. G. Wilson and H. H. Cheng, *J. Environ. Qual.*, **7**(2), 281 (1978).

14.   J. F. Lutz, G. E. Byers, and T. J. Sheets, *J. Environ. Qual.*, **2**(4), 485 (1973).

15.   E. L. Bjerke, unpublished method of The Dow Chemical Company, Midland, MI, (1973).

16.   G. Yip and R. E. Ney, Jr., *Weeds*, **14**, 167 (1966).

17.   H. E. Munro, *Pestic. Sci.*, **3**, 371 (1972).

18.   H. Løkke, *Bull. Environ. Contam. Toxicol.*, **13**(6), 730 (1975).

19.   C. Chow, M. L. Montgomery, and T. C. Yu, *Bull. Environ. Contam. Toxicol.*, **6**(6), 576 (1971).

20.   D. E. Clark, *J. Agr. Food Chem.*, **17**(6), 1168 (1969).

21. K. Erne, *Acta Vet. Scand.*, **7**, 77 (1966).
22. M. T. Shafik, H. C. Sullivan, and H. F. Enos, *Int. J. Environ. Anal. Chem.*, **1**, 23 (1971).
23. D. E. Clark, J. S. Palmer, R. D. Radeleff, H. R. Crookshank, and F. M. Farr, *J. Agric. Food Chem.*, **23**(3), 573 (1975).
24. A. H. Kutschinski and V. Riley, *J. Agr. Food Chem.*, **17**(2), 283 (1969).
25. E. L. Bjerke, A. H. Kutschinski, and J. C. Ramsey, *J. Agr. Food Chem.*, **15**(3), 469 (1967).
26. E. A. Woolson and C. I. Harris, *Weeds*, **15**, 168 (1967).
27. W. M. Edwards and B. L. Glass, *Bull. Environ. Contam. Toxicol.*, **6**, 81 (1971).
28. D. W. Woodham, W. G. Mitchell, C. D. Loftis, and C. W. Collier, *J. Agr. Food Chem.*, **19**(1), 186 (1971).
29. N. C. Glaze and M. Wilcox, *Proc. Soil Crop Sci. Soc. Fla.*, **26**, 271 (1966).
30. W. R. Meagher, *J. Agr. Food Chem.*, **14**, 374 (1966).
31. C. A. Bache, L. E. St. John, Jr., and D. T. Lisk, *Anal Chem.*, **40**(8), 1241 (1968).
32. E. L. Bjerke, J. L. Herman, P. W. Miller, and J. H. Wetters, *J. Agr. Food Chem.*, **20**(5), 963 (1972).
33. A. H. Kutchinski, *J. Agr. Food Chem.*, **17**(2), 288 (1969).
34. G. Yip, *J. Assoc. Offic. Anal. Chem.*, **47**, 343 (1962).
35. G. Yip, *J. Assoc. Offic. Anal. Chem.*, **54**, 966 (1971).
36. S. O. Farwell, F. W. Bowes, and D. F. Adams, *Anal. Chem.*, **48**(2), 420 (1976).
37. R. Albright et al., *Bull. Environ. Contam. Toxicol.*, **12**(3), 378 (1974).
38. M. C. Bowman and M. Beroza, *J. Assoc. Offic. Anal. Chem.*, **48**, 943 (1965).
39. I. H. Suffet, *J. Agr. Food Chem.*, **21**(4), 591 (1973).
40. Th. J. BeBoer and H. G. Backer, *Org. Synth. Coll.*, **4**, 250 (1963).

ANALYSIS OF ORGANOPHOSPHORUS PESTICIDES

MALCOLM C. BOWMAN

Department of Health, Education, and Welfare, Food and Drug Administration,
National Center for Toxicological Research, Jefferson, Arkansas

CONTENTS

7.1. INTRODUCTION

The introduction of DDT as a contact insecticide in the early 1940s marked the beginning of the synthetic organic pesticide era and a large family of organochlorine pesticides soon emerged. However, through widespread use and misuse their effectiveness began to diminish in some instances and by the early 1950s new synthetic pesticides of the organophosphorus (OP) and carbamate groups were often used as supplements or substitutes.

This article was written by Malcolm C. Bowman in his private capacity. No official support or endorsement by the Food and Drug Administration is intended or should be inferred.

Early analytical methods for the OP compounds were based on cholinesterase inhibition (ChEI) (1), total phosphorus determinations (2) and colorimetric procedures such as the classical Averell–Norris method (3). By today's standards, these methods lack speed, sensitivity and/or specificity; however, the basic analytical concepts are still widely used. For example, ChEI is used to signal and measure biological response in animals exposed to the pesticides and for ultrasensitive detection of residues in thin layer chromatography (TLC); colorimetric procedures for functional groups or total phosphorus are used in automated residue analyses.

During the late 1950s and the 1960s, research on the development of new pesticides of the three major classes (organochlorine, organophosphorus, and carbamates) as well as analytical methodology for the parent compounds, analogs, and breakdown products proceeded rapidly. Through the use of sophisticated analytical techniques developed during this period questions were raised concerning the safety of pesticide residues remaining in the environment that led to essentially the complete curtailment of organochlorine pesticides usage in the Western World and their replacement with OP and carbamate compounds.

The OP pesticides, which may be considered our second generation of synthetic pesticides, have gained favor because they are generally of low persistence, biodegradable, and do not tend to accumulate in the food chain, or pollute the environment. They are generally more toxic (potent cholinesterase inhibitors) than the organochlorine or carbamate compounds however, through *proper usage*, they continue to serve us well.

During the past 3 to 5 years, research on the development of new pesticides and analytical methodology has been severely curtailed in favor of alternate means of pest control. Nevertheless, the use of pesticides continues to be our major weapon for controlling pests, and the development of new and improved techniques of analyzing these agents and their breakdown products has continued at a reduced pace.

Many reviews pertaining to the analysis of OP compounds have appeared in the literature during the past several years and the most recent one was an excellent state-of-the-art overview by Burchfield and Storrs (4). The biyearly pesticide reviews of Thornburg (5–8) are also valuable sources of information on the analysis of these compounds. The *Pesticide Analytical Manual* (*PAM*) (9) and the *Manual of Analytical Methods* (*MAM*) (10) prepared by the U.S. Food and Drug Administration (FDA) and the U.S. Environment Protection Agency (EPA), respectively, describe proven methods for analyzing many of the compounds in various substrates. Manuals and methods are also available from many of the pesticide manufacturers. An excellent additional reference is the Pesticide Index by Frear (11), which contains about 1500 entries for pesticides of all classes and includes their chemical and common names, structural formulas, physical properties, toxicities and usages.

The purpose of this chapter is to present the salient elements of analytical procedures currently used for the analysis of OP pesticide residues and to provide a convenient reference to some of their key analytical properties as well as procedures for multiple or individual residue determinations. Common names (or trade names), chemical names, the phosphorus (P) and sulfur (S) content of the molecule, and references to analytical methods for 132 OP pesticides, metabolites and break-down products (including 6 insect chemosterilants) are presented in Table 7.1 (11–163).

7.2. SAMPLE PREPARATION, EXTRACTION, AND CLEANUP

7.2.1. Preparation of the Sample for Extraction

Excellent procedures for preparing samples for extraction are described in the *PAM* (8), therefore, only a brief overview of these procedures is presented here. The main objective of the preparation step is to produce a final product that is representative of the entire amount from which the sample(s) was taken and in a form that is amenable to extraction without sustaining a loss of pesticide residues or a change in their chemical nature during the process.

Samples are processed in various ways depending on their physical characteristics. For example, dried beans, peas, grains, and so on are ground to about 20-mesh in a Wiley mill while vegetables, fruits, meats, and so on are finely chopped in a Hobart Vertical Cutter. Liquids or solid samples consisting of fine and uniform particles require essentially no preparation. Often several portions of a commodity sampled at random are processed and then composited to obtain a representative sample. The compositing of samples such as eggs and cheese is best accomplished by using a Waring blendor to obtain a uniform mixture. In some cases only the edible portion of a product is subjected to residue analysis, therefore the peel, husk, shell, stem, rind, seed, and so on, is discarded prior to commutation. It is a good practice to perform a moisture determination (percentage of weight lost in a 110°C oven overnight) on a portion of the final sample for use in the event that it becomes necessary to express the residues on a dry weight basis. In the event that residues of volatile or labile compounds are sought, special precautions must be taken in transporting and processing the samples. For example, a sample might be quick-frozen and transported under dry ice then chopped in the presence of dry ice; the dry ice can then be allowed to sublime from the sample in a freezer prior to the extraction process.

TABLE 7.1. Common Names and/or Trade Names, Chemical Names, Phosphorus (P) and Sulfur (S) Content of the Molecule and References to Analytical Methods for 132 Organophosphorus Pesticides, Metabolites, Breakdown Products, and Insect Chemosterilants

Pesticide (or related compound)[a]	PS in Molecule	Chemical Name	References for Analytical Methods
Abate® (Temephos)	P_2S_3	O,O'-(Thiodi-4,1-phenylene) bis(O,O-dimethyl phosphorothioate)	11–16
Accothion®, see fenitrothion			
Amidithion	PS_2	S-{[(2-Methoxyethyl)carbamoyl] methyl} O,O-dimethyl phosphorodithioate	17–18
Apholate	P_3	2,2,4,4,6,6-Hexakis(1-aziridinyl) 2,2,4,4,6,6-hexahydro-1,3,5,2,4,6-triazatriphosphorine	17–21
Azinphosethyl	PS_2	O,O-Diethyl phosphorodithioate S-ester with 3-(mercaptomethyl)-1,2,3-benzotriazin-4(3H)-one	17, 18, 22, 23
Azinphosmethyl	PS_2	O,O-Dimethyl phosphorodithioate S-ester with 3-(mercaptomethyl)-1,2,3-benzotriazin-4(3H)-one	17, 18, 22, 24–28
Azodrin®, see Monocrotophos			
Bay 30911	PS	O-(2,4-Dichlorophenyl) O-methyl methylphosphonothioate	17, 18
Bay 37289 (Trichloronate)	PS	O-Ethyl O-(2,4,5-trichlorophenyl) ethylphosphonothioate	17, 18
Bay 37342	PS_2	O,O-Dimethyl O-{4-(methylthio)-3,5-xylyl} phosphorothioate	17, 18
Bay 68138 (Phenamiphos)	PS	Ethyl 4-(methylthio)-m-tolyl isopropylphosphoramidate	29–31

Name	Type	Chemical name	References
Bay 68138 sulfoxide	PS	Ethyl 4-(methylsulfinyl)-*m*-tolyl isopropylphosphoramidate	29
Bay 68138 sulfone	PS	Ethyl 4-(methylsulfonyl)-*m*-tolyl isopropylphosphoramidate	29
Bay 77488, see Phoxim			
Bay 93820	PS	Isopropyl salicylate *O*-ester with *O*-methyl phosphoramidothioate	32–34
Bay 93820 *O*-analog	P	Isopropyl salicylate *O*-ester with *O*-methyl phosphoramidate	32–34
Bidrin©, see Dicrotophos			
Bromophos	PS	*O*-(4-Bromo-2,5-chlorophenyl) *O*,*O*-di–ethyl phosphorothioate	35, 37
Carbophenothion	PS₃	*S*-{[(*p*-Chorophenyl)thio]methyl} *O*,*O*-diethyl phosphorodithioate	17, 18
Carbophenothion *O*-analog	PS₂	*S*-{[(*p*-Chlorophenyl) thio]methyl} *O*,*O*-diethyl phosphorothioate	17, 18
Chevron Ortho® 9006	PS	*O*,*S*-Dimethyl phosphoramidothioate	38, 39
Chipman RP-11783	PS₂	*O*,*O*-Dimethyl phosphorodithioate *S*-ester with 3-(mercaptomethyl)-2- benzoxazolinone	17, 18
Clorphoxim	PS	(*o*-Chlorophenyl)glyoxylonitrile oxime *O*,*O*-diethyl phosphorothioate	40
Chlorphoxim *O*-analog	P	(*o*-Chlorophenyl)glyoxylonitrile oxime diethyl phosphate	40
Chlorpyrifos	PS	*O*,*O*-Diethyl *O*-(3,5,6-trichloro-2-pyridyl) phosphorothioate	37, 41–51
Chlorpyrifos *O*-analog	P	Diethyl 3,5,6-trichloro-2-pyridyl phosphate	41, 43 48, 50–52
Chlorpyrifos-methyl	PS	*O*,*O*-Dimethyl *O*-(3,5,6-trichloro-2-pyridyl) phosphorothioate	53, 54
Chlorpyrifos-methyl *O*-analog (Fospirate)	P	Dimethyl 3,5,6-trichloro-2-pyridyl phosphate	53, 54

267

TABLE 7.1. (Continued)

Pesticide (or related compound)[a]	PS in Molecule	Chemical Name	References for Analytical Methods
Chlorpyrifos, Chlorpyrifos-methyl and their O-analogs (hydrolysis product)	—	3,5,6-trichloro-2-pyridinol	41, 51, 53, 54
Chlorthion®	PS	O-(3-Chloro-4-nitrophenyl)O,O-dimethyl phosphorothioate	17, 18
Ciba C-2307	P	Dimethyl phosphate ester with (E)-3-hydroxy-N-methoxy-N-methylcrotonamide	17, 18
Ciba C-8874	PS	O-(2,5-Dichloro-4-iodophenyl)O,O-diethyl phosphorothioate	17, 18
Ciba C-9491 (Iodofenphos)	PS	O-(2,5-Dichloro-4-iodophenyl)O,O-dimethyl phosphorothioate	55–57
Ciba C-9491 O-analog	P	2,5-Dichloro-4-iodophenyl dimethyl phosphate	55–57
Ciba C-8874, Ciba C-9491, and their O-analogs (hydrolysis product)	—	2,5-Dichloro-4-iodophenol	55–57
Ciodrin®, see Crotoxyphos			
Compound 4072	P	2-Chloro-1-(2,4-dichlorophenyl) vinyl diethyl phosphate	58–60
Compound 4072 (hydrolysis product)	—	2,2′,4′-Trichloroacetophenone	60
Coumaphos	PS	O,O-Diethyl O-(3-chloro-4-methyl-2-oxo-2H-1-benzopyran-7-yl) phosphorothioate	61–63
Coumaphos O-analog	P	Diethyl 3-chloro-4-methyl-2-oxo-2H-1-benzopyran-7-yl phosphate	61, 62
Crotoxyphos	P	α-Methylbenzyl (E)-3-hydroxycrotonate dimethyl phosphate	64–66
Crufomate	P	4-tert-Butyl-2-chlorophenyl methyl methylphosphoroamidate	67

268

Compound	Type	Chemical name	Ref
Dasanit® (Fensulfothion)	PS₂	O,O-Diethyl O-{p-(methylsulfinyl)phenyl} phosporothioate	68–70
Dasanit sulfone	PS₂	O,O-Diethyl O-{p-(methylsulfonyl)phenyl} phosphorothioate	68,70
Dansanit O-analog	PS	Diethyl p-(methylsulfinyl)phenyl phosphate	69, 70
Dansanit O-analog, sulfone	PS	Diethyl p-(methylsulfonyl)phenyl phosphate	69, 70
DEF®	PS₃	S,S,S-Tributyl phosphorotrithioate	17, 18
Demeton	PS₂	Mixture of O,O-diethyl S(and O)-{2-(ethylthio)ethyl} phosphorothioates	17, 18
Dialifor	PS₂	O,O-Diethyl phosphorodithioate S-ester with N-(2-chloro-1-mercaptoethyl) phthalimide	71
Diazinon	PS	O,O-Diethyl O-(2-isopropyl-6-methyl-4-pyrimidinyl) phosphorothioate	72–78
Diazoxon	P	Diethyl 2-isopropyl-6-methyl-4-pyrimidinyl phosphate	72, 74
Dicapthon	PS	O-(2-Chloro-4-nitrophenyl)O,O-dimethyl phosphorothioate	17, 18
Dichlorvos	P	2,2-Dichlorovinyl dimethyl phosphate	79–84
Dicrotophos	P	Dimethyl phosphate ester with (E)-3-hydroxy-N,N-dimethylcrotonamide	84–87
Dimethoate	PS₂	O,O-Dimethyl phosphorodithioate S-ester with 2-mercapto-N-methylacetamide	84, 88–94
Dimethoate O-analaog	PS	Dimethyl phosphorothioate S-ester with 2-mercapto-N-methylacetamide	88–91, 94
Dioxathion	P₂P₄	S,S'-p-Dioxane-2,3-diyl O,O,O',O'-tetraethyl bis(phosphorodithioate)	17, 18 95
Disulfoton	PS₃	O,O-Diethyl S-{2-(ethylthio)ethyl} phosphorodithioate	96–100
Disulfoton sulfoxide	PS₃	O,O-Diethyl S-{2-(ethylsulfinyl)ethyl} phosphorodithioate	98

TABLE 7.1. (Continued)

Pesticide (or related compound)[a]	PS in Molecule	Chemical Name	References for Analytical Methods
Disulfoton sulfone	PS₃	O,O-Diethyl S-{2-(ethylsulfonyl)ethyl} phosphorodithioate	98
Disulfoton O-analog	PS₂	Diethyl S-{2-(ethylthio)ethyl} phosphorothioate	98
Disulfoton O-analog Sulfoxide	PS₂	Diethyl S-{2-(ethylsulfinyl)ethyl} phosphorothioate	98
Disulfoton O-analog sulfone	PS₂	Diethyl S-{2-(ethylsulfonyl)ethyl} phosphorothioate	98
Dition®	PS	O,O-Diethyl O-(7,8,9,10-tetrahydro-6-oxo-6H-dibenzo[b,d]pyran-3-yl)phosphorothioate	17, 18
Dowco. 214, see Chlorpyrifos-methyl			
Dursban®, see Chlorpyrifos			
Dyfonate® (Fonofos)	PS₂	O-Ethyl S-phenyl ethylphosphonodithioate	101, 102
Dyfonate O-analog	PS	O-Ethyl S-phenyl ethylphosphonothioate	17, 18
EPN	PS	O-Ethyl O-(p-nitrophenyl) phenyl-phosphonothioate	17, 18
Ethion	P₂S₄	S,S-Methylene O,O,O',O'-tetraethyl phosphorodithioate	17, 18, 103
Etrimfos	PS	O,O-Dimethyl O-(6-ethoxy-2-ethyl-4-pyrimidinyl) phosphorothioate	104
Etrimfos O-analog	P	Dimethyl 6-ethoxy-2-ethyl-4-pyrimidinyl phosphate	104
Etrimfos and its O-analog (hydrolysis product)	—	6-Ethoxy-2-ethyl-4-hydroxypyrimidine	104
Famphur	PS₂	O-[p-(Dimethylsulfamoyl)phenyl] O,O-dimethyl phosphorothioate	17, 18, 105

270

Name	Chemical name		Reference
Fenitrothion	O,O-Dimethyl O-(4-nitro-m-tolyl) phosphorothioate	PS	106–108
Fenitrothion O-analog	Dimethyl 4-nitro-m-tolyl phosphate	P	108
Fenitrothion an its O-analog (hydrolysis product)	4-Nitro-m-cresol	—	108
Fenthion	O,O-Dimethyl O-{4-methylthio)-m-tolyl} phosphorothioate	PS_2	97, 109–111
Fenthion sulfoxide	O,O-Dimethyl O-{4-(methylsulfinyl)-m-tolyl} phosphorothioate	PS_2	109, 110
Fenthion sulfone	O,O-Dimethyl O-{4-(methylsulfonyl)-m-tolyl} phosphorothioate	PS_2	109, 110
Fenthion O-analog	Dimethyl 4-(methylthio)-m-tolyl phosphate	PS	109, 110
Fenthion O-analog sulfoxide	Dimethyl 4-(methysulfinyl)-m-tolyl phosphate	PS	109, 110
Fenthion O-analog sulfone	Dimethyl 4-(methylsulfonyl)-m-tolyl phosphate	PS	109, 110
Formothion	O,O-Dimethyl phosphorodithioate S-ester with N-formyl-2-mercapto-N-methylacetamide	PS_2	112
Gardona® (Stirofos)	2-Chloro-1-(2,4,5-trichlorophenyl)vinyl dimethyl phosphate	P	60, 113, 114
Gardona (hydrolysis product)	2,2′,4′,5′-Tetrachloroacetaphenone	—	60
Geigy GS-13005	S-{(2-Methoxy-5-oxo-Δ^2-1,3,4-thiadiazolin-4-yl)methyl}-O,O-dimethyl phosphoro-dithioate	PS_2	115, 116
Geigy G-28029	S-{{(2,5-Dichlorophenyl)thio}methyl} O,O-diethyl phosphorodithioate	PS_3	17, 18
Hempa	Hexamethylphosphoric triamide	P	17–19, 117
Imidan® (Phosmet)	O,O-Dimethyl phosphorodithioate S-ester with N-(mercaptomethyl)- phthalimide	PS_2	118–121
Imidoxon	Dimethyl phosphorothioate S-ester with N-(mercaptomethyl)phthalimide	PS	120

271

TABLE 7.1. (Continued)

Pesticide (or related compound)[a]	PS in Molecule	Chemical Name	References for Analytical Methods
Leptophos	PS	O-(4-Bromo-2,5-dichlorophenyl) O-methyl phenylphosphonothioate	122–127
Leptophos O-analog	P	4-Bromo-2,5-dichlorophenyl methyl phenylphosphonate	122–125
Leptophos and its O-analog (hydrolysis product)	—	4-Bromo-2,5-dichlorophenol	122–125
Malathion	PS₂	Diethyl mercaptosuccinate S-ester with O,O-dimethyl phosphorodithioate	78, 128–132
Malaoxon	PS	Diethyl mercaptosuccinate S-ester with O,O-dimethyl phosphorothioate	129, 131
Menazon	PS₂	S-{(4,6-Diamino-s-triazin-2-yl)methyl} O,O-dimethyl phosphorodithioate	17, 18
Merphos®	PS₃	S,S,S-Tributyl phosphorotrithioite	17, 18
Metepa	P	Tris(2-methyl-1-aziridinyl)phosphine oxide	17–19, 21
Methiotepa	PS	Tris(2-methyl-1-aziridinyl)phosphine sulfide	17–19, 21
Methyl Parathion	PS	O,O-Dimethyl O-(p-nitrophenyl)phosphorothioate	78, 133–135
Methyl Trithion®	PS₃	S-{[(p-Chlorophenyl)thio]methyl} O,O-dimethyl phosphorodithioate	17, 18
Mevinphos	P	Methyl (E)-3-hydroxycrotonate dimethyl phosphate	136, 137
Monitor®, see Chevron Ortho® 9006			
Monocrotophos	P	Dimethyl phosphate ester with (E)-3-hydroxy-N-methylcrotonamide	84–87, 138, 139

Common name	Type	Chemical name	References
Naled	P	1,2-Dibromo-2,2-dichloroethyl dimethyl phosphate	17, 18
Nemacur®, see Bay 68138			
Nemacide® (Dichlofenthion)	PS	O-(2,4-Dichlorophenyl) O,O-diethyl phosphorothioate	140
Oxydemetonmethyl sulfone	PS_2	S-{2-(Ethylsulfonyl)ethyl} O,O-dimethyl phosphorothioate	98
Parathion	PS	O,O-Diethyl O-(p-nitrophenyl) phosphorothioate	141–149
Paraoxon	P	Diethyl p-nitrophenyl phosphate	146
Phorate	PS_3	O,O-Diethyl S-{(ethylthio)methyl} phosphorodithioate	97, 150–152
Phorate sulfoxide	PS_3	O,O-Diethyl S-{(ethylsulfinyl)methyl} phosphorodithioate	151
Phorate sulfone	PS_3	O,O-Diethyl S-{(ethylsulfonyl)methyl} phosphorodithioate	151
Phorate O-analog	PS_2	O,O-Diethyl S-{(ethylthio)methyl} phosphorothioate	151
Phorate O-analog sulfoxide	PS_2	O,O-Diethyl S-{(ethylsulfinyl)methyl} phosphorothioate	151
Phorate O-analog sulfone	PS_2	O,O-Diethyl S-{(ethylsulfonyl)methyl} phosphorothioate	151
Phosalone	PS_2	O,O-Diethyl S-{(6-chloro-2-oxo-benz-oxazolin-3-yl)methyl} phosphorodithioate	153, 154
Phosfon®	P	Tributyl(2,4-dichlorobenzyl)phosphonium chloride	17, 18
Phosphamidon	P	Dimethyl phosphate ester with 2-chloro-N,N-diethyl-3-hydroxycrotonamide	84, 137, 155
Phoxim	PS	Phenylglyoxylonitrile oxime O,O-diethyl phosphorothioate	156, 157

273

TABLE 7.1. (Continued)

Pesticide (or related compound)[a]	PS in Molecule	Chemical Name	References for Analytical Methods
Phoxim O-analog	P	Phenylglyoxylonitrile oxime diethyl phosphate	156, 157
Pirazinon®	PS	O,O-Diethyl O-(6-methyl-2-propyl-4-pyridiminyl) phosphorothioate	17, 18
Potasan®	PS	7-Hydroxy-4-methylcoumarin O-ester with O,O-diethyl phosphorothioate	61
Ronnel	PS	O,O-Dimethyl O-(2,4,5-trichlorophenyl) phosphorothioate	17, 18, 158
Ruelene®, see Crufomate			
Schradan	P_2	Octamethylpyrophosphoramide	17, 18
Shell SD-8280	P	2-Chloro-1-(2,4-dichlorophenyl)vinyl dimethyl phosphate	17, 18, 159
Shell SD-8436	P	2-Chloro-1-(2,4,5-tribromophenyl)vinyl dimethyl phosphate	17, 18
Shell SD-8448	P	2-Chloro-1-(2,4,5-trichlorophenyl)vinyl diethyl phosphate	17, 18
Stauffer N-2788	PS_2	O-Ethyl S-p-tolyl ethylphosphonodithioate	17, 18
Stauffer R-3828	PS_2	S-(p-Chloro-α-phenylbenzyl)-O,O-dimethyl phosphorodithioate	160

274

Common name		Chemical name	Ref.
Tepp	P	Tetraethyl pyrophosphate	161
Tepa	P	Tris(1-aziridinyl)phosphine oxide	17–19, 21
Thiometon	PS$_3$	S-{2-(Ethylthio)ethyl} O,O-dimethyl phosphorodithioate	17, 18
Thiometon sulfoxide	PS$_3$	S-{2-(Ethylsulfinyl)ethyl} O,O-dimethyl phosphorodithioate	17, 18
Thiometon sulfone	PS$_3$	S-{2-(Ethylsulfonyl)ethyl} O,O-dimethyl phosphorodithioate	17, 18
Thiotepa	PS	Tris(1-aziridinyl)phosphine sulfide	17–19, 21
Trichlorfon	P	Dimethyl (2,2,2,-trichloro-1-hydroxyethyl) phosphonate	83, 111, 162
Union Carbide UC-8305	PS	2-Chlorohexahydro-4-methyl-4H-benzodioxaphosphorin 2-sulfide	17, 18
Velsicol VCS-506, see Leptophos			
Zinophos®	PS	O,O-Diethyl O-pyrazinyl phosphorothioate)	163
Zytron®	PS	O-(2,4-Dichlorophenyl) O-methyl isopropylphosphoramidothioate	17, 18

[a] Common names that have been recently approved are shown in parentheses.

Conditions used by this author in preparing a variety of foods for OP residue analysis are presented in Table 7.2 (164). The entire analytical process (preparation, extraction, and analysis) should be performed without delay. If this is not possible, information should be available concerning the stability of residues in storage. Generally, residues are more stable in an extract than in the unextracted commodity.

7.2.2. Extraction Efficiency

The "completeness of extracting pesticide residues from a substrate" (extraction efficiency) is an important procedural step that has been given only minimal attention until the past few years; even today, the concern of the pesticide analysts for this matter is probably not commensurate with its importance. Good recoveries obtained for substrates spiked in the laboratory indicate good analytical techniques but not necessarily a complete recovery of residues deposited via a biological process or even from biological surfaces after they are subjected to weathering and/or aging. The substitution of a radiolabeled pesticide in performance tests designed to simulate the intended use of the unlabeled compound is probably the best means of optimizing extraction conditions if metabolites can be identified and distinguished from the parent. However, in the absence of a radioisotope study, extraction trials with mechanical, reflux, and Soxhlet procedures employing a variety of solvents or mixtures possessing a range of polarity usually reveal a procedure that yields higher residue levels than the others. It is apparent that all of the cleanup and detection procedures at our disposal are useless if we fail to remove the residue from the sample.

Data from some of the earlier procedures (including some of my own) employing "surface stripping" with nonpolar solvents or blending with solvents of medium polarity are undoubtedly biased toward low residues. A good example of such data was reported in a study involving the field persistence of Ciba C-9491, its O-analog, and phenolic hydrolysis product in sweet corn (56). Identical corn samples were extracted by two methods: (1) equal weights of sample and anhydrous sodium sulfate were blended at high speed with benzene (3 ml/g of sample) in a Waring blender, and (2) the sample was Soxhlet extracted (under N_2) with chloroform–10% methanol (7.5 ml/g of sample) for 4 hr at the rate of about 6 solvent exchanges per hour. Residues of the three compounds recovered via Soxhlet extraction were 1.5–3.3 times higher than from the blend. Also, the differences between the extraction efficiencies of the two methods were proportional to the period of time the sample had weathered in the field. The failure to optimize extraction efficiency could therefore yield data that are erroneously low by a factor of 2 or more.

Wide differences in efficiencies of extracting residues of six OP insecticides and their metabolites from field-treated crops by using nine different extraction procedures were reported in 1968 (165) and the pitfalls of inadequate extraction procedures were discussed. Variables considered in that study were extraction solvent and method, time of extraction, crop extracted, temperature of extraction, type of compound extracted, effect of weathering, and formulation of insecticide. Recoveries were best when Soxhlet extraction with chloroform-10% methanol was continued until appreciable amounts of the residues were no longer removed.

7.2.3. Extraction Methods

Procedures for extracting residues of OP pesticides may vary widely depending on the nature of the substrate and the residues sought. Generally, the procedure of choice should satisfy as many of the following conditions as possible: (1) provides high efficiency of extracting the pesticide residues while removing minimal amounts of extraneous material, (2) maintains the chemical integrity of the residues (e.g., no changes due to oxidation, hydrolysis, conjugation, etc.), (3) holds losses of residues via volatilization and other means to a minimum, (4) provides a rapid extraction or one that operates essentially unattended, (5) employs solvents that minimize hazards of fire, explosion, and toxic effects and that are also easily removed from the residual extractives, and (6) employs apparatus that is relatively inexpensive and easily cleaned.

Various procedures for extracting several OP pesticides and their metabolites from field-weathered crops were evaluated by using blending, Soxhlet, Soxhlet at ambient temperature, reflux and blending followed by Soxhlet or reflux employing various solvents or mixtures (165). As previously stated, Soxhlet extraction with chloroform-10% MeOH for the period of time appropriate for the particular substrate and pesticide emerged as the method of choice. An extraction and partitioning scheme for determining OP residues in a variety of foods was reported by this author (164) and it will be presented under a separate heading.

Selection of the proper extraction procedure for a given commodity from the *PAM* is based on its fat and moisture content. Procedures for extracting nonfatty samples (< 10% fat) employ high-speed blending with acetonitrile and varying amounts of water for periods up to 5 min followed by filtration through coarse fluted paper. High moisture products (fruits and vegetables) are extracted by blending 100 g of sample with 200 ml of acetonitrile for 1 to 2 min. For dry products (hays, grains, feedstuffs) and products of intermediate water content (fish, shrimp, silages), a 20- to 50-g sample is blended with 350 ml of acetonitrile–35% water. In the extraction of dry products

TABLE 7.2. Conditions Used in Preparing Foods for GLC Analysis and Dry Matter and Fat Content of Foods (164)

Crop or Product (Food)	Sample Preparation for Extraction[a]	Extrac-tion—Parti-tioning Method	Gram Equiv. of Food/ Analysis	Total Wt of Extrac-tive (mg)	Final Vol. (ml)	Wt. Extrac-tive Injected in 5 μl (μg)	Dry Matter Content[b] (%)	Fat Content[c] (%)
Apples, Winesap	Washed, chopped	C-1	100	31.4	10	16	16.7	—
Bacon, sugar-cured	Chopped	B-2	12.5	89.9	5	90	75.4	63.0
Bananas	Peeled, chopped	C-1	100	33.5	10	17	22.3	—
Beans, lima, dry	Ground	A-1	50	66.7	10	33	87.7	—
Beans, snap, fresh	Washed, chopped	A-1	50	18.3	10	9.2	10.1	—
Brazil nuts	Shelled, chopped	B-2	12.5	471.3	5	471	96.2	70.2
Butter	—	D-2	12.5	402.6	5	403	83.5	82.2
Cabbage	Washed, chopped	A-1	50	7.6	10	3.8	7.45	—
Carrots	Washed and brushed, chopped	A-1	50	10.0	10	5.0	10.9	—
Cauliflower	Washed, chopped	A-1	50	13.2	10	6.6	6.28	—
Celery	Washed, chopped	C-1	100	47.7	10	24	3.67	—
Cheese, cheddar	—	C-2	12.5	111	5	111	42.4	9.62
Chicken, fresh, whole	Dressed chicken chopped whole	A-1	50	9.4	10	4.7	33.3	—
Corn silage	Chopped	A-1	50	262.3	10	131	26.4	—
Cucumbers, fresh	Washed, chopped	C-1	100	36.4	10	18	4.40	—
Flour, white enriched	—	A-1	50	59.5	10	30	89.1	—
Grapes, fresh (purple)	Washed, chopped	C-1	100	19.4	10	9.7	12.1	—
Lard	—	D-2	12.5	120.5	5	120	100	100

278

Lemons	Washed, chopped	C-1	100	92.8	10	46	16.1	—
Lettuce	Washed, chopped	C-1	100	24.3	10	12	4.47	—
Margarine	—	D-2	12.5	82.8	5	83	85.1	82.0
Melon, honeydew	Washed, peeled, chopped (seed and peel discarded)	C-1	100	8.6	10	4.3	5.82	—
Milk, fresh, whole	—	C-1	100	27.4	10	14	15.1	5.72
Oil, cooking	—	D-2	12.5	196.2	5	196	100	100
Onions, cured	Peeled, chopped	C-1	100	45.1	10	23	8.9	—
Oranges	Washed, chopped	C-1	100	73.3	10	37	17.4	—
Peanuts	Shelled, chopped	B-2	12.5	92.5	5	92	93.5	47.8
Peas, blackeyed, dry	Ground	A-1	50	58.5	10	29	88.6	—
Peas, English, fresh	Shelled, chopped	A-1	50	30.7	10	15	22.6	—
Pineapple, fresh	Peeled, cored, chopped (peeling and core discarded)	C-1	100	24.2	10	12	11.8	—
Potatoes, Irish, cured	Washed and brushed, chopped	C-1	100	35.3	10	18	20.9	—
Rutabagas, waxed	Peeled, chopped	A-1	50	36.0	10	18	13.5	—
Spinach, fresh	Washed, chopped	A-1	50	39.1	10	20	7.38	—
Steak, sirloin	Chopped (including bone)	B-2	12.5	31.0	5	31	41.1	14.7
Strawberries, fresh	Washed, chopped	C-1	100	48.4	10	24	5.37	—
Turnip greens, fresh	Washed, chopped	A-1	50	19.8	10	9.9	6.33	—
Turnip roots, fresh	Washed and brushed, chopped	A-1	50	4.3	10	2.2	6.04	—
Yams, cured	Washed and brushed, chopped	A-1	50	78.5	10	39	28.0	—

[a] Chopping was done in a Hobart cutter and grinding was done mechanically until samples were 20 mesh or smaller.

[b] Material not volatile at 110°C overnight.

[c] Material dissolved during extraction and not volatile at 110°C overnight.

where hydration is required prior to extraction, 80 ml of water is mixed with 20- to 50-g of sample in a blender, allowed to stand for 15 min, then blended with 200 ml of acetonitrile for 5 min. *PAM* procedures for fatty samples vary widely depending on the nature of the sample. Generally, the fat is isolated from the product and 3 g dissolved in petroleum ether; the residues are then partitioned from the petroleum ether into acetonitrile. Animal tissues are blended with anhydrous sodium sulfate to absorb the water present then are extracted three times with petroleum ether in a centrifuge bottle. Butter is simply warmed to 50°C until the fat separates, then about 3 g is taken for the fat cleanup. Cheese is blended with sodium or potassium oxalate and methyl or ethyl alcohol for 2 or 3 min then extracted three times in a centrifuge bottle with mixed ethers consisting of equal volumes of petroleum ether and diethyl ether; milk is extracted in centrifuge bottles in a similar manner. Samples slated for cleanup via sweep codistillation are usually extracted by a single blend with ethyl acetate. Direct reference to *PAM* should be made for details pertaining to the extraction of specific commodities and to which OP pesticides are recovered through the analytical scheme.

Another extraction procedure that should be mentioned was reported by Luke et al. (167). The samples are extracted with acetone and partitioned with dichloromethane–petroleum ether to remove water. Although the procedure is essentially identical to many others already described in the literature (69, 97, 108, 125, 164, 166) for extracting residues of OP compounds and carbamates from milk and fruits, the authors have expanded it to include four classes of pesticide residues in a variety of substrates. The fact that residues were quantitatively recovered for 15 organophosphorus, 9 organochlorine, 5 organonitrogen, and 2 hydrocarbon compounds indicates that the procedure may be valuable for use in multiresidue procedures for pesticides having widely differing polarities.

The analyst should be aware of the fact that many of the OP pesticides and especially their metabolites are quite polar and that their residues may be associated with both the fatty and aqueous portions of a substrate. For example, good recoveries of the nonpolar chlorinated pesticides from milk are achieved only by rupturing the fat globules (addition of alcohol), which allows complete extraction of the fat. On the other hand, better recoveries of the highly polar OP pesticides are obtained when no alcohol is added and the fat recovery is incomplete (ca. 30%) (121, 168).

The extraction of conjugates and breakdown products of OP pesticides involves complex procedures that are beyond the scope of this chapter.

7.2.4. Solvent Partitioning and the Use of p Values

In the development of analytical methods for new OP pesticides, it has been the practice for this author to routinely determine the p values of the compound in a variety of solvent systems immediately after the solubilities of the compound have been approximated and the appropriate detection system selected. The p value (partition value) which is defined as the fractional amount of solute partitioning into the nonpolar phase of an equivolume immiscible binary solvent system is very useful in developing extraction and partition cleanup procedures as well as confirming identities of GC peaks (169–175).

The p value is easily determined as follows: A 5-ml aliquot of the nonpolar phase (preequilibrated with the polar phase) containing the compound of interest is analyzed by GC or any other means. To a second 5-ml aliquot in a graduated glass-stoppered 10-ml centrifuge tube an equal volume of polar phase (preequilibrated with the nonpolar phase) is added; the tube is shaken for about one minute and the nonpolar phase is analyzed exactly as before. The ratio of the second analysis to the first is the p value—that is, the amount of compound in the nonpolar phase (second analysis) divided by the total amount (first analysis). Some of the attractive features of the p value technique are (1) neither the identity nor the actual amount of pesticide are required to determine p values, merely the analytical responses before and after distribution between the phases, (2) the integrity of the p value is maintained even in the presence of relatively large amounts of extractives or other pesticides, and (3) values may be determined with subnanogram amounts by using smaller volumes of phases and sensitive GC detectors. Although, by definition, the p values are based on equivolume equilibrated phases, they may also be determined in unequal volumes of unequilibrated phases (173).

One of the most widely used binary solvent systems is hexane–acetonitrile; p values for 59 OP pesticides and related compounds in this system are presented in Table 7.3. These values provide valuable information concerning the relative polarities of the pesticides and the number of extractions of one phase by the other that is required to remove a known percentage of the residue. A knowledge of the polarity of the pesticide based on its p-value also aids in the selection of an appropriate solvent for extracting residues from the samples. The p values listed in Table 7.3 illustrate the fact that polarities of OP pesticides vary widely; the least polar compound listed has a value of 0.54, while the most polar ones are <0.01. Many of the compounds contain either a P=S or P=O (O analog) moiety; the O analogs as well as the hydrolysis products are more polar (lower p value) than the parent pesticide.

Some OP pesticides contain a thioether linkage that can be oxidized from the sulfide to the sulfoxide and further to the sulfone. The sulfoxides are more polar than the sulfones which are more polar than the sulfides. The p values of several thioether compounds and their oxidation products in seven solvent systems are presented in Table 7.4. In cases where the solubility of

TABLE 7.3. *p* **Values of 59 Organophosphorus Pesticides and Related Compounds in Hexane—Acetonitrile at 25°C**

Apholate	0.031	Etrimfos (hyd. prod.)	0.014
Azinphosmethyl	0.008	Fenitrothion	0.032
Bay 30911	0.23	Fenitrothion *O*-analog	0.011
Bay 37289	0.54	Fenitrothion (hyd. prod.)	0.012
Bay 68138	0.038	Gardona®	0.051
Bay 68138 Sulfoxide	0.004	Gardona (hyd. prod.)	0.18
Bay 68138 Sulfone	0.004	Geigy G-28029	0.29
Bay 93820	0.016	Hempa	0.068
Bay 93820 *O*-analog	0.041	Imidan®	0.007
Carbophenothion	0.21	Leptophos	0.35
Chipman RP-11783	0.019	Leptophos *O*-analog	0.10
Chlorphoxim	0.047	Leptophos (hyd. prod.)	0.10
Chlorphoxim *O*-analog	0.011	Malathion	0.042
Chlorpyrifos	0.28	Metepa	0.035
Chlorthion®	0.026	Methiotepa	0.14
Ciba C-9491	0.26	Methyl Parathion	0.019
Compound 4072	0.058	Methyl Paraoxon	0.010
Compound 4072 (hyd. prod.)	0.12	Methyl Trithion	0.075
Coumaphos	0.006	Naled	0.043
Crufomate	0.031	Parathion	0.048
Diazinon	0.28	Paraoxon	0.010
Dicapthon	0.031	Phorate	0.21
Dichlorvos	0.043	Phoxim	0.056
Dicrotophos	0.008	Phoxim *O*-analog	0.012
Disulfoton	0.20	Stauffer N-2788	0.22
Dyfonate	0.21	Tepa	0.012
EPN	0.038	Thiotepa	0.063
Ethion	0.079	Zinophos®	0.058
Etrimfos	0.200	Zytron®	0.12
Etrimfos *O*-analog	0.048		

the compound in hexane is not the limiting factor, the *p* value increases with the addition of water to the acetonitrile phase. This method of forcing pesticide residues into the nonpolar phase is commonly used. In *PAM*, the acetonitrile phase is adjusted to contain about 90% water then extracted with petroleum ether to recover the residue, thus the problems associated with the evaporation of acetonitrile are eliminated and some of the polar interfering sample extractives are allowed to remain in the aqueous phase. It should be noted, however, that the *p* values of some of the highly polar compounds (Table 7.4) in hexane–water are very low and minimal recoveries would be expected from the method just described. The *p* values in benzene–water are much higher, therefore benzene is preferred for extracting these compounds. However, a few of the compounds are also too polar to extract with benzene and alternate methods such as the use of more polar solvents (e.g., chloroform or dichloromethane), evaporation of the acetonitrile, or direct analysis of the acetonitrile phase are required.

The p-value technique has been used with more than twenty binary solvent systems (171, 174). Some of the systems were hexane–acetonitrile, –dimethyl sulfoxide, –oxydipropionitrile, –90% acetic acid; cyclohexane–methanol; heptane–90% ethanol; isooctane–dimethylformamide, and isooctane–80% acetone. Extensive research on p values led to the development of apparatus and procedures for the rapid extraction and identification of pesticides by single and multiple distribution in binary solvent systems (172).

Since solvent partitioning is a simple, rapid, and effective means of separating organic compounds, it has played an important role in the development of most analytical procedures for pesticide residues. It usually appears as a procedural step immediately following the extraction of the substrate. Through the use of hexane–acetonitrile, large amounts of nonpolar compounds (e.g., fats and oils) are separated from the pesticide residues in extraction and discarded with the hexane phase (s) thus achieving a good cleanup from the standpoint of nonpolar interferences. Unfortunately, comparable progress with partitioning systems for the removal of polar interferences has not been recorded and other methods such as adsorption chromatography have been required to supplement the partitioning procedure.

7.2.5. An Extraction and Partitioning Procedure for a Variety of Food Types

A general procedure for preparing a variety of foods for multicomponent analysis of OP pesticides via GC with flame photometric detection (FPD) was reported by this author (164). In Table 7.3, extraction and partitioning methods for 38 foods are designated by extraction methods A, B, C, or D and partitioning methods 1 or 2; these methods are described as follows.

7.2.5.1. Extraction Methods

A. Fifty grams of the comminuted sample is placed in a Soxhlet extraction apparatus containing a plug of glass wool to prevent insoluble matter from siphoning over and extracted under a slow flow of nitrogen with 150 ml of chloroform–10% methanol for 4 hr at a rate of about 6 solvent exchanges per hour. If an aqueous layer is present in the extraction flask it

TABLE 7.4. p Values of OP Pesticides and Metabolites Containing a Thioether Linkage in Seven Solvent Systems

Compound	Hexane–Water	Hexane–20% MeCN (80% Water)	Hexane–40% MeCN (60% Water)	Hexane–60% MeCN (40% Water)	Hexane–80% MeCN (20% Water)	Hexane–MeCN	Benzene–Water
Bay 68138	0.94	0.65	0.20	0.051	0.023	0.038	0.98
Bay 68138 sulfoxide	0.002	0.002	0.002	0.002	0.002	0.004	0.52
Bay 68138 sulfone	0.009	0.002	0.002	0.002	0.002	0.004	0.93
Dasanit	0.75	0.34	0.056	0.012	0.006	—	1.00
Dasanit sulfone	0.98	0.72	0.18	0.027	0.008	—	1.00
Dasanit O-analog	0.002	0.001	0.001	<0.001	<0.001	—	0.62
Dasanit O-analog sulfone	0.021	0.018	0.009	0.003	0.002	—	0.96
Disulfoton	1.00	1.00	1.00	0.81	0.52	0.20	—
Disulfoton sulfoxide	0.50	0.23	0.06	0.01	0.00	0.00	—
Disulfoton sulfone	0.82	0.52	0.13	0.01	0.00	0.00	—
Disulfoton O-analog	0.83	0.75	0.39	0.11	0.05	0.02	—
Disulfoton O-analog sulfoxide	0.00	0.00	—	—	—	—	0.18

Compound							
Disulfoton O-analog sulfone	0.01	0.00	0.00	—	—	—	0.78
Fenthion	1.00	0.98	0.92	—	—	—	—
Fenthion sulfoxide	0.50	0.18	0.03	—	—	—	—
Fenthion sulfone	0.94	0.61	0.12	—	—	—	—
Fenthion O-analog	0.92	0.65	0.18	—	—	—	0.35
Fenthion O-analog sulfoxide	0.00	0.00	0.00	—	—	—	
Fenthion O-analog sulfone	0.01	0.00	0.00	—	—	—	1.00
Oxydemetonmethyl	0.00	0.00	—	—	—	—	0.03
Oxydemetonmethyl sulfone	0.00	0.00	—	—	—	—	0.10
Phorate	1.00	1.00	1.00	0.81	0.54	0.21	—
Phorate sulfoxide	0.54	0.35	0.10	0.02	0.01	0.00	—
Phorate sulfone	0.98	0.79	0.30	0.05	0.02	0.01	—
Phorate O-analog	0.89	0.73	0.36	0.11	0.05	0.02	0.21
Phorate O-analog sulfoxide	0.00	0.00	0.00	—	—	—	—
Phorate O-analog sulfone	0.01	0.00	0.00	—	—	—	0.78

is separated and the organic layer is percolated through a plug of sodium sulfate (ca. 25 mm diam. × 30 mm thick). Any aqueous layer is extracted twice with 25-ml portions of chloroform which are successively percolated through the plug. Otherwise, the plug is washed twice with 25-ml portions of chloroform. The combined extracts and washings are then evaporated to dryness on a 60°C water bath under water pump vacuum and reserved for the appropriate partitioning step.

B.   The extraction is performed as described in method *A* except that the water bath is set at 100°C to remove last traces of solvent from the fatty residue. The residue is then dissolved in hexane, diluted to 200 ml, and reserved for the partitioning step.

C.   The sample (100 g) is blended at high speed in a Waring Blendor for 3 min with 200 ml of acetone; the slurry is then filtered by suction through Whatman No. 40 paper on a Buchner funnel. The blender and the filter are washed twice with 25-ml portions of acetone, and the combined acetone filtrates are extracted with 250 ml of dichloromethane and the organic (lower) layer percolated through a plug of sodium sulfate (ca. 3 cm diam. × 15 cm thick). The aqueous layer is again extracted with 100 ml of dichloromethane, which is also percolated through the plug. The combined dichloromethane extracts are then evaporated to near dryness on a steam bath under a 3-ball Snyder column and finally to dryness with a jet of dry nitrogen. The residue is reserved for the partitioning step. (Exception: Cheese residue is diluted to 400 ml with hexane for the partitioning step.)

D.   The sample (100 g) is dissolved in 150 ml of hexane (warming if necessary), filtered through a plug of sodium sulfate (ca. 25 mm diam. × 50 mm thick), the plug washed twice with 25-ml portions of hexane then the filtrate diluted to 400 ml with hexane (Exception: The butter solution in 150 ml of hexane is passed through a plug of sodium sulfate (ca. 3 cm diam. × 15 cm thick) then filtered through Whatman No. 40 paper on a Buchner funnel. The plug and filter are washed twice with 25-ml portions of hexane and the combined filtrate adjusted to 400 ml). The hexane solutions are reserved for the partitioning step.

7.2.5.2.   *Partitioning Methods*

1.   The residue is transferred to a 15-ml glass-stoppered separatory funnel by using 5 ml of preequilibrated hexane and acetonitrile. This mixture is shaken for 1 min, the acetonitrile layer withdrawn, and the hexane phase extracted again with 5 ml of the acetonitrile. The hexane layer is discarded and the combined acetonitrile extracts reserved for direct injection into the FPD-GC.

2.   50-ml portions of the hexane solution are extracted twice with 50-ml portions of acetonitrile and the hexane layer discarded. The combined ace-

tonitrile extracts are mixed in a separatory funnel with 1 liter of 20% aqueous sodium chloride then extracted twice with 50-ml portions of benzene which were successively passed through a plug of sodium sulfate (ca. 25 mm diam. × 50 mm thick). The combined extracts are evaporated to near dryness on a 60°C water bath under water pump vacuum and the volume adjusted to 5 ml with benzene for injection into the FPD-GC.

The specificity and sensitivity of FPD-GC operated either in the phosphorus or sulfur mode permits the analysis of most OP pesticides in these extracts with only the solvent partitioning cleanup. Details concerning the GC analysis will be discussed later. Relative efficiencies of the partitioning cleanup can be obtained by comparing the gram equivalents of food per analysis (original weight of sample) with the total weight of extractives (weight after extraction and partitioning) in Table 7.2. The most efficient was fresh turnip roots with only 8.6 mg extractives/100 g sample and the least efficient was Brazil nuts with 3.77 g/100 g sample.

7.2.6. Column Cleanup Procedures

7.2.6.1. Florisil

Cleanup on Florisil as described in the *PAM* (8) is performed on a glass column (22 mm i.d. × 300 mm long) containing about 10 cm of activated Florisil (after settling) topped with about 1 cm of sodium sulfate and prewet with about 50 ml of petroleum ether. The petroleum ether extract, after the acetonitrile partitioning step, containing the pesticide residues is percolated through the column at a rate not to exceed 5 ml/min and the container and column are rinsed twice with 5-ml portions of petroleum ether. The column is then sequentially eluted with 200-ml portions of 6, 15, and 50% diethyl ether in petroleum ether. Carbophenothion, ethion, and ronnel are found in the 6% fraction, diazinon, parathion, and methyl parathion in the 15% eluate, and malathion is eluted with the 50% mixture.

The behavior of 60 OP compounds through the Florisil cleanup procedure was evaluated by Pardue (176) in the absence of sample extractives. Recoveries of 14 of the compounds in the various eluates were 80 to 133% while 9 others were partially recovered (33–76%); the remaining 37 compounds, many of them highly polar, were not recovered.

7.2.6.2. Charcoal

Storherr and coworkers (177) demonstrated that losses of OP pesticides occurred both in the extraction of residues from aqueous acetonitrile with petroleum ether and on the Florisil column; the following procedure was then devised to reduce these losses. One third of the aqeous acetonitrile phase (after hexane–acetonitrile partitioning) was extracted three times with dichloromethane, and the extracts were successively percolated through a column containing a mixture of acid-treated charcoal, hydrated SeaSorb 43, and Celite 545 (1:2:4). The pesticides were then eluted with a mixture containing equal volumes of benzene and acetonitrile. Kale spiked at various levels (0.01–0.80 ppm) with 41 OP pesticides, and their analogs yielded 85–104% recoveries. Nine of the compounds were also added to apples, lettuce, carrots, strawberries, and green beans, and satisfactory recoveries were obtained at levels from 0.02 to 0.40 ppm.

Recoveries of OP pesticides spiked into apple and lettuce extracts were compared by using various cleanup procedures employing charcoal, Florisil, attaclay, silica gel, and so on in combination with a variety of solvents and/or mixtures (178).

7.2.6.3. Silica Gel

Columns of silica gel (partially deactivated in various ways) have been employed by this author for cleaning up sample extracts and separating residues of many OP pesticides and their analogs for analysis via FPD-GC (29, 40, 55, 61, 69, 83, 98, 108, 109, 125, 151, 157). The columns (ca. 1 cm diam.) usually consisted of about 5 g of the silica gel deactivated with 3–15% water. In some cases deactivation of the adsorbent with buffer (pH 7) was required to prevent degradation of the residues on the column. A benzene solution of the sample extractive is added to the column (prewet with benzene) and the column eluted with portions of benzene containing increasing amounts of acetone (e.g., 1, 2, 5, 10, 20, 50, 75, and 100%). Under these conditions the less polar OP pesticides (e.g., parent compounds) are eluted first, followed by the analogs in an order of increasing polarity. Alumina columns have also been used in tandem with silica gel to separate the phenolic hydrolysis products of certain OP pesticides (55, 125). In these procedures the alumina column is placed below and in tandem with a silica gel column; the parent pesticide passes through both columns with benzene while the O-analog is retained by the silica gel and the phenol by the alumina. The columns are then separated and the O-analog and phenol eluted with acetone and benzene–2% acetic acid, respectively. These procedures will be illustrated in the discussion of metabolite analysis.

7.2.6.4. Gel Permeation

Cleanup of sample extracts via gel permeation chromatography (GPC) is accomplished by separating the constituents on the basis of their molecular size. Most pesticides are of low molecular weights (e.g., 400 or less), while the bulk of the interfering extractives have high molecular weights of 500 or more. The constituents of high molecular size are eluted from the column first followed by those of decreasing size; the interfering substrates would therefore be eluted and discarded prior to the emergence of pesticide residues.

Pflugmacher and Ebing (179) described a combination partitioning–GPC cleanup procedure for 22 OP pesticides in 12 vegetable extracts spiked at levels of 0.05–0.50 ppm; recoveries were 68–98%. Samples were blended with acetone and a portion of the extract diluted with water and extracted with dichloromethane. The extract was concentrated to 1 or 2 ml, diluted with 5 ml of ethanol, then reduced to a 1-ml volume. The concentrated extract was then added to a column consisting of 118 g of Sephadex LH-20 and the pesticides eluted with ethanol at a flow rate of 45 ml/hr. Elution of the various OP pesticides from the column occurred between 305 and 428 ml.

Stalling et al. described a GPC cleanup system for the analysis of chlorinated pesticide and polychlorinated biphenyl (PCB) residues in fish extracts (180), and later Tindle and Stalling developed the system into an automated cleanup apparatus which appears promising for cleaning up fatty samples for pesticide residue analysis (181).

Griffitt and Craun (182) evaluated the automated GPC apparatus of Tindle and Stalling by using about 30 pesticides including 8 OP compounds in milk fat to compare recoveries via GPC before and after Florisil with those from the partitioning and Florisil procedure in the PAM; elution patterns and cleanup efficiencies were also determined. The GPC system utilized Bio Beads SX-2 with cyclohexane as a solvent. Cleanup efficiency was better than that obtained with acetonitrile partitioning, and recoveries generally were better than those obtained with acetonitrile partitioning followed by Florisil cleanup. Currently, Bio Beads SX-3 with toluene–ethyl acetate or dichloromethane are also being used with the GPC system (183).

An automated GPC system employing Bio Beads SX-3 gel and a toluene–ethyl acetate $(1 + 3)$ elution solvent was employed by Johnson et al. (184) to clean up residues of disulfoton, diazinon, methyl parathion, malathion, parathion, dichlorvos, and dasanit in chicken and turkey fat.

7.2.7. Other Cleanup Procedures

7.2.7.1. Sweep Codistillation

The apparatus of Storherr and Watts (185, 186) is approved in the PAM for the cleanup of parent OP residues of carbophenothion, diazinon, ethion, parathion, methyl parathion and malathion in extracts of kale, endive, carrots, lettuce, potatoes, apples, and strawberries. The sample (25 g) is blended with 125 ml of ethyl acetate and 25 g of sodium sulfate for 5 min and a 10-g equivalent of the extract is concentrated to 5 ml. One ml (2 g equiv.) of the concentrate is cleaned up by injecting it into a heated (ca. 185°C) glass column (24 cm long) packed with glass wool, followed by injections of ethyl acetate (0.5 ml) every 3 min until a total of 8 injections are made. A nitrogen carrier gas (ca. 600 ml/min) sweeps the volatile components through the packed tube to a condensing bath and through an Anakrom scrubber tube to a collection tube. The collected sample is analyzed by GC employing a FPD or alkali flame ionization detector (AFID). The cleanup is not adequate for analysis by GC with the electron capture detector (ECD). In a collaborative evaluation of the sweep codistillation cleanup with 6 OP pesticides in 6 crops, recoveries at the 0.5 and 1.0 ppm levels were better than 90% (187).

7.2.7.2. Low-Temperature Precipitation of Lipids

Procedures for cleaning up tissue extracts by freezing out the lipid interferences were reported by Grussendorf et al. (188) and McLeod and Wales (189). In the latter method the acetone-5% benzene extract of the tissue is held at −78°C for 30 min to precipitate the lipids; the suspension is then filtered through cellulose and the extract assayed by ECD- or FPD-GC. The method has been applied to the analysis of malathion, parathion, fenitrothion, their O-analogs, phosphamidon, and other classes of pesticides. Recoveries from carrots and wheat were 98 ± 10%.

7.2.7.3. Channel Chromatography

Matherne and Bathalter (190) reported a method employing a square glass plate containing recessed parallel channels filled with aluminum oxide G. The plates were developed with acetonitrile:tetrahydrofuran (1:1, v/v) followed by acetonitrile. Plant extractives remained near the origin, while the pesticide residues moved with the solvent front. However, recoveries of OP pesticides were low. Later, Hetherington and Parouchais (191) modified the procedure to obtain good recoveries of malathion, parathion, diazinon, carbophenothion, and azinphosmethyl from grain, forage, and poultry feed.

Determinative procedures such as TLC, GC, and high-pressure liquid chromatography (HPLC) have also been employed to resolve special cleanup problems that might have otherwise been insurmountable via conventional cleanup methods.

7.3. MEASUREMENT OF ORGANOPHOSPHORUS PESTICIDE RESIDUES

7.3.1. Gas Chromatography

GC is unquestionably the most widely used determinative procedure for the analysis of OP pesticides and many of their metabolites. The procedure is rapid, provides good resolution for determining multicomponent residues, and through the use of highly sensitive and specific detectors, trace level residues are quantified with a high degree of precision and accuracy.

7.3.1.1. Detectors

The two most widely used detectors for GC analysis of OP pesticides are the alkali flame ionization detector (AFID) and the flame photometric detector (FPD). The AFID is based on the phenomenon that the flame ionization detector (FID) yields enhanced response to heteroatoms in the presence of alkali metal salts (192). Many arrangements for applying salts to the flame have been reported and the most common salts used for OP compounds are K_2SO_4, KCl, and KBr (4). Generally the response of the AFID is proportional to the amount of heteroatom introduced and typical OP compounds can be detected in the lower picogram range with discrimination against other compounds by three to five decades (193). However, the sensitivity of the detector varies widely with the nature of the alkali salt, the gas flow rate and the positioning of the electrodes (4). Although the detector is somewhat temperamental and must be carefully optimized by the analyst, it is well accepted and widely used for OP pesticide assays.

Another AFID system currently employed for the analysis of pesticide residues is the nitrogen/phosphorus (N/P) detector. The commercial systems usually employ an electrically heated bead containing a rubidium salt; flow rates of gases are low and usually critical. For example, under conditions that detect both nitrogen and phosphorus compounds, hydrogen flow rates are precisely controlled at rates of 1 to 3 ml/min. Confirmation of whether a GC response is due to phosphorus, nitrogen, or both is usually accomplished by (1) increasing the hydrogen flow (flame temperature) and reversing the polarity of the electrical potential applied to the jet which renders the detector sensitive only to phosphorus (194) or (2) by lowering the applied polarizing potential, which reduces the phosphorus response by 30% while maintaining normal response for nitrogen (195). When compared with a conventional flame ionization detector, the N/P system is about 50 and 500 times more sensitive to nitrogen and phosphorus compounds, respectively.

The FPD (196) operates with a cool, hydrogen-rich flame for the detection of P- and S-containing compounds whose product species POH and S_2 emit band spectra that are monitored by using interference filters and a photomultiplier. Phosphorus compounds (526-nm filter) respond linearly from about 0.1 ng amounts over about 4 decades, while the response from sulfur compounds (394-nm filter) is roughly a square function of the concentration, unless SO_2 makeup gas is used. Most of the research on OP pesticides reported by this author was performed with the FPD. The detector is easy to operate, reproducible from day to day, relatively insensitive to moderate changes in operating conditions and easily temperature-programmed; however, it is somewhat less sensitive than the AFID. The detector is highly specific with the response (peak height) of 100 ng of parathion being 132,500 and 48,750 times greater than those of equal amounts of aldrin and lindane in the P mode, respectively. Under the same conditions, the response of parathion in the S mode was 10,750 times more than aldrin (197). Another attractive feature of the detector is that any solvent may be used for injecting the extract into the GC (e.g., hexane, benzene, MeCN, MeOH or even water if the OP residue is stable under the GC conditions employed). Injections of 25 μl or more of concentrated extracts into the GC are often made in the S mode of operation to compensate for the lower sensitivity of sulfur. In the earlier versions of the detector each injection of the sample to be chromatographed extinguished the flame and reignition was required. Many analysts regarded this as a disadvantage and devised methods to divert the injection solvent from the detector. Moye (198) modified the burner tip to produce a nonextinguishing flame. Others have reversed the hydrogen and oxygen–air inlets to the detector which is said to prevent the flame from being extinguished. On the other hand, this analyst has not regarded the reignition requirement as a disadvantage but more of an asset, since benzene is commonly used as an injection solvent and a nonextinguishing flame would rapidly deposit a film of carbon on the detector window, which would decrease the sensitivity and reproducibility of the analysis. In certain analyses, however, a means of diverting the solvent from the detector would be advantageous. The operating temperature of the original FPD was limited to about 165°C; however, with the use of water-cooled adapters (15, 199), this analyst has routinely operated at temperatures as high as 280°C. The P mode of operation is the method of choice for analyzing residues of OP pesticides, however by also using the S mode, even with its nonlinearity and 10-fold lesser sensitivity, valuable information concerning residues of OP pesticides that contain both phosphorus and sulfur can be obtained. A dual FPD that simultaneously monitors GC effluents containing phosphorus on one channel and sulfur on the other was developed by Bowman and Beroza (200). The response ratio, defined as the phosphorus response divided by the square root of the sulfur response, provides a means of estimating the atomic ratio of P to S in a molecule. The use of this detector for confirmatory identification of OP residues will be discussed later.

The electron capture detector (ECD) is sometimes used for OP pesticides. Generally, it responds well to those compounds containing the P=S moiety, nitro groups, or halogens, and in the hands of an experienced analyst good results can be obtained with those compounds. Although the ECD is a rather poor choice for the general analysis of OP pesticides, because of its insensitive response to many of the compounds (especially those with a P=O moiety), the requirement for extensive cleanup, and its lack of specificity; it has been invaluable in the analysis of their hydrolysis products (53, 55–57, 108, 122–125). The ECD has been greatly improved recently; the use of ^{63}Ni or scandium tritide sources allows operation at temperatures in excess of 300°C and a linear dynamic range of five decades is obtained in the variable frequency mode by using analog conversion of the detector signal to a linear function of the sample concentration.

The microcoulometric titration cell was used by Burchfield et al. to measure OP pesticides after their reduction to phosphine in an open quartz tube (201). Hydrogen halides were subtracted from the reaction products on a short alumina column and H_2S and PH_3 were separated on a short gas–solid chromatographic (GSC) column packed with silica gel. Both sulfur and phosphorus were determined by microcoulometry. Berck et al. (202) determined the minimum amounts of phosphine detectable by three different detectors as follows: FPD, 5 pg; AFID, 20 pg; and microcoulometer, 5 ng.

7.3.1.2. Columns and Operating Parameters

Solid supports are usually of the Chromosorb W or Gas Chrom Q types (80–100 mesh) that are acid-washed and silanized prior to the application of the stationary liquid phase. The use of SE-30, DC-200, QF-1, and other silicones either alone or in admixture is common, and the total amount of liquid phase is generally in the range of 3–10% by weight. Compounds of the diethyleneglycol succinate (DEGS) type are sometimes used in instances where a more polar liquid phase is required to achieve separations not possible with the silicones. However, some of the polar OP pesticides do not elute, and these phases also have lower temperature limitations than the silicones. The analyst is also cautioned against the use of DEGS-type phases stabilized with phosphoric acid in temperature-programmed analyses for traces of OP pesticides; the reasons are obvious.

This author chose to standardize GC analysis of OP pesticides on the newer liquid phases which are listed in order of increasing polarity as follows: Dexsil 300 (polycarboranesiloxane), OV-101 (dimethyl silicone), OV-17 (phenylmethyl silicone, 50% phenyl), OV-210 (trifluoropropylmethyl silicone, 50% trifluoropropyl), and OV-225 (cyanopropylphenylmethyl silicone, 25% cyanopropyl, 25% phenyl). The use of AN-600 (also known as XF-1150, a cyanoethylmethyl silicone, 50% cyanoethyl) has also been very satisfactory where a more polar silicone phase was required (83, 203). The retention times (t_R) of 108 OP pesticides and related compounds relative to parathion with five of these phases on 240 cm glass columns (4 mm, i.d.) temperature-programmed from 150 to 300°C at the rate of 10°C/min then held at 300°C until the last peak emerged are presented in Table 7.5 (17, 18). These data provide valuable information concerning GC separation of multicomponent residues as well as a reference for the identification of unknown GC peaks obtained under similar analytical conditions. All of the GC column packings are 5% (w/w) on Gas Chrom Q (80–100 mesh) except Dexsil 300 which is 5% on specially washed Chromosorb W (80–100 mesh).

The primary advantage of using Dexsil 300 is its exceptionally high thermal stability (204). It was found that interfering materials that accumulated on the column could be removed by purging at 400°C without appreciable loss of the liquid phase. One exception to the use of a silanized support for pesticide analysis is that of Dexsil 300. A column of 5% Dexsil 300 on Gas Chrom Q was tried by this analyst but chromatograms of OP pesticides (Fig. 7.1) contained peaks that were lower in response and broader (decreased resolution) than those obtained with 5% OV-101 on the same support (17). This probably stemmed from uneven wetting and spreading of the phase on the silanized support. A column of Dexsil 300 on specially washed unsilanized Chromosorb W (Fig. 7.2) gave results that were comparable to those with 5% OV-101 on Gas Chrom Q. Despite the use of 4 and 5 times more pesticides (P and S modes, respectively) with the Dexsil-silanized packing than with the Dexsil-unsilanized packing, responses were still better with the latter preparation (compare Fig. 7.1 with Fig. 7.2).

Temperature-programmed operation of the GC is highly desirable for multiresidue determinations of OP pesticides, since compounds with a wide range of retention times can be analyzed with a single injection of sample. Isothermal operation is used for a specific compound or a mixture having similar t_R values. Before the GC column is connected to the detector it is usually conditioned in the instrument overnight at temperatures 20–50°C higher than the maximum operating conditions for the assays. On-column injection into the packed portion of the GC column generally gives the best results. Injection port and detector temperatures are usually about 10–30°C higher than the column oven; any transfer lines connecting the column oven with the detector are set at a temperature intermediate to those of the

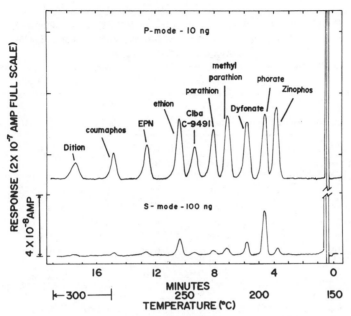

Figure 7.1 Chromatograms of pesticide standards in the P and S modes with 5% Dexsil on Gas Chrom Q (18).

column and the detector. Flow rates of carrier gases are most often in the range of 60–160 ml/min, while those of other gases are optimized to produce the most desirable detector response.

7.3.1.3.   Multiresidue Determinations

The utility of temperature-programmed FPD-GC for trace-level multicomponent residue analysis of OP pesticides is well documented (17, 18, 164, 205). Chromatograms of the 39 foods, with no pesticides added, prepared as previously described (see Table 7.2) and temperature-programmed on a 240 cm glass column (4 mm i.d.) containing 5% OV-101 on Gas Chrom Q using FPD-GC (P mode) are presented in Figure 7.3a,b. Chromatograms of OP pesticide standards analyzed under the same conditions in both the P and S modes and on a similar column packed with 5% OV-210 and analyzed in the P mode are presented in Figure 7.4. Based on the amount of sample extract injected and a reasonable assumption that 0.5 ng of P-containing pesticides or analog is detectable, then the limit of detection would be about 0.01–0.05 ppm. An examination of the chromatograms in Figure 7.3a,b indicates that very few foods generate serious interference. Interference of variable magnitude was observed at t_R 9.5 min which cor-

TABLE 7.5. Relative Retention Times (t_R) of 108 Organophosphorus Pesticides, Metabolites, Break-Down Products and Insect Chemosterilants Temperature-Programmed on 5 Columns[a]

Pesticide (or related compound)	Ratio of t_R of Component to t_R Parathion (1.00)				
	Dexsil 300	OV-101	OV-17	OV-210	OV-225
Abate®	2.57	2.67	3.06	2.43	[b]
Amidithion	0.94	0.95	1.04	1.01	1.15
Apholate	1.60	1.83	1.83	1.41	[d]
Azinphosethyl	1.68	1.85	1.85	1.75	1.72
Azinphosmethyl	1.62	1.75	1.79	1.70	[b]
Bay 30911	0.64	0.65	0.67	0.42	0.61
Bay 37289	0.98	1.08	0.96	0.68	0.80
Bay 37342	0.97	1.01	1.05	0.73	0.92
Carbophenothion	1.37	1.48	1.41	1.08	1.22
Carbophenothion O-analog	1.26	1.35	1.33	1.18	1.21
Chipman RP-11783	1.40	1.42	1.49	1.45	1.45
Chlorpyrifos	0.92	1.00	0.98	0.65	0.82
Chlorpyrifos O-analog	0.93	0.97	1.00	0.95	0.93
Chlorthion	1.04	1.00	1.05	1.03	1.08
Ciba C-2307	0.55	0.53	0.64	0.69	0.70
Ciba C-8874	1.25	1.39	1.30	0.93	1.08
Ciba C-9491	1.16	1.25	1.23	0.86	1.06
Ciba C-9491 O-analog	1.10	1.15	1.18	1.01	1.09
Compound 4072	1.04	1.13	1.10	0.98	1.00
Coumaphos	1.88	1.97	1.88	2.10	1.81
Coumaphos O-analog	1.80	1.90	1.83	2.29	1.86
Crotoxyphos	1.17	1.14	1.16	1.14	1.07
Crufomate	0.98	1.02	1.04	1.00	1.03
Dasanit	1.34	1.36	1.43	1.56	1.42
Dasanit sulfone	1.38	1.38	—	1.62	—
Dasanit O-analog	1.28	1.27	1.36	1.72	1.43
Dasanit O-analog sulfone	1.31	1.28	—	1.73	—
DEF®	1.16	1.32	1.16	0.89	0.95
Demeton	0.48	0.48	0.50	0.31	—
	0.64	0.62	0.67	0.55	0.63
Diazinon	0.66	0.73	0.71	0.41	0.58
Diazoxon	0.64	0.69	0.70	0.60	0.63
Dicapthon	1.02	1.01	1.03	0.98	1.03
Dichlorvos	0.17	0.17	0.18	0.17	0.21
Dicrotophos	0.60	0.55	0.67	0.81	0.78
Dimethoate	0.68	0.61	0.78	0.72	0.96

296

TABLE 7.5. (Continued)

Pesticide (or related compound)	Ratio of t_R of Component to t_R Parathion (1.00)				
	Dexsil 300	OV-101	OV-17	OV-210	OV-225
Dimethoate O-analog	0.51	0.49	—	0.71	—
Dioxathion	0.15	0.23	0.23	0.16	0.20
	0.66	0.67	0.76	0.51	0.71
	1.44	2.10	—	1.67	—
Disulfoton	0.71	0.75	0.74	0.47	0.66
Disulfoton sulfoxide	1.19	1.18	1.25	1.42	1.36
Disulfoton sulfone	1.19	1.18	1.24	1.43	1.36
Disulfoton O-analog	0.63	0.63	0.66	0.55	0.65
Disulfoton O-analog sulfoxide	1.08	1.02	—[b]	—[b]	—[b]
Disulfoton O-analog sulfone	1.08	1.01	1.16	1.46	[b]
Dition®	2.25	2.34	2.23	2.40	2.16
Dyfonate®	0.72	0.72	0.75	0.46	0.66
Dyfonate O-analog	0.61	0.60	0.65	0.54	0.64
EPN	1.57	1.66	1.59	1.58	1.46
Ethion	1.29	1.41	1.36	1.12	1.19
Famphur	1.44	1.46	1.50	1.75	1.55
Fenitrothion	0.92	0.93	1.00	0.93	1.00
Fenitrothion O-analog	0.83	0.81	0.91	1.08	1.00
Fenthion	0.93	1.00	1.02	0.72	0.93
Fenthion sulfoxide	1.36	1.36	1.47	1.60	1.44
Fenthion sulfone	1.35	1.36	1.47	1.66	1.50
Fenthion O-analog	0.85	0.89	0.99	0.88	0.95
Fenthion O-analog sulfoxide	1.27	1.27	1.42	1.76	1.45
Fenthion O-analog sulfone	1.27	1.27	1.42	1.80	1.54
Formothion	0.81	0.77	—	0.88	—
Gardona®	1.11	1.21	1.19	1.05	1.09
Geigy G-28029	1.53	1.69	1.58	1.27	1.37
Hempa	0.24	0.23	0.22	0.31	0.25
Imidan®	1.53	1.64	1.68	1.60	1.60
Imidoxon	1.41	1.51	1.59	1.70	[b]
Leptophos	1.58	1.79	1.66	1.32	1.40
Leptophos O-analog	1.48	1.66	1.59	1.40	1.39
Malathion	0.89	0.98	0.97	0.87	0.92
Malaoxon	0.82	0.85	0.88	0.99	0.92
Menazon	1.43	1.63	1.75	1.39	—[b]

TABLE 7.5. (Continued)

Pesticide (or related compound)	Ratio of t_R of Component to t_R Parathion (1.00)				
	Dexsil 300	OV-101	OV-17	OV-210	OV-225
Merphos®	1.00	1.13	0.97	0.51	0.68
	1.16	1.29	1.17	0.88	0.95
Metepa	0.39	0.41	0.44	0.44	0.54
Methiotepa	0.41	0.43	0.43	0.28	0.39
Methyl parathion	0.88	0.85	0.93	0.90	0.97
Methyl trithion®	1.29	1.36	1.36	1.00	1.21
Mevinphos	0.30	0.29	0.34	0.34	0.38
Monocrotophos	0.59	0.55	0.73	0.82	0.95
Naled	0.52	0.55	0.61	0.43	0.57
Nemacide®	0.81	0.84	0.80	0.54	0.69
Oxydemetonmethyl sulfone	0.95	0.88	1.08	1.38	—[b]
Parathion	1.00	1.00	1.00	1.00	1.00
Paraoxon	0.90	0.90	0.95	1.14	1.00
Phorate	0.57	0.60	0.60	0.35	0.53
Phorate sulfoxide	0.98	0.96	1.05	1.05	1.16
Phorate sulfone	0.99	0.97	1.05	1.14	1.16
Phorate O-analog	0.48	0.50	0.54	0.43	0.51
Phorate O-analog sulfoxide	0.87	0.83	0.97	1.17	1.14
Phorate O-analog sulfone	0.87	0.83	0.97	1.18	1.14
Phosalone	1.66	1.77	1.68	1.72	1.58
Phosfon®	0.70	0.75	0.60	0.81	0.64
Phosphamidon	0.84	0.85	0.89	1.12	0.97
Phoxim	1.14	—	—	—	—
Phoxim O-analog	0.92	0.94	—	1.16	—
Pirazinon®	0.48	0.50	0.56	0.52	0.57
Potasan®	1.70	1.73	1.70	1.98	1.70
Ronnel	0.85	0.93	0.88	0.60	0.76
Schradan	0.73	0.70	0.73	0.31	0.81
Shell SD-8280	1.07	1.00	1.04	0.89	0.97
Shell SD-8436	1.15	1.24	1.29	1.09	1.18
Shell SD-8448	1.19	1.33	1.25	1.14	1.11
Stauffer N-2788	0.84	0.86	0.86	0.57	0.75
Tepa	0.37	0.33	0.46	0.40	0.58
Tepp	0.12	0.12	0.12	0.14	0.12
Thiometon	0.61	0.63	—	0.43	—
Thiometon sulfoxide	—[b]	—[b]	—	—[b]	—
Thiometon sulfone	1.10	1.05	—	1.32	—

TABLE 7.5. (Continued)

Pesticide (or related compound)	Ratio of t_R of Component to t_R Parathion (1.00)				
	Dexsil 300	OV-101	OV-17	OV-210	OV-225
Thiotepa	0.56	0.41	0.48	0.29	0.48
Union Carbide UC-8305	0.69	0.68	0.79	0.70	0.87
Zinophos®	0.46	0.48	0.54	0.33	0.51
Zytron®	0.90	0.97	0.94	0.68	0.88
Parathion-t_R (min)	8.80	7.50	10.3	7.10	10.8

[a] Temperature programming: started at 150°C, 10°/min for 15 min, held at 300°C until last peak emerged.
[b] No peak appeared despite several 5-μg injections of compound.

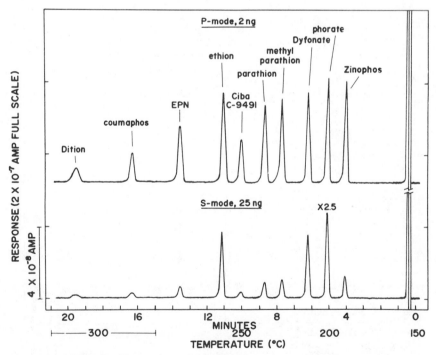

Figure 7.2 Chromatograms of pesticide standards in the P and S modes with 5% Dexsil on Chromosorb W specially washed with HCl (18).

299

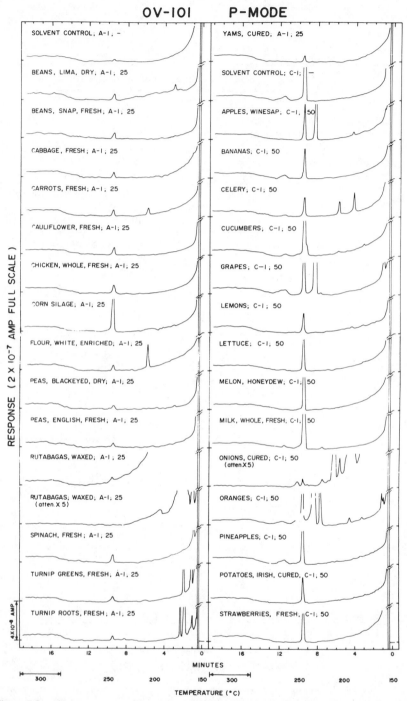

Figure 7.3a Chromatograms of extracts of foods, using a 240 cm glass column (4 mm, I.D.) containing 5% OV-101 on Gas Chrom Q and FPD-GC (P mode). Listed above chromatogram are the food, extraction-partitioning method, and mg equivalents of food in extract injected (164).

300

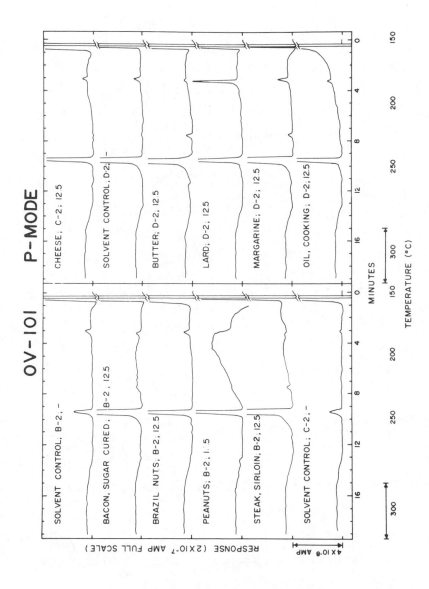

Figure 7.3b Chromatograms of extracts of foods, using the OV-101 packing and FPD-GC, P mode (continuation of Figure 7.3*a*) (164).

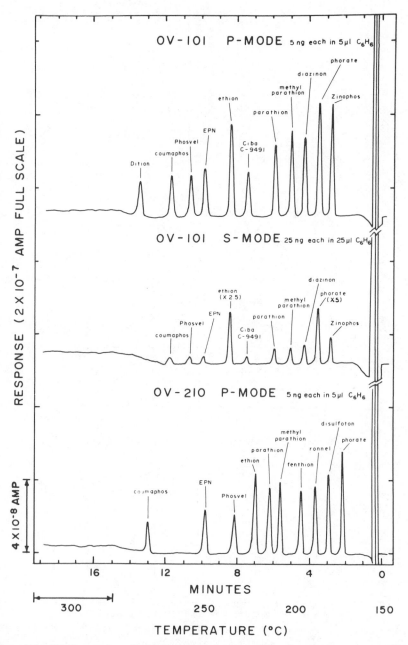

Figure 7.4 Chromatograms of pesticide standards on two GC columns using same conditions as used for chromatograms of extracts of foods. [*Note:* Phosvel is an earlier name for leptophos (164).]

responds to a t_R value of 1.62 relative to parathion. The interference appeared to arise in part from the acetone used; all reagents and solvents, as well as detergents used in cleaning glassware were checked, but means of eliminating the interference could not be devised. Other liquid phases could be of benefit if this interference causes difficulty. Significant interference in the P mode was encountered only with onions and peanuts and minor interference with oranges and waxed rutabagas. Since the treatment history of the commodities was not known, it is possible that some of the peaks noted may arise from pesticides on the crop and not from natural interference. Nevertheless, it is apparent that very little interference is present in the chromatograms of about 90% of the foods with the P detector and that minimal cleanup (solvent partitioning) is adequate for these analyses (164).

Quantitative aspects of the multicomponent residue procedure were subjected to test by adding 10 OP pesticides to milk at five levels (0.01–0.25 ppm). The temperature-programmed chromatograms (P mode) presented in Figure 7.5 of milk (unspiked), the extract spiked with 0.025 ppm and milk spiked with 0.01 and 0.025 ppm prior to extraction were obtained with a 240-cm column of 5% Dexsil 300 supported by specially washed Chromosorb W. Steady baselines and no interferences (except the peak at about 12 minutes already mentioned) were observed despite the minimum cleanup given the milk (i.e., the hexane-acetonitrile partitioning). Recoveries of the pesticides added to the milk prior to extraction, determined at 5 levels between 0.01 and 0.25 ppm are presented in Table 7.6 (18). Corrections in recoveries were made for losses in the hexane-acetonitrile partitions based on the p-values of the pesticides. The GC responses (P mode) obtained in these analyses plotted in Figure 7.6 are linearly related to concentration; that is, peak height is directly proportional to ppm added to the milk for all of the pesticides.

In another type of multiresidue procedure where several metabolites of a pesticide are likely to be present and only the total residue resulting from the pesticide is sought, the metabolites are converted to the same compound for analysis. For example, residues of fenthion, disulfoton, and phorate may each consist of the parent compound and five metabolites formed by oxidation of the thionophosphate and sulfide groups in each molecule. The analysis of those residues was simplified and speeded by oxidizing the pesticide and its metabolites to the O-analog sulfone with m-chloroperbenzoic acid. After removal of the acid on an alumina column the O-analog sulfone(s) was analyzed by FPD-GC (P mode) on a 90-cm glass column of 10% OV-101 operated isothermally at 215°C. Recoveries of total residues from corn, milk, grass, and feces were satisfactory and limits of detection were about 0.001 ppm based on twice noise (97).

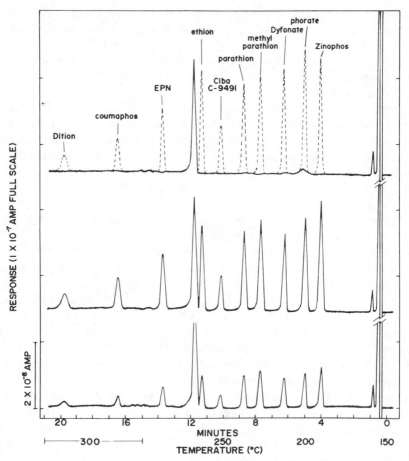

Figure 7.5 Chromatograms of pesticides (P mode) added to milk or to its extract, using 5% Dexsil 300 on Chromosorb W specially washed with HCl. 5-μl samples of extract, each equivalent to 50 mg milk, were injected. *Top*: solid line = blank milk extract; broken line = blank milk extract to which 0.025 ppm of each pesticide was added just before injected (1.25 ng of each pesticide in 5-μl injection); *Center*: milk extract obtained by adding 0.025 ppm of each pesticide to milk sample and then extracting; *Bottom*: same as center, but 0.010 ppm of each pesticide added (18).

7.3.1.4. Analysis for Specific Compounds

The analysis for traces of a specific OP pesticide alone or combined with a few related compounds is generally best accomplished isothermally once the GC characteristics of the compound(s) are determined and a column with suitable resolution is selected. However, in any trace-level OP pesticide analysis via GC it may be necessary to condition the column to the com-

TABLE 7.6.   Recoveries (%) of 10 OP Pesticides Added to 100 g Portions of Milk at 5 Levels, then Extracted and Analyzed by FPD-GC, P-mode (18)

Pesticide	Pesticide Added (ppm)					p Value
	0.25	0.10	0.05	0.025	0.010	
Zinophos	95	94	95	93	87	0.058
Phorate[a]	83	83	84	84	76	0.21
Dyfonate[a]	85	84	82	80	75	0.21
Methyl parathion	95	95	97	95	94	0.022
Parathion	91	91	91	89	89	0.044
Ciba C-9491[a]	79	80	78	75	73	0.26
Ethion	83	84	83	82	78	0.079
EPN	88	86	87	84	80	0.038
Coumaphos	98	98	96	94	91	0.006
Dition	96	95	97	95	93	0.027

[a] Recoveries corrected for loss in hexane–acetonitrile partitioning at 25°C. Losses of pesticides having p values of 0.08 or less were not appreciable (163).

pound and substrate being analyzed in order to obtain maximum sensitivity and reproducibility, particularly for those compounds containing a P=O moiety or other polar groups. The conditioning requirement is usually accomplished by repeatedly injecting 250-ng amounts of the compound in the extract until 5-ng amounts give a constant response. Standards are then injected alternately with extracts of unknown pesticide content to ensure the integrity of the analysis. Residues are satisfactorily quantified on the basis of peak height. The use of short columns operated isothermally that produce short retention times provides high sensitivity and ease of conditioning the column; therefore this mode of operation is recommended for routine analysis of specific OP pesticides (206). Analyses for trace-level residues of most of the OP pesticides are accomplished without difficulty by using the highly specific and sensitive FPD-GC system. Nevertheless, problems associated with intereference, thermal instability, etc. invariably arise and must be solved on an individual basis.

For example, in the analysis of Bay 68138 (an *O*-analog sulfide) and two of its metabolites (the sulfoxide and the sulfone) in turf grass, the fraction from the silica gel column that contained the parent compound and the sulfone also contained a naturally occurring interference peak that emerged under the sulfone peak in GC on a 45-cm column of 10% OV-101 (methyl silicone); with a column of 5% OV-25 (methyl phenyl silicone, 75% phenyl) of the same length it emerged under the parent pesticide peak. Therefore, a 5% OV-17 (methyl phenyl silicone, 50% phenyl) column was selected for the

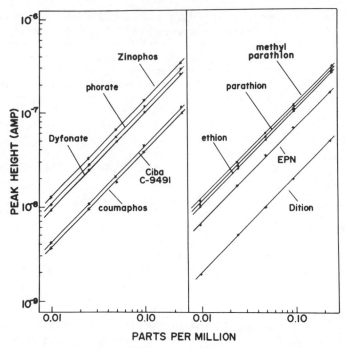

Figure 7.6 Responses obtained by FPD-GC (P mode) from extracts of milk; 10 OP pesticides added at 5 levels prior to extraction (18).

analysis because it shifted the peak to a portion of the chromatogram where it did not interfere; see chromatogram A of Figure 7.7 (29). Selection of a methyl silicone liquid phase containing an appropriate amount of phenyl groups (e.g., OV-3, 10% phenyl; OV-7, 20% phenyl; OV-61, 33% phenyl; OV-17, 50% phenyl; OV-22, 65% phenyl; or OV-25, 75% phenyl) may be invaluable for resolving interference problems.

Another rather unique analytical problem was encountered in the analysis of dichlorvos in the presence of large amounts of its precursor, trichlorfon (83). Trichlorfon is not chromatographed intact, but it is reproducibly broken down to dimethyl phosphite and dichlorvos (only a small amount) in the heated injection port of the GC. However, in analyzing for trichlorfon, the quantitation is based only on the dimethyl phosphite peak and ignored on the small and late-emerging dichlorvos peak. Then since the small dichlorvos peak always results from injecting trichlorfon into the GC, the injection of a single extract for analysis of both trichlorfon and dichlorvos could give an erroneously high result for dichlorvos. Thornton (207) developed a procedure for separating the two compounds by solvent partitioning prior to GC analysis. The separation, though adequate for conventional residue analysis, failed in the presence of large amounts of trichlorfon. A procedure was therefore developed for the absolute separation of the two compounds on a silica gel column (Figure 7.8) (83). A benzene extract containing the two compounds was added to a glass column (12 mm, i.d.) containing 5 g of silica gel as received (5.2% moisture) and washed with benzene to obtain a 50-ml eluate which was discarded. The dichlorvos was then eluted with 50 ml of benzene–5% acetone followed by 50 ml acetone to elute the trichlorfon. The separate fractions were concentrated and the trichlorfon assayed as dimethyl phosphite at 120°C and the dichlorvos at 165°C. Chromatograms of the two compounds on a 100 cm glass column of 16% XF-1150 on Chromosorb W (AW) at 120°C and of the dichlorvos at 165°C are presented in Figure 7.9.

Some other analytical problems pertaining to OP pesticide residue analysis encountered by this author during the past few years were as follows. (1) In the analysis of fenitrothion, its O-analog and its cresol hydrolysis product, in corn, grass and milk; a conventional silica gel column was found to hydrolyze the O-analog to the cresol (easily detected by the characteristic yellow color). The problem was corrected by deactivating the

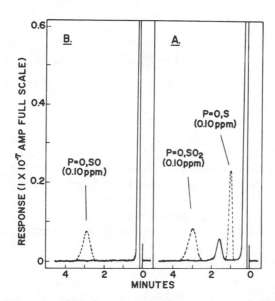

Figure 7.7 Chromatograms (FPD, P-mode) of Bay 68138 and 2 of its metabolites in fractions A and B obtained from a silica gel column. Solid lines are unspiked controls of turf grass equivalent to 20 mg; broken lines are the same fractions spiked with 2 ng (0.10 ppm) of the 3 compounds (29).

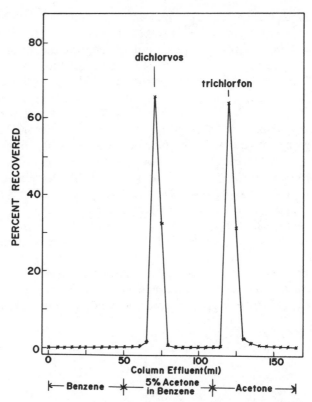

Figure 7.8 Separation of dichlorvos and trichlorfon on a column containing 5 g of silica gel (83).

silica gel with 20% water (108). (2) Residues of Bay 68138 and its sulfoxide in turf grass were partially oxidized to the sulfone during a Soxhlet extraction under a blanket of nitrogen; the use of carbon dioxide prevented the oxidation (29). (3) Phoxim was found to be thermally unstable in the GC while its O-analog was not; this was surprising since O-analogs are generally more unstable than the parent phosphorothioates (157). Unusual analytical behavior and problems that may challenge our ingenuity can be expected occasionally; however, few are insurmountable and they help to make our chosen specialty an interesting one.

7.3.2. Thin Layer Chromatography

TLC procedures for OP pesticides were reviewed by Watts (208) and recently MacNeil and Frei discussed the quantitative aspects of TLC for several classes of pesticides (209). Since this subject is also discussed in depth in another chapter, only a brief overview of the technique as it is applied to OP pesticides is presented here. The thin layers most frequently used are the silica gels (H, G, and G-HR), neutral alumina G, and Adsorbosil G-1. Some of the mobile phases that have been reported for TLC of OP pesticides on silica gel are hexane:acetone, 1:4 (210); benzene:cyclohexane, 3:1 (211); benzene:hexane, 1:1; benzene:dichloromethane, 4:1, and benzene:acetone, 3:1 (212); acetone:hexane, 15:85 (213), and benzene (214). An immobile liquid phase such as dimethylformamide is sometimes used. Two-dimensional TLC has also been used with silica gel plates to obtain better resolution of the OP pesticides. The plates were first developed with toluene or heptane:ethyl acetate, 1:3, then turned 90° and developed with ethyl acetate. Exposure of the plates to bromine vapors (converts thions to oxons) prior to the second development was useful for identification purposes (215).

Some of the procedures for visualizing the OP pesticide zones on the developed TLC plates include the use of fluorescent silica gel GF_{254} and viewing the plate under uv light (213, 216), exposure to bromine vapors then spraying with ethanolic Congo Red or flavone robinetin (211, 217), spraying with rhodamine B (214), treatment with silver nitrate and bromophenol blue (218), spraying with 4-(4-nitrobenzyl)pyridine then with tetraethylenepentamine in acetone (219), and spraying with strong acids or bases followed by heating the plates (220). The method that is probably the

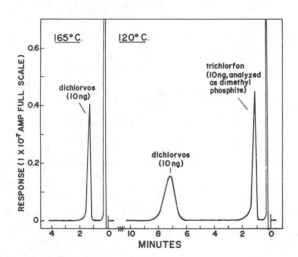

Figure 7.9 Chromatograms (FPD, P mode) of trichlorfon (as dimethyl phosphite) the dichlorvos at 120°C and dichlorvos at 165°C (83).

most specific and sensitive one available for visualizing OP pesticides on thin layers is based on the cholinesterase inhibiting properties of the compounds. Since the oxons are more powerful inhibitors than the thions, the OP pesticide zones on the developed plate are first converted to oxons by exposure to bromine vapors (221), then the plate is sprayed with a solution containing cholinesterase which is inhibited by the phosphate zones. Finally, the plate is sprayed with a substrate such as indoxyl acetate, which is converted to the fluorescent indoxyl and indigo, except in the zones where the inhibiting phosphates are present and the phosphate zones are visualized under uv light as dark spots against a fluorescent background. The indigo white background, upon exposure to the air, slowly oxidizes to visible indigo blue. Zones containing as little as 1 ng of some OP pesticides have been detected by this procedure.

TLC has been used for many years as a valuable and inexpensive technique for the cleanup, separation, identification, and semiquantitative estimation of pesticide residues. However, it was necessary to elute the zones from the TLC plates for analysis by alternate methods (e.g., GC, colorimetry, fluorescence, etc.) in order to obtain reliable quantitative data. Although quantitative TLC is still considered to be in the developmental stage, the advent of instruments dedicated to in situ analysis directly on the plates such as the one described by Beroza et al. (222) has inspired extensive research in this area. The use of such devices was later discussed by Lefar and Lewis (223), Getz and Hill (224), Hurtubise et al. (225), and Frei and MacNeil (226). The availability of suitable instruments and sensitive methods is expected to lead to wide acceptance of TLC as an attractive alternative to GC and other methods for the quantitative determination of pesticide residues, especially from the standpoints of economy and simplicity.

7.3.3. Other Determinative Procedures

7.3.3.1. High-Pressure Liquid Chromatography (HPLC)

This subject was recently reviewed by Moye (227) for all classes of pesticides and it will also be discussed in a separate chapter, therefore, only brief comments pertaining to the OP pesticides are presented here. Although, the use of HPLC for pesticide residue analysis is still in its infancy, theory and technology have developed rapidly and many applications of the technique are sure to follow. The main advantages of HPLC are in its capability of analyzing compounds that are heat labile or of such low volatility or high polarity that they cannot be eluted from a GC column. Therefore, it is not surprising that much of the pesticide residue work with

HPLC has pertained to the carbamates. However, Abate (an OP pesticide difficult to assay via GC) was analyzed in pond water with a sensitivity of 0.05 ppm by using a 1-m column (2.1 mm i.d.) packed with 1% β,β'-oxydipropionitrile on Zipax and a 254-nm uv detector (228). A polarographic detector for HPLC was used by Koen et al. to determine residues of parathion and methyl parathion in lettuce (229). Wade and Moye metered buffered solutions of cholinesterase enzyme into the effluent of a reverse phase HPLC column, incubated the mixture at 40°C, then metered in a substrate (N-methylindoxyl acetate) which was hydrolyzed by the enzyme to form a fluorescent background (230). The presence of cholinesterase inhibiting pesticides reduced the background fluorescence and yielded negative peaks. Several carbamates and 4 OP pesticides were included in their experiments. The EC detector commonly used in GC analysis has also been adapted for use with HPLC (231) and it may be used for the detection of some OP pesticides. The prospect of using HPLC for pesticide residue analysis is a very attractive one. However, the development of more sensitive and specific detection systems will be required before the technique can become competitive with GC for the analysis of most OP pesticides.

7.3.3.2. Plasma Chromatography (PC)

The term plasma chromatography (ion-drift spectrometry), originated by Franklin GNO Corp., West Palm Beach, Fl., is based on the generation of ions at atmospheric pressure via ion-molecule reactions and their analysis in an ion mobility drift tube. The time required for the ions to traverse the length of the drift tube is a function, among other parameters, of the mass and charge of the ion (232). Moye (233) reported ion mobility spectra for 14 OP pesticides in 1975 and no other work on this class of pesticides via PC could be found in the literature. An attractive feature of the PC is its high sensitivity (low picogram range) under certain conditions. The technique also provides a relatively specific "fingerprint" and has the advantage of operating at atmospheric pressure. Nevertheless, much research with the PC and probably the coupling of it with other instruments such as the GC or HPLC will be required with many pesticides and substrates to evaluate its applicability to trace-level residue analysis.

7.3.3.3. Optical Methods

Most of the colorimetric methods once used for specific OP pesticides have been replaced by chromatographic procedures, however, optical methods are still widely used for the purpose of screening commodities for compliance with established pesticide tolerances. Some of these procedures are summarized as follows: (1) total organic phosphorus is determined by digesting the sample with a mixture of perchloric and sulfuric acids to yield the inorganic phosphate followed by the addition of ammonium molybdate and 1-amino-2-naphthol-4-sulfonic acid reagents to develop the characteristic phosphomolybdenum blue color. The absorbance of the colored product is measured at 815 nm; (2) in an anticholinesterase method based on acid-phenol red, a buffered mixture of sample extract and cholinesterase are incubated, then combined with buffer and acetylcholine iodide and further incubated to generate acetic acid which is proportional to the amount of remaining active enzyme. The acetic acid is removed by dialysis, combined with phenol red and the absorbance measured at 550 nm (234); (3) in another anticholinesterase method the remaining active cholinesterase hydrolyzes acetylthiocholine to form thiocholine which is then reacted with 5,5'-dithio-bis-2-nitrobenzoic acid (DTNB). The resulting compound is measured at 420 nm (235); (4) the Schoenemann reaction which is based on the fact that aromatic amines react in the presence of organophosphates and alkaline peroxides to form fluorescent or colored products has also been adapted to OP pesticide analysis. Indole as a substrate for this reaction is oxidized to the highly fluorescent indoxyl or to the blue indigo dye (236) and (5) 4-(4-nitrobenzyl) pyridine reagent after being heated with OP pesticides in aqueous alcohol, cooling, and the addition of tetraethylenepentamine reagent yields a colored product that is measured at 580 nm. Twenty-three OP pesticides responded to this test with good sensitivity (237).

7.3.3.4. Automated Analysis

An excellent review on the development of automated systems for pesticide residue analysis from 1960 through 1974 was recently presented by McLeod (238). A dual AutoAnalyzer® system for the simultaneous determination of OP pesticide residues via anticholinesterase (DTNB method) and total phosphorus (molybdenum blue) reactions was described by Ott (239). Limits of sensitivity were about 0.2 ppm and analyses were performed at the rate of about 20 samples per hour. Several other researchers have reported similar performance by using single colorimetric detection systems. McLeod et al. described a semiautomated system consisting of an automatic sample injector in tandem with a five detector GC system with data recording facilities for offline computer processing (240). During an overnight operation the system is capable of handling 48 solutions and it detects compounds containing N, P, or S; that capture electrons, or that respond to a flame ionization detector (reducing mode). With the availability of automated systems for sample injection (Autosamplers), gel permeation cleanup, sweep codistillation, extraction and evaporation, rapid progress in the area of automated analysis is expected. A completely automated analysis includes all steps from the preparation of the sample to the typed report and several laboratories have demonstrated that most of these steps can be automated. However, problems pertaining to the automation of adsorption chromatographic cleanup, concentrating of eluates, quantitative transfers, adjustments to final volumes and the loading of automatic samplers for GC analysis have not been fully resolved (238).

7.4. CONFIRMATORY IDENTIFICATION OF RESIDUES

A tentative identification of a pesticide is often made by a single determinative procedure, however, any such identification should be either confirmed or rejected by using several alternate procedures, particularly where the history of the sample is not known. For example, several years ago this analyst demonstrated the presence of EC-GC peaks corresponding to various pesticides in soil held in a sealed container since before the advent of synthetic pesticides in the United States (241); subsequently several other investigators have made similar reports. The pitfalls of incorrect identification of pesticide residues, or the failure to detect residues, are apparent.

The method most commonly used for confirming the identity of a pesticide GC peak is the simple comparison of the retention time of the unknown vs. peaks of standard pesticides on several different types of column packings such as those already mentioned. An identical retention time of a standard and unknown peak on two or more columns of differing polarities constitutes good supplementary evidence that the compounds are the same.

The p value, which was discussed earlier, is another popular means of confirming the identity of GC peaks at the nanogram level. Its use for this purpose is probably exceeded only by the technique of comparing retention times on different columns. The technique is simple and rapid and it is especially useful at the low levels commonly encountered for quantitative analysis via EC-, AFID-, and FPD-GC. Each pesticide has a characteristic p value with a given solvent pair (Tables 7.3 and 7.4) and one of the most attractive features of the technique is that an exact determination of the amounts of substance present is not necessary; only the relative amount present in the original and the extracted solutions is required. This is especially useful in dealing with an unknown compound for which the response factor is not known. The p values are sufficiently reproducible and different from one another to allow their use in confirming the identity of pesticides to be meaningful. The integrity of a *p* value confirmation is greatly enhanced by using several different solvent systems.

An excellent review concerning the use of chemical derivatization techniques for the identification of pesticide residues was recently presented by Cochrane (242) and the salient elements of his presentation pertaining to OP pesticides follow. The various reactions used for the confirmation of OP pesticides are illustrated in Figure 7.10; parathion is used as an example. The three general approaches are (1) alkaline hydrolysis and derivatization of the P moiety, (2) hydrolysis and derivatization of the alkyl or aryl moiety, or (3) derivatization of the intact pesticide. In the first approach NaOH is generally used to perform the hydrolysis and the interference from inorganic phosphate is overcome by splitting the extract and preparing both the methyl and ethyl derivatives. The test suffers from the fact that the hydrolysis products (i.e., phosphoric acid derivatives) are so soluble in water that quantitative extraction is difficult. Also, if more than one compound yields the same product no distinction can be made between them. Analysis of the P moiety is therefore best suited for characterization and screening procedures. In the second approach, if the hydrolysis of the OP pesticide

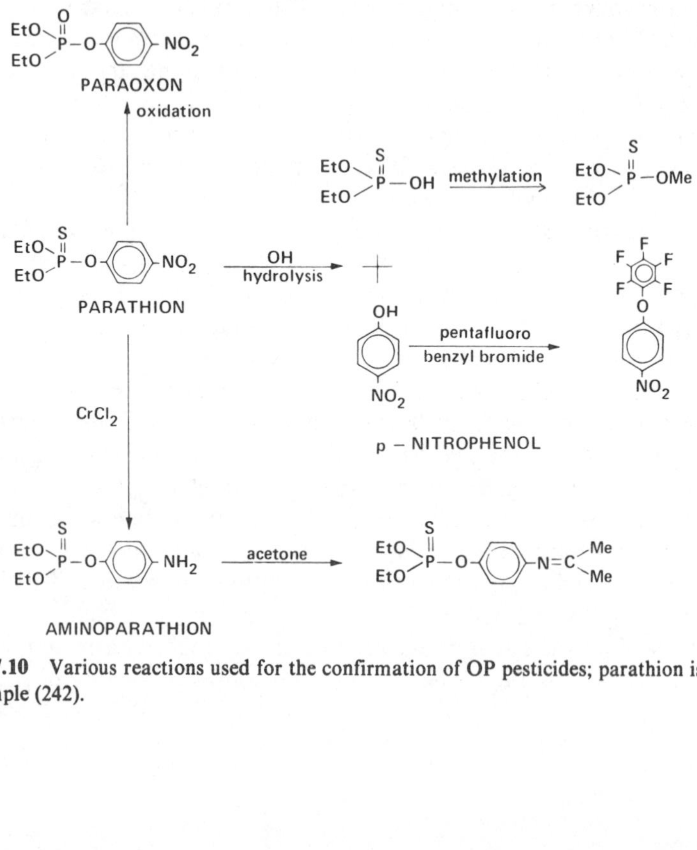

Figure 7.10 Various reactions used for the confirmation of OP pesticides; parathion is used as an example (242).

also forms a phenol, this moiety can be derivatized to an electron-capturing product. Many derivatizing agents have been tested but recent trends are toward the use of perfluorinated reagents that form trifluoroacetyl, heptafluorobutyryl, pentafluoroaryl, etc. derivatives and the nitroaromatic reagents to produce 2,4-dinitrophenyl (DNP)- and 2,6-dinitro-4-trifluoromethyl phenyl (DNT)-ethers. The pentafluoroaryl derivatives generally give the best sensitivity for EC detection. The confirmation of p-nitrophenol, which is a hydrolysis product of both parathion and methyl parathion, can be done by using the pentafluorobenzyl (PFB), DNP, or DNT derivatives. In cases where the phenol already possesses sufficient EC properties, the methyl ether, formed by reaction with diazomethane, or the trimethylsilyl ether, formed by reaction with N,O-Bis(trimethylsilyl)acetamide, may be employed. The third approach, based on the derivatization of the intact molecule, is by far the most desirable. In one of the methods $CrCl_2$, $PdCl_2$, or Zn-HCl is used to reduce compounds containing aryl nitro or aryl cyano groups to their corresponding amines (243). OP pesticides containing the P=S moiety have also been oxidized to the corresponding oxons by using neutralized sodium hypochlorite (244). A base-catalyzed alkylation technique employing a reagent mixture consisting of NaH, MeI, and DMSO has been employed for the derivatization of compounds possessing the NH or NH_2 moiety (245). The uv irradiation of crufomate in a hexane solution to form the deschloro derivative has also been used as a confirmatory test (246).

Bioassays with test organisms such as mosquito larvae or Drosophila have been employed through the years as sensitive procedures for approximating trace levels of pesticides. Although these procedures provide actual tests for toxicity, their application has been limited probably because of the requirement of maintaining a colony of test organisms and the masking effects obtained in tests with sample extractions. The biological response (LD_{50}) is influenced by the test compound, nature of the extractive, extent of cleanup, etc. and response curves for a given pesticide in a specific sample extractive must be determined under standarized conditions before valid data can be obtained. However, a bioassay procedure that is properly designed is extremely valuable as a determinative or confirmatory test for OP pesticides. Procedures based on the cholinesterase inhibitive (ChEI) properties of OP pesticides have been more widely used for determinative and confirmatory tests. Phillips et al. (247) combined these two procedures into a parallel screening method employing a mathematical treatment of $ChEI_{50}$ and LD_{50} values from the same extract to obtain good selectivity. The parallel screening procedure was never widely used probably because commercial GC equipment with sensitive and specific detectors soon appeared; nevertheless the basic concept is sound and it offers an alternate means of confirming identities of OP pesticides.

The dual channel FPD (200), which simultaneously monitors GC effluents for compounds containing P (526-nm filter) and S (394-nm filter) provides good confirmatory information concerning the presence (or absence) of P and/or S as well as an estimation of the atomic ratio of P to S in the molecule are obtained. A typical temperature-programmed analysis of 9 pesticides is presented in Figure 7.11. Gardona, which contains P but no S, is detected on the P channel but not on the S channel. A similar analysis with 13 pesticides is presented in Figure 7.12. Sulphenone, which contains S but no P, responds only on the S channel. In compounds containing both P and S, the atomic ratio of these elements in the molecule may be approximated by dividing the P response by the square root of the S response. Response ratios for 21 pesticides at three different levels of concentration (25, 50, and 100 ng) are presented in Table 7.7; these ratios remain fairly constant for any given compound and they are related to the PS contents of the molecules. For example, the ratios at three concentration levels for Zinophos, a PS compound, range from 5.5 to 6.0 as do the ratios of all other pesticides containing one P and one S. Ratios of the PS_2 pesticides are about 3 (ca. half that of the PS compounds), and those with a PS_3 content are about 2 (ca. one third of the PS compounds). By using a

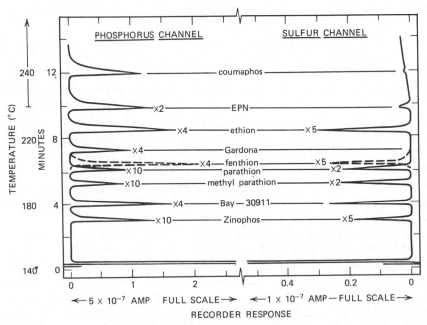

Figure 7.11   Dual-channel recording of P and S response to 50-ng amounts of 9 pesticides in 5-μl benzene in temperature-programmed GC (200).

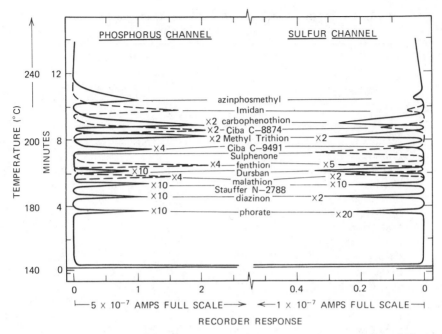

Figure 7.12 Dual-channel recording of P and S response to 50-ng amounts of 13 pesticides in 5-μl benzene in temperature-programmed GC (200).

reverse application of this process, the analyst can deduce the atomic ratio of P to S in an unknown compound. This type of information is invaluable for use in confirmatory tests for OP pesticides.

The use of determinative procedures such as spectrophotometry, mass spectrometry, TLC, NMR, etc. or combinations of these techniques for confirmatory tests should not be minimized. For example, GC-mass spectrometry is a powerful confirmatory technique; Vander Velde and Ryan recently reported on its use as applied to pesticide analysis (248). In any event the analytical sample generally consists of no more than a few nanograms of the compound for confirmation in combination with large amounts of extraneous material, therefore the analyst must exercise sound judgement in the choice of techniques that possess adequate sensitivity and specificity for the task.

7.5. ANALYSIS OF METABOLITES

Metabolism studies with pesticides are best accomplished by using radiolabeled compounds with the active atoms (usually ^{14}C) properly posi-

TABLE 7.7. Response Ratios $(R_p/\sqrt{R_s})$ of Compounds Containing Phosphorus and Sulfur (200)

Pesticide	PS Content	Response ratio × 10^3 at concentration (ng) indicated		
		100	50	25
Zinophos	PS	5.5	5.8	6.0
Phorate	PS$_3$	2.0	2.1	2.2
Bay 30911	PS	5.7	5.5	5.3
Diazinon	PS	5.5	5.6	5.3
Stauffer N-2788	PS$_2$	2.8	2.9	2.8
Methyl parathion	PS	5.8	5.7	5.6
Malathion	PS$_2$	3.0	3.2	3.1
Chlorpyrifos	PS	5.6	5.6	5.5
Parathion	PS	5.6	5.5	5.5
Fenthion	PS$_2$	3.0	3.0	2.9
Ciba C-9491	PS	5.3	5.2	5.2
Methyl trithion	PS$_3$	1.9	2.0	1.9
Ethion	P$_s$S$_4$	2.8	2.9	2.8
Ciba C-8874	PS	5.4	5.2	5.3
Carbophenothion	PS$_3$	1.9	2.0	1.8
Imidan	PS$_2$	3.3	3.2	3.3
EPN	PS	6.0	5.8	—
Azinphosmethyl	PS$_2$	3.2	3.3	—
Coumaphos	PS	5.9	—	—

tioned within the molecule to serve as tracers in the event that the molecule is fragmented. Then by using various radioassay techniques, the analyst can isolate and account for all of the test material and determine whether it is intact, fragmented, conjugated, extractable or unextractable. Once the metabolic products are determined in this manner, conventional methods may be used in many instances for subsequent studies. The metabolites of OP pesticides are generally regarded as either ionic (formed by the cleavage of a linkage), or nonionic (formed by the oxidation or reduction of a sulfur, nitrogen or alkyl groups). Some of these reactions were illustrated in Figure 7.10. A few examples of procedures for assaying metabolites of OP pesticides via conventional means follow.

Bay 68138 (P=O,S) is a systemic OP insecticide and nematocide containing a P=O (O-analog) moiety and also a thioether linkage (sulfide) which oxidizes to the sulfoxide (P=O,SO) and further to the sulfone (P=O,SO$_2$).

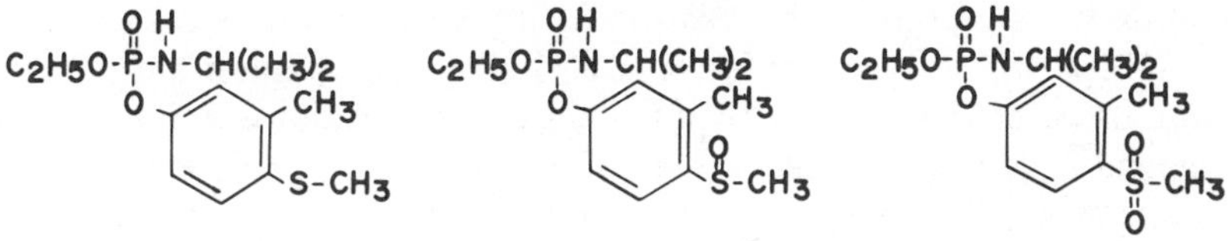

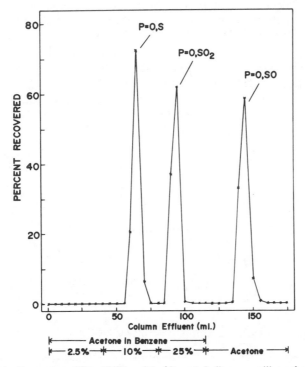

Figure 7.13 Formulas of Bay 68138 (I) and its sulfoxide (II) and sulfone (III) metabolites (29).

Formulas of the three compounds are shown in Figure 7.13. The three compounds may be completely separated on a 5 g column (12 mm i.d.) of silica gel (15% water) by elution with portions of increasingly polar mixtures of acetone in benzene then with acetone alone (Figure 7.14). Oxidation of about 10 and 2% of the parent compound to the sulfoxide was observed on silica gel containing 4.8 and 10% water, respectively, however no detectable

Figure 7.14 Separation of Bay 68138 and 2 of its metabolites on a silica gel column (29).

oxidation was observed with 15% water. In the analysis for residues in turf grass (29) all three compounds could easily be separated on the silica gel column however this was not required since good separations of the parent compound from the sulfoxide or the sulfone were obtained via FPD-GC. On the other hand, the sulfoxide and the sulfone were not resolved by the GC column employed (Figure 7.7). The extract of turf grass containing residues of the 3 compounds was therefore separated and partly cleaned-up as shown in Figure 7.15. The benzene and 2.5% acetone eluate from the column were discarded then fraction A consisting of the parent compound and its sulfone were collected by eluting the column with 45 ml of benzene–25% acetone; fraction B containing the sulfoxide (most polar of the three residues) was eluted with 50 ml of acetone. The two fractions were analyzed separately via FPD-GC (Figure 7.7).

Fenthion is an excellent example of a phosphorothioate insecticide containing a thioether linkage that oxidizes to form a family of metabolites. The structure of fenthion and 5 of its metabolites is shown in Figure 7.16. The separation of all 6 compounds on a 4 g column (12 mm i.d.) of silica gel (3.5% water) by elution with portions of benzene, benzene-acetone, then acetone is illustrated in Figure 7.17. In the analysis of corn, grass or milk for residues of these 6 compounds, only three fractions from the column were required because of the additional resolution obtained in the FPD-GC analysis. The sample extract in 10 ml of benzene was added to the column and eluted with 50 ml of benzene-1% acetone to obtain the parent and its sulfone (P=S,S; P=S,SO$_2$). Next the column was eluted with 50 ml of benzene-10% acetone to remove the O-analog sulfide, parent sulfoxide and O-analog sulfone (P=O,S; P=S,SO; P=O,SO$_2$); finally 50 ml of acetone was

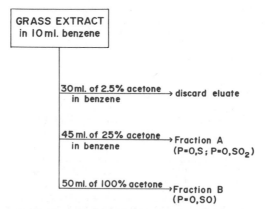

Figure 7.15 Summary of operations of silica gel column for the assay of Bay 68138 and 2 of its metabolites via FPD-GC (29).

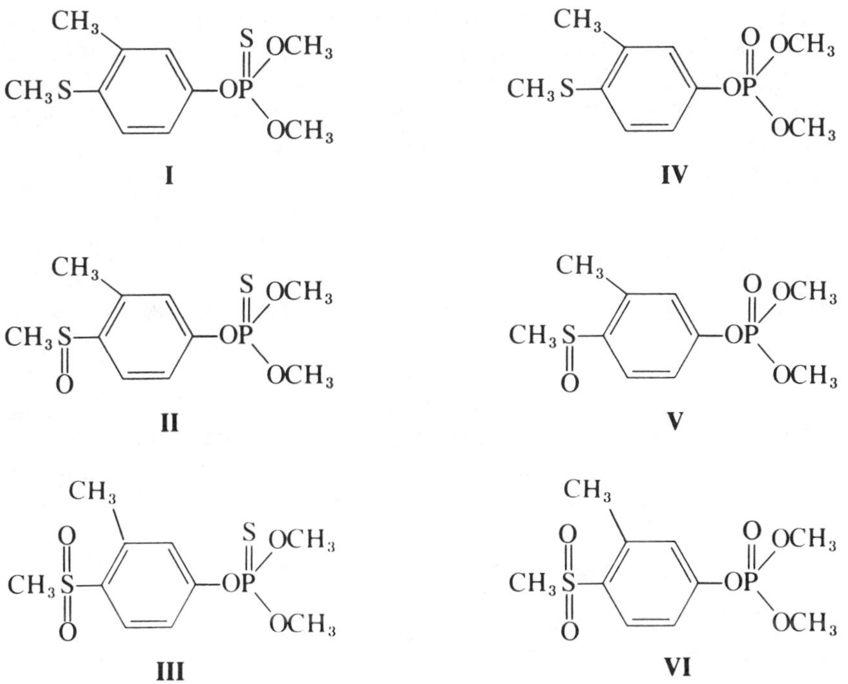

CH₃ / CH₃S— —OP S OCH₃ / OCH₃

I

CH₃ / CH₃S— —OP O OCH₃ / OCH₃

IV

CH₃ / CH₃S(O)— —OP S OCH₃ / OCH₃

II

CH₃ / CH₃S(O)— —OP O OCH₃ / OCH₃

V

CH₃ / CH₃S(O₂)— —OP S OCH₃ / OCH₃

III

CH₃ / CH₃S(O₂)— —OP O OCH₃ / OCH₃

VI

Fig. 7.16 Fenthion (I) and five of its metabolites. (I) P=S,S; (II) P=S,SO; (III) P=S,SO₂; (IV) P=O,S; (V) P=O,SO; (VI) P=O,SO₂.

used to elute the O-analog sulfoxide (P=O,SO) which is the most polar of the 6 compounds. The three fractions were then assayed by FPD-GC (109). Similar separations on silica gel and analysis via FPD-GC in various substrates were described for dasanit and 3 of its metabolites (69), phorate and 5 metabolites (151), disulfoton and 5 metabolites (98) as well as many phosphorothioates and their O-analogs (40, 50, 108,131, 156).

OP pesticides such as leptophos, Ciba C-9491 and many others oxidize to form O-analogs and both the parent compound and the O-analog may be hydrolyzed to form a phenol; assays for residues of all three compounds are often required. An analytical scheme for the separation of leptophos, its O-analog, and its phenolic hydrolysis product (4-bromo-2,5-dichlorophenol) employing the use of silica gel and alumina columns operated in tandem is presented in Figure 7.18 (125). A 10 g column (2 cm i.d.) of silica gel deactivated with 10% buffer (pH 7) is positioned on top and in tandem with a 5 g column (12 mm i.d.) of alumina (3% water). The sample extract of corn or milk in 10 ml of benzene followed by 60 ml of benzene is added to the silica gel and allowed to percolate through both columns. The eluate from both columns contains the parent leptophos which is concentrated and assayed via FPD-GC. The columns are then separated and the *O*-analog adsorbed on the silica gel is eluted with 50 ml of acetone, concentrated and also assayed by FPD-GC. The phenol adsorbed on the alumina is eluted with 50 ml of benzene-2% acetic acid and neutralized in a separatory funnel with 50 ml of 5% aqueous sodium bicarbonate. The benzene extract is dried, concentrated, methylated with diazomethane, and the methyl ether derivative assayed by EC-GC. This analyst has been successful in assaying some phenols without preparing the derivative, however the process is difficult and therefore not recommended. Methanol has also been used to elute the phenol from alumina (55); however, the use of a benzene–acetic acid mixture yields better recoveries.

The analysis of metabolites of OP pesticides is a formidable undertaking, particularly where highly polar or conjugated products are sought. Although the procedures just described are useful for analyzing residues of a few of these products each fragment and conjugate of each pesticide should be studied individually.

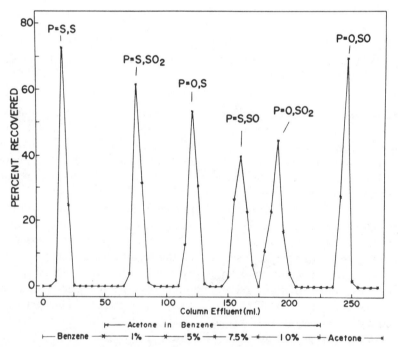

Figure 7.17 Separation of fenthion and 5 of its metabolites on a silica gel column (109).

I. ADD EXTRACT, IN 10 ML. BENZENE,
   THEN ELUTE SILICA GEL AND ALUMINA
   WITH 60 ML. BENZENE

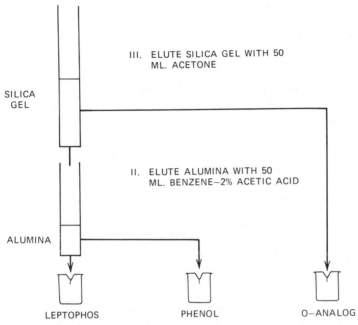

Figure 7.18   Separation of residues of leptophos, its *O*-analog and its phenol on silica gel and alumina columns.

REFERENCES

1. P. A. Giang and S. A. Hall, *Anal. Chem.*, **23,** 1830 (1951).

2. W. I. H. Holman, *Biochem. J.*, **87,** 256 (1943).

3. P. R. Averell and M. V. Norris, *Anal. Chem.*, **20,** 753 (1948).

4. H. P. Burchfield and E. E. Storrs, *J. Chromatogr. Sci.*, **13,** 202 (1975).

5. W. Thornburg, *Anal. Chem.*, **43,** 145 R (1971).

6. W. Thornburg, *Anal. Chem.*, **45,** 151 R (1973).

7. W. Thornburg, *Anal. Chem.*, **47,** 157 R (1975).

8. W. Thornburg, *Anal. Chem.*, **49,** 98 R (1977).

9. *Pesticide Analytical Manual*, Vols. I and II, Food and Drug Administration, Washington, D.C., 1971.

10. J. F. Thompson (Ed.), *Manual of Analytical Methods*, U.S. Environmental Protection Agency, Research Triangle Park, N.C., 1974.

11. D. E. H. Frear (Ed.), *Pesticide Index*, 4th ed., College Science Publishers, State College, Pa., 1969.

12. C. W. Miller and A. F. M. Funes, *J. Chromatogr.*, **59**, 161 (1971).

13. R. A. Henry, J. A. Schmit, J. A. Dieckman, and F. J. Murphey, *Anal. Chem.*, **43**, 1053 (1971).

14. R. Blinn, *J. Agric. Food Chem.*, **17**, 118 (1969).

15. M. C. Bowman, H. R. Ford, C. S. Lofgren, and D. E. Weidhass, *J. Econ. Entomol.*, **61**, 1586 (1968).

16. W. E. Dale and J. W. Miles, *J. Agric. Food Chem.*, **17**, 60 (1969).

17. M. C. Bowman and M. Beroza, *J. Assoc. Off. Anal. Chem.*, **53**, 499 (1970).

18. M. C. Bowman and M. Beroza, *J. Assoc. Off. Anal. Chem.*, **54**, 1086 (1971).

19. J. A. Seawright, M. C. Bowman, and C. S. Lofgren, *J. Econ. Entomol.*, **66**, 613 (1973).

20. J. A. Seawright and M. C. Bowman, *Mosq. News*, **35**, 41 (1975).

21. M. C. Bowman and M. Beroza, *J. Assoc. Off. Anal. Chem.*, **49**, 1046 (1966).

22. L. M. Hunt and B. N. Gilbert, *Bull. Environ. Contam. Toxicol.*, **5**, 42 (1970).

23. V. B. Stein and K. A. Pittman, *J. Assoc. Off. Anal. Chem.*, **59**, 1094 (1976).

24. J. Wieneke and W. Steffens, *J. Agric. Food Chem.*, **24**, 416 (1976).

25. D. C. Staiff, S. W. Comer, J. F. Armstrong, and H. R. Wolfe, *Bull. Environ. Contam. Toxicol.*, **13**, 362 (1975).

26. T. T. Liang and E. P. Lichtenstein, *J. Agric. Food Chem.*, **24**, 1205 (1976).

27. K. M. Al-Adil, E. R. White, W. L. Winterlin, and W. W. Kilgore, *J. Agric. Food Chem.*, **21**, 376 (1973).

28. R. J. Kuhr, A. C. Davis, and J. B. Bourke, *Bull. Environ. Contam. Toxicol.*, **11**, 244 (1974).

29. M. C. Bowman, *Int. J. Environ. Anal. Chem.*, **1**, 307 (1972).

30. J. S. Thornton, *J. Agric. Food Chem.*, **19**, 890 (1971).

31. T. B. Waggoner, *J. Agric. Food Chem.*, **20**, 157 (1972).

32. J. S. Thornton and C. W. Stanley, *J. Agric. Food Chem.*, **19**, 73 (1971).

33. D. L. Bull and J. J. Whitten, *J. Econ. Entomol.*, **65**, 973 (1972).

34. M. C. Bowman and D. B. Leuck, *Int. J. Environ. Anal. Chem.*, **1**, 35 (1971).

35. J. Stenersen, *Bull. Environ. Contam. Toxicol.*, **4**, 104 (1969).

36. M. Stiasni, W. Deckers, K. Schmidt, and H. Simon, *J. Agric. Food Chem.*, **17**, 1017 (1969).

37. B. G. Luke and C. J. Dahl, *J. Assoc. Off. Anal. Chem.*, **59**, 1081 (1976).

38. J. B. Leary, *J. Assoc. Off. Anal. Chem.*, **57**, 189 (1974).

39. J. A. Lubkowitz, J. Baruel, A. P. de Revilla, and M. M. Cermeli, *J. Agric. Food Chem.*, **21**, 143 (1973).

40. D. B. Leuck and M. C. Bowman, *J. Econ. Entomol.*, **66,** 798 (1973).

41. H. E. Braun, *J. Assoc. Off. Anal. Chem.*, **57,** 182 (1974).

42. P. Maini and A. Collina, *J. Assoc. Off. Anal. Chem.*, **55,** 1265 (1972).

43. D. L. Struble, *J. Assoc. Off. Anal. Chem.*, **56,** 49 (1973).

44. P. Maini, A. Collina, and I. Passarini, *Pest. Sci.*, **3,** 533 (1972).

45. S. Uk and C. M. Himel, *J. Agric. Food Chem.*, **20,** 638 (1972).

46. H. V. Claborn, H. D. Mann, and D. D. Oehler, *J. Assoc. Off. Anal. Chem.*, **51,** 1243 (1969).

47. M. E. Dusch, W. E. Westlake, and F. A. Gunther, *J. Agric. Food Chem.*, **18,** 178 (1970).

48. R. L. McKellar, H. J. Dishburger, J. R. Rice, L. F. Craig, and J. Pennington, *J. Agric. Food Chem.*, **24,** 283 (1976).

49. L. M. Hunt, B. N. Gilbert, and J. C. Schlinke, *J. Agric. Food Chem.*, **17,** 1166 (1969).

50. M. C. Bowman and M. Beroza, *J. Agric. Food Chem.*, **15,** 651 (1967).

51. H. D. Mann, M. C. Ivey, S. E. Kunz, and B. F. Hogan, *J. Econ. Entomol.*, **66,** 715 (1973).

52. M. C. Ivey and H. V. Claborn, *J. Assoc. Off. Anal. Chem.*, **51,** 1245 (1968).

53. D. B. Leuck, R. L. Jones, and M. C. Bowman, *J. Econ. Entomol.*, **68,** 287 (1975).

54. J. C. Johnson, Jr., R. L. Jones, D. B. Leuck, M. C. Bowman, and F. E. Knox, *J. Dairy Sci.*, **57,** 1467 (1974).

55. M. C. Bowman and M. Beroza, *J. Agric. Food Chem.*, **16,** 280 (1968).

56. M. C. Bowman and J. R. Young, *J. Econ. Entomol.*, **52,** 1468 (1969).

57. R. E. Bry, M. C. Bowman, H. J. Lang, and F. G. Crumley, *J. Econ. Entomol.*, **64,** 177 (1971).

58. M. C. Ivey, D. D. Oehler, and H. V. Claborn, *J. Agric. Food Chem.*, **21,** 822 (1973).

59. D. L. Suett, *Pestic. Sci.*, **5,** 57 (1974).

60. M. Beroza and M. C. Bowman, *J. Agric. Food Chem.*, **14,** 625 (1966).

61. M. C. Bowman, M. Beroza, C. H. Gordon, R. W. Miller, and N. O. Morgan, *J. Econ. Entomol.*, **61,** 358 (1968).

62. J-G Zakrevsky and V. N. Mallet, *J. Assoc. Off. Anal. Chem.*, **58,** 554 (1975).

63. Y. Volpe and V. N. Mallet, *Anal. Chim. Acta.*, **81,** 111 (1976).

64. J. G. Konrad and G. Chesters, *J. Agric. Food Chem.*, **17,** 226 (1969).

65. D. D. Oehler and H. V. Claborn, *J. Assoc. Off. Anal. Chem.*, **53,** 1045 (1970).

66. A. Westlake, F. A. Gunther, and W. E. Westlake, *J. Agric. Food Chem.*, **17,** 1157 (1969).

67. R. Greenhalgh, J. Dokladalova, and W. O. Haufe, *Bull. Environ. Contam. Toxicol.*, **7,** 237 (1972).

68. I. H. Williams, M. J. Brown, and D. G. Finlayson, *J. Agric. Food Chem.*, **20**, 1219 (1972).

69. M. C. Bowman and K. R. Hill, *J. Agric. Food Chem.*, **19**, 342 (1971).

70. I. H. Williams, R. Kore and D. G. Finlayson, *J. Agric. Food Chem.*, **19**, 456 (1971).

71. W. E. Westlake, M. E. Dusch, F. A. Gunther, and L. R. Jeppson, *J. Agric. Food Chem.*, **19**, 1191 (1971).

72. N. F. Janes, A. F. Mackin, M. P. Quick, H. Rogers, D. E. Mundy, and A. J. Cross, *J. Agric. Food Chem.*, **21**, 121 (1973).

73. A. F. Machin and M. P. Quick, *Analyst*, **94**, 221 (1969).

74. A. F. Machin, M. P. Quick, and D. F. Waddell, *Analyst*, **98**, 176 (1973).

75. J. W. Hogan and C. O. Knowles, *Bull. Environ. Contam. Toxicol.*, **8**, 61 (1972).

76. R. S. H. Yang, E. Hodgson, and W. C. Dauterman, *J. Agric. Food Chem.*, **19**, 10 (1971).

77. D. Eberle and D. Novak, *J. Assoc. Off. Anal. Chem.*, **52**, 1067 (1969).

78. N. M. Randolph, H. W. Dorough, and G. L. Teetes, *J. Econ. Entomol.*, **62**, 462 (1969).

79. M. C. Ivey and H. V. Claborn, *J. Assoc. Off. Anal. Chem.*, **52**, 1248 (1969).

80. W. Crisp and K. R. Tarrant, *Analyst*, **96**, 310 (1971).

81. D. R. Schultz, R. L. Marxmiller, and B. A. Koos, *J. Agric. Food Chem.*, **19**, 1238 (1971).

82. A. F. Machin, M. P. Quick, and D. F. Waddell, *Analyst*, **98**, 176 (1973).

83. M. C. Bowman, M. M. Cole, P. H. Clark, and D. E. Weidhaas, *J. Med. Entomol.*, **10**, 405 (1973).

84. R. E. Menzer and W. C. Dauterman, *J. Agric. Food Chem.*, **18**, 1031 (1970).

85. B. Y. Giang and H. Beckman, *J. Econ. Entomol.*, **16**, 899 (1968).

86. B. Y. Giang and H. Beckman, *J. Econ. Entomol.*, **17**, 63 (1969).

87. M. C. Bowman and M. Beroza, *J. Agric. Food Chem.*, **15**, 465 (1967).

88. W. A. Steller and N. R. Pasarela, *J. Assoc. Off. Anal. Chem.*, **55**, 1280 (1972).

89. D. W. Woodham, J. C. Hatchett, and C. A. Bond, *J. Agric. Food Chem.*, **22**, 239 (1974).

90. D. W. Woodham, R. G. Reeves, C. B. Williams, H. Richardson, and C. A. Bond, *J. Agric. Food Chem.*, **22**, 731 (1974).

91. R. W. Storherr and R. R. Watts, *J. Assoc. Off. Anal. Chem.*, **52**, 511 (1969).

92. J. D. MacNeil, R. W. Frei, S. Safe, and O. Hutzinger, *J. Assoc. Off. Anal. Chem.*, **55**, 1270 (1972).

93. L. P. Sarna, G. J. Howe, G. M. Findlay, and G. R. B. Webster, *J. Agric. Food Chem.*, **24**, 1046 (1976).

94. J. D. MacNeil, M. Hikichi, and F. L. Banham, *J. Agric. Food Chem.*, **23**, 758 (1975).

95. W. H. Harned and J. E. Casida, *J. Agric. Food Chem.*, **24**, 689 (1976).

96. M. C. Kleinschmidt, *J. Agric. Food Chem.*, **19**, 1196 (1971).

97. M. C. Bowman and M. Beroza, *J. Assoc. Off. Anal. Chem.*, **52**, 1231 (1969).

98. M. C. Bowman, M. Beroza and C. R. Gentry, *J. Assoc. Off. Anal. Chem.*, **52**, 157 (1969).

99. F. B. Ibrahim, J. M. Gilbert, R. T. Evans, and J. C. Cavagnol, *J. Agric. Food Chem.*, **17**, 300 (1969).

100. J. S. Thornton and C. A. Anderson, *J. Agric. Food Chem.*, **17**, 895 (1968).

101. N. S. Talekar and E. P. Lichtenstein, *J. Agric. Food Chem.*, **21**, 851 (1973).

102. J. B. McBain, L. J. Hoffman, and J. J. Menn, *J. Agric. Food Chem.*, **18**, 1189 (1970).

103. M. C. Ivey and H. D. Mann, *J. Agric. Food Chem.*, **23**, 319 (1975).

104. M. C. Bowman, C. L. Holder, and L. G. Rushing, *J. Agric. Food Chem.*, **26**, 35 (1978).

105. M. C. Ivey, *J. Assoc. Off. Anal. Chem.*, **59**, 261 (1976).

106. W. N. Yule and I. W. Varty, *Bull. Environ. Contam. Toxicol.*, **13**, 678 (1975).

107. R. L. Mundy, M. C. Bowman, J. H. Farmer, and T. J. Haley, *Arch. Toxicol.*, **41**, 111 (1978).

108. M. C. Bowman and M. Beroza, *J. Agric. Food Chem.*, **17**, 271 (1969).

109. M. C. Bowman and M. Beroza, *J. Agric. Food Chem.*, **16**, 399 (1968).

110. F. C. Wright and J. C. Riner, *J. Agric. Food Chem.*, **26**, 1258 (1978).

111. E. Mollhoff, *Pestic. Sci.*, **2**, 179 (1971).

112. H. H. Sauer, *J. Agric. Food Chem.*, **20**, 578 (1972).

113. W. H. Gutenmann and D. J. Lisk, *J. Agric. Food Chem.*, **19**, 200 (1971).

114. M. C. Ivey, R. A. Hoffman, and H. V. Claborn, *J. Econ. Entomol.*, **61**, 1647 (1968).

115. C. E. Polan and P. T. Chandler, *J. Dairy Sci.*, **54**, 847 (1971).

116. D. O. Eberle and W. D. Hormann, *J. Assoc. Off. Anal. Chem.*, **54**, 150 (1971).

117. D. L. Bull and A. B. Borkovec, *Arch. Environ. Contam. Toxicol.*, **1**, 148 (1973).

118. M. Tanabe, R. L. Dehn and R. R. Bramhall, *J. Agric. Food Chem.*, **22**, 54 (1974).

119. M. C. Bowman and M. Beroza, *J. Assoc. Off. Anal. Chem.*, **48**, 922 (1965).

120. M. C. Bowman and M. Beroza, *J. Assoc. Off. Anal. Chem.*, **49**, 1154 (1966).

121. M. C. Bowman and M. Beroza, *J. Assoc. Off. Anal. Chem.*, **50**, 940 (1967).

122. H. E. Braun, *J. Assoc. Off. Anal. Chem.*, **57**, 182 (1974).

123. N. Aharonson and A. Ben-Aziz, *J. Agric. Food Chem.*, **22**, 704 (1974).

124. R. A. Currie, *J. Assoc. Off. Anal. Chem.*, **57**, 1056 (1974).
125. M. C. Bowman and M. Beroza, *J. Agric. Food Chem.*, **17**, 1054 (1969).
126. R. L. Holmstead, T. R. Fukuto, and R. B. March, *Arch. Environ. Contam. Toxicol.*, **1**, 133 (1973).
127. H. E. Braun, F. L. McEwen, R. Frank, and G. Ritcey, *J. Agric. Food Chem.*, **23**, 90 (1975).
128. D. Gegiou, *Anal. Chem.*, **46**, 742 (1974).
129. A. El-Ratei and T. L. Hopkins, *J. Assoc. Off. Anal. Chem.*, **55**, 526 (1972).
130. W. Crisp and K. R. Tarrant, *Analyst*, **96**, 310 (1971).
131. P. B. Morgan, M. C. Bowman, and G. C. LaBrecque, *J. Econ. Entomol.*, **65**, 1269 (1972).
132. G. H. Cook and J. C. Moore, *J. Agric. Food Chem.*, **24**, 631 (1976).
133. P. S. Jaglan, R. B. March, T. R. Fukuto, and F. A. Gunther, *J. Agric. Food Chem.*, **18**, 809 (1970).
134. J. Gabica, J. Wyllie, M. Watson, and W. W. Benson, *Anal. Chem.*, **43**, 1102 (1971).
135. F. M. Kishk, T. El-Essawi, S. Abdel-Ghafar, and M. B. Abov-Donia, *J. Agric. Food Chem.*, **24**, 305 (1976).
136. J. W. Seiber and J. C. Markle, *Bull. Environ. Contam. Toxicol.*, **1**, 72 (1972).
137. S. J. Yu and F. O. Morrison, *J. Econ. Entomol.*, **62**, 1296 (1969).
138. K. I. Beynon, K. E. Elgar, B. L. Mathews, and A. N. Wright, *Analyst*, **98**, 194 (1973).
139. J. F.Lawrence and H. A. McLeod, *J. Assoc. Off. Anal. Chem.*, **59**, 637 (1976).
140. M. Sherman, J. Beck and R. B. Herrick, *J. Agric. Food Chem.*, **20**, 617 (1972).
141. Y. Iwata, W. E. Westlake, and F. A. Gunther, *Arch. Environ. Contam. Toxicol.*, **1**, 84 (1973).
142. R. L. Joiner and K. P. Baetcke, *J. Agric. Food Chem.*, **21**, 391 (1973).
143. R. L. Joiner and K. P. Baetcke, *J. Assoc. Off. Anal. Chem.*, **56**, 338 (1973).
144. R. L. Joiner and K. P. Baetcke, *J. Assoc. Off. Anal. Chem.*, **57**, 508 (1974).
145. E. R. White, K. M. Al-Adil, W. L. Winterlin, and W. W. Kilgore, *Bull. Environ. Contam. Toxicol.*, **10**, 140 (1973).
146. P. S. Jaglan and F. A. Gunther, *J. Chromatogr.*, **46**, 79 (1970).
147. T. E. Archer, *J. Agr. Food Chem.*, **22**, 974 (1974).
148. T. E. Archer, *J. Agr. Food Chem.*, **23**, 858 (1975).
149. L. Kliger and B. Yaron, Bull. *Environ. Contam. Toxicol.*, **13**, 714 (1975).
150. D. L. Suett, *Pestic. Sci.*, **5**, 57 (1974).
151. M. C. Bowman, M. Beroza, and J. H. Harding, *J. Agr. Food Chem.*, **17**, 138 (1969).

152. M. J. Brown, *J. Agr. Food Chem.*, **23**, 334 (1975).

153. W. O. Gauer and J. N. Seiber, *Bull. Environ. Contam. Toxicol.*, **6**, 183 (1971).

154. A. Guardigli, W. Chow, P. M. Martwinski, and M. S. Lefar *J. Agr. Food Chem.*, **19**, 742 (1971).

155. W. E. Westlake, M. Ittig, D. E. Ott, and F. A. Gunther, *J. Agr. Food Chem.*, **21**, 846 (1973).

156. W. A. Manson and C. E. Melozn, *J. Agr. Food Chem.*, **24**, 299 (1976).

157. M. C. Bowman and D. B. Leuck, *J. Agr. Food Chem.*, **19**, 1215 (1971).

158. F. C. Wright, *J. Agr. Food Chem.*, **23**, 820 (1975).

159. T. R. Roberts and G. Stoydin, *Pestic. Sci.*, **7**, 135 (1976).

160. H. V. Claborn, R. A. Hoffman, H. D. Mann, and D. D. Oehler, *J. Econ. Entomol.*, **63**, 1560 (1970).

161. J. Crossley, *J. Assoc. Off. Anal. Chem.*, **53**, 1036 (1970).

162. D. L. Bull and R. L. Ridgway, *J. Agr. Food Chem.*, **17**, 837 (1969).

163. L. W. Getzin, *J. Econ. Entomol.*, **61**, 1560 (1968).

164. M. C. Bowman, M. Beroza, and K. R. Hill, *J. Assoc. Off. Anal. Chem.*, **54**, 346 (1971).

165. M. C. Bowman, M. Beroza, and D. B. Leuck, *J. Agr. Food Chem.*, **16**, 796 (1968).

166. M. C. Bowman and M. Beroza, *J. Assoc. Off. Anal. Chem.*, **52**, 1054 (1969).

167. M. A. Luke, J. E. Froberg, and H. T. Masumoto, *J. Assoc. Off. Anal. Chem.*, **58**, 1220 (1975).

168. M. Beroza and M. C. Bowman, *J. Assoc. Off. Anal. Chem.*, **49**, 1007 (1966).

169. M. Beroza and M. C. Bowman, *Anal. Chem.*, **37**, 291 (1965).

170. M. Beroza and M. C. Bowman, *J. Assoc. Off. Anal. Chem.*, **43**, 358 (1965).

171. M. C. Bowman and M. Beroza, *J. Assoc. Off. Anal. Chem.*, **48**, 943 (1965).

172. M. Beroza and M. C. Bowman, *Anal. Chem.*, **38**, 837 (1966).

173. M. C. Bowman and W. Beroza, *Anal. Chem.*, **38**, 1427 (1966).

174. M. C. Bowman and M. Beroza, *Anal. Chem.*, **38**, 1544 (1966).

175. M. Beroza, M. N. Inscoe, and M. C. Bowman, *Res. Rev.*, **30**, 1 (1969).

176. J. R. Pardue, *J. Assoc. Off. Anal. Chem.*, **54**, 359 (1971).

177. R. W. Storherr, P. Ott, and R. R. Watts, *J. Assoc. Off. Anal. Chem.*, **54**, 513, (1971).

178. B. Versino, M. Th. Van der Venne, and H. Vissers, *J. Assoc. Off. Anal. Chem.*, **54**, 147 (1971).

179. J. Pflugmacher and W. Ebing, *J. Chromatogr.*, **93**, 457 (1974).

180. D. L. Stalling, R. C. Tindle, and J. L. Johnson, *J. Assoc. Off. Anal. Chem.*, **55**, 28 (1972).

181. R. C. Tindle and D. L. Stalling, *Anal. Chem.*, **44**, 1768 (1972).

182. K. R. Griffitt and J. C. Craun, *J. Assoc. Off. Anal. Chem.*, **57**, 168 (1974).
183. R. H. Waltz, *12th Ann. Pest. Conf.*, *Fla. Dept. Agr.*, *Palm Beach, Fla.*, *July 15, 1975.*
184. L. D. Johnson, R. H. Waltz, J. P. Ussary, and F. E. Kaiser, *J. Assoc. Offic. Anal. Chem.*, **59**, 174 (1976).
185. R. W. Storherr and R. R. Watts, *J. Assoc. Off. Anal. Chem.*, **48**, 1154 (1965).
186. R. R. Watts and R. W. Storherr, *J. Assoc. Off. Anal. Chem.*, **48**, 1158 (1965).
187. R. W. Storherr and R. R. Watts, *J. Assoc. Off. Anal. Chem.*, **51**, 662 (1968).
188. O. W. Grussendorf, A. J. McGinnis, and J. Solomon, *J. Assoc. Off. Anal. Chem.*, **53**, 1048 (1970).
189. H. A. McLeod and P. J. Wales, *J. Agr. Food Chem.*, **20**, 624 (1972).
190. M. J. Matherne and W. H. Bathalter, *J. Assoc. Off. Anal. Chem.*, **49**, 1012 (1966).
191. R. M. Hetherington and C. J. Parouchais, *J. Assoc. Off. Anal. Chem.*, **53**, 146 (1970).
192. A. Karmen and L. Guiffrida, *Nature*, **201**, 1204 (1964).
193. W. A. Aue, *J. Cromatogr. Sci.*, **13**, 329 (1975).
194. J. E. Baudean, *Can. Res. Dev.* May-June **1975.**
195. Anonymous, *Retention Times*, Austin, TX, Vol. 4, Tracor, Inc. (August, 1977).
196. S. S. Brody and J. E. Chaney, *J. Gas Chromatogr.*, **4**, 42 (1966).
197. M. Beroza and M. C. Bowman, *Environ. Sci. Technol.*, **2**, 450 (1968).
198. H. A. Moye, *Anal. Chem.*, **41**, 1717 (1969).
199. W. E. Dale and C. C. Hughes, *J. Gas Chromatogr.*, **6**, 603 (1968).
200. M. C. Bowman and M. Beroza, *Anal. Chem.*, **40**, 1448 (1968).
201. H. P. Burchfield, J. W. Rhoades, and R. J. Wheeler, *J. Agr. Food Chem.*, **13**, 513 (1965).
202. B. Berck, W. E. Westlake, and F. A. Gunther, *J. Agr. Food Chem.*, **18**, 143 (1970).
203. J. E. Wright and M. C. Bowman, *J. Econ. Entomol.*, **66**, 707 (1973).
204. R. W. Finch, *Analabs Res. Notes*, **10**, 1 (1970).
205. M. C. Bowman and M. Beroza, *J. Assoc. Off. Anal. Chem.*, **50**, 1228 (1967).
206. M. C. Bowman, *J. Chromatogr. Sci.*, **13**, 307 (1975).
207. J. S. Thornton, *Chemagro Corp. Rept. No. 30, 993* (1971).
208. R. R. Watts, *Res. Rev.*, **18**, 105 (1967).
209. J. D. MacNeil and R. W. Frei, *J. Chromatogr. Sci.*, **13**, 279 (1975).
210. C. E. Mendoza and J. B. Shields, *J. Assoc. Off. Anal. Chem.*, **54**, 508 (1971).
211. K. Cywinska-Smoter, *Rocz. Panstw. Zakl. Hig.*, **23**, 505 (1972); *Chem. Abstr.*, **78**, 39090q (1973).

212. G. Ligeti and A. Katona, *Gyogyszereszet*, **18**, 11 (1974); *Chem. Abstr.*, **81**, 34327z (1974).

213. D. C. Villeneuve, A. G. Butterfield, D. L. Grant, and K. A. McCully, *J. Chromatogr.*, **48**, 567 (1970).

214. S. N. Tewari and S. P. Harplani, *Mikrochim. Acta*, **2**, 321 (1973).

215. A. M. Gardner, *J. Assoc. Off. Anal. Chem.*, **54**, 517 (1971).

216. L. P. Manner, *J. Chromatogr.*, **21**, 430 (1966).

217. R. W. Frei, V. Mallet, and C. Pothier, *J. Chromatogr.*, **59**, 135 (1971).

218. A. El-Refai and T. L. Hopkins, *J. Agr. Food Chem.*, **13**, 479 (1965).

219. R. R. Watts, *J. Assoc. Offic. Anal. Chem.*, **48**, 1161 (1965).

220. G. L. Brun and V. Mallet, *J. Chromatogr.*, **80**, 117 (1973).

221. C. E. Mendoza, *Res. Rev.*, **43**, 105 (1972).

222. M. Beroza, K. R. Hill, and K. H. Morris, *Anal. Chem.*, **40**, 1611 (1968).

223. M. S. Lefar and A. D. Lewis, *Anal. Chem.*, **42**, 79A (1970).

224. M. E. Getz and K. R. Hill, Paper No. 54, ACS Div. Pestic. Chem., 163rd ACS Natl. Mtg., Boston, April (1972).

225. R. G. Hurtubise, P. F. Lott, and J. R. Dias, *J. Chromatogr. Sci.*, **11**, 476 (1973).

226. R. W. Frei and J. D. MacNeil, *Diffuse Reflectance Spectroscopy in Environmental Problem-Solving*, CRC Press, Cleveland, OH, 220 pp (1973).

227. H. A. Moye, *J. Chromatogr. Sci.*, **13**, 268 (1975).

228. R. A. Henry, J. A. Schmit, J. F. Dieckman, and F. J. Murphey, *Anal. Chem.*, **43**, 1053 (1971).

229. J. G. Koen, J. F. K. Huber, H. Poppe, and G. den Boef, *J. Chromatogr. Sci.*, **8**, 192 (1970).

230. T. C. Wade and H. A. Moye, Paper No. 7, Pestic. Div., 166th ACS Natl. Mtg. Chicago, IL, August (1973).

231. F. W. Willmott and R. J. Dolphin, *J. Chromatogr. Sci.*, **12**, 695 (1974).

232. M. J. Cohen and F. W. Karasek, *J. Chromatogr. Sci.*, **8**, 330 (1970).

233. H. A. Moye, *J. Chromatogr. Sci.*, **13**, 285 (1975).

234. G. D. Winter, *N.Y. Acad. Sci.*, **87**, 629 (1960).

235. P. J. Gary and J. I. Roth, *Clin. Chem.*, **11**, 91 (1965).

236. W. Rusiecki, J. Brzezinski, and M. Szutowski, *Acta Pol. Pharm.*, **28**, 385 (1971); *Health Aspects of Pesticides Abstr. Bull.*, **6**, 0461 (1973).

237. C. R. Turner, *Analyst*, **99**, 431 (1974).

238. H. A. McLeod, *J. Chromatogr. Sci.*, **13**, 302 (1975).

239. D. E. Ott, *J. Agr. Food Chem.*, **16**, 874 (1968).

240. H. A. McLeod, A. G. Butterfield, D. Lewis, W. E. J. Phillips, and D. E. Coffin, *Anal. Chem.* **47**, 674 (1975).

241. M. C. Bowman, H. C. Young, and W. F. Barthel, *J. Econ. Entomol.*, **58**, 896 (1965).

242. W. P. Cochrane, *J. Chromatogr. Sci.*, **13**, 246 (1975).

243. M. A. Forbes, B. P. Wilson, R. Greenhalgh, and W. P. Cochrane, *Bull. Environ. Contam. Toxicol.*, **13**, 141 (1975).

244. J. Singh and M. R. Lapointe, *J. Assoc. Off. Anal. Chem.*, **57**, 1285 (1974).

245. R. Greenhalgh and J. Kavacicova, *J. Agr. Food Chem.* **23**, 325 (1975).

246. R. Greenhalgh, J. Dokladolova, and W. O. Haufe, *Bull. Environ. Contam. Toxicol.*, **7**, 237 (1972).

247. W. F. Phillips, M. C. Bowman, and R. J. Schultheisz, *J. Agr. Food Chem.*, **10**, 486 (1962).

248. G. Vander Velde and J. F. Ryan, *J. Chromatogr. Sci.*, **13**, 322 (1975).

CARBAMATE INSECTICIDE RESIDUE ANALYSIS BY GAS–LIQUID CHROMATOGRAPHY

JAMES N. SEIBER

Department of Environmental Toxicology,
University of California, Davis

CONTENTS

8.1. INTRODUCTION

The group of commercial insecticides known as the carbamates (Table 8.1) have increased substantially in popularity in recent years as a consequence of their (generally) selective insecticidal properties, low mammalian toxicity, and lack of undue persistence. A relatively low chemical and biochemical stability accounts for these desirable properties, but also introduces analytical difficulties. For example, some carbamates may be unstable to column chromatography on highly active adsorbents, and some may undergo at least partial thermal decomposition under relatively common gas–liquid chromatographic (GLC) conditions. Furthermore, many of the carbamates lack electron-capturing properties or heteroatoms amenable to the use of highly sensitive and selective detectors which have so greatly facilitated the analysis for residues of other pesticide classes. As a result methodology basic to the analysis of organochlorine and organophosphorus

TABLE 8.1. Nomenclature for Carbamate Insecticides Mentioned in Text

Common Name	Trade Name (or other)	Chemical Designation
Aldicarb	Temik	2-Methyl-2-(methylthio)propionalde-hyde O-(methylcarbamoyl)oxime
Aminocarb	Matacil	4-Dimethylamino-3-methylphenyl N-methylcarbamate
	Bay 32651	4-(Methylthio)-2-methylphenyl N-methylcarbamate
	Bay 78537	2,3-Dihydro-2,2-dimethyl-7-benzofura-nyl N-acetyl-N-methylcarbamate
	Butacarb	3,5-Di-*tert*-butylphenyl N-methylcarbamate
Carbanolate	Banol	2-Chloro-4,5-dimethylphenyl N-methylcarbamate
Carbaryl	Sevin	1-Naphthyl N-methylcarbamate
Carbofuran	Furadan	2,2-Dimethyl-2,3-dihydro-7-benzofura-nyl N-methylcarbamate
	Hopcide (CPMC)	2-Chlorophenyl N-methylcarbamate
	Ciba C-9643	O-(4-Methyl-1,3-dioxolan-2-yl)phenyl N-methylcarbamate
	Cosban, Macbal	3,5-Dimethylphenyl N-methylcarbamate
Decarbofuran	Bay 62863	2,3-Dihydro-2-methyl-7-benzofuranyl N-methylcarbamate
Dimetilan	Dimetilan	1-(Dimethylcarbamoyl)-5-methyl-3-py-razolyl N,N-dimethylcarbamate
Formetanate	Carzol	m-{[(Dimethylamino)methylene]amino} phenyl N-methylcarbamate
Isoprocarb	Etrofolan, Mipcin, MIPC	2-Isopropylphenyl N-methylcarbamate
	Hercules 5727	3-Isopropylphenyl N-methylcarbamate
	Hercules 9007	3-Isopropylphenyl N-(chloroacetyl)-N-methylcarbamate
	Landrin	Mixture 3,4,5-trimethylphenyl and 2,3,5-trimethylphenyl N-methylcar-bamate
	Meobal	3,4-Dimethylphenyl N-methylcarba-mate
Metalkamate, bufencarb	Bux	3:1 Mixture 3-(1-methylbutyl)phenyl and 3-(1-ethylpropyl)phenyl N-methylcarbamate
Methiocarb, metmercapturon	Mesurol	4-(Methylthio)-3,5-dimethylphenyl N-methylcarbamate

TABLE 8.1.   (Continued)

Common Name	Trade Name (or other)	Chemical Designation
Methomyl	Lannate, Nudrin	Methyl N-[(methylcarbamoyl) oxy]thioacetamidate
Mexacarbate	Zectran	4-(Dimethylamino)-3,5-dimethylphenyl N-methylcarbamate
	Mobam	Benzo[b]thien-4-yl N-methylcarbamate
Oxamyl	Vydate	Methyl 2-(dimethylamino)-N-{[(methylamino)carbonyl]oxy}-2-oxo-ethanimidothioate
Pirimicarb	PP062	2-(Dimethylamino)-5,6-dimethyl-4-py-rimidinyl N,N-dimethylcarbamate
Promecarb	Minacide	3-Methyl-5-isopropylphenyl N-methylcarbamate
Propoxur	Baygon	2-Isopropoxyphenyl N-methylcarba-mate
Pyrolan		1-Phenyl-3-methyl-5-pyrazolyl N,N-dimethylcarbamate
Terbutol	Azak	2,6-Di-tert-butyl-4-methylphenyl N-methylcarbamate
Thiofanox	Dacamox	3,3-Dimethyl-1-(methylthio)-2-butanone O-(methylcarbamoyl) oxime
	Tranid	2-exo-Chloro-6-endo-cyano-2-norborna-none O-(methylcarbamoyl)oxime
	Tsumacide (MTMC)	3-methylphenyl N-methylcarbamate

insecticides embodied in multiresidue schemes such as that of the FDA (1) is inappropriate for the carbamates as a class. There have thus evolved GLC methods for carbamates which are characterized by relatively mild cleanup conditions, the use of deactivated columns or derivatization for improved GLC resolution, and detection via nitrogen-selective detectors or others responding to a derivatized fragment.

Proper selection of pesticide residue methods depends to a considerable extent on the commodity or substrate of interest, the required analytical sensitivity, accuracy, and precision, the need for accounting for metabolites and other conversion products, and whether or not a single compound or group of compounds must be considered. These variables account for the seemingly endless proliferation of methods and their modifications for the carbamates, as well as other pesticide classes. Detailed methods suitable for particular carbamates and their metabolites (compound-specific methods), often applicable to only one or a few commodities, are particularly useful in studies of field dissipation, tissue clearance, and residue tolerance adherence. However, for samples of unknown history, "screening" or "multiresidue" procedures, which can account simultaneously for compounds of several classes, are more appropriate. These latter include methods capable of signaling the presence (or absence) of classes of chemicals—an example being cholinesterase inhibition screens for the carbamates and organophosphorus insecticides—and those quantitative for each of several compounds from analysis of a single sample set.

This review will focus primarily on compound-specific methods for the carbamates from necessity, since these are most adequately documented in the literature. As other chapters in this book deal with thin-layer and high-performance liquid chromatographic determinations, only those methods based on GLC will be considered. Extraction and cleanup steps will be mentioned in connection with the appropriate GLC resolution/detection technique, the latter providing the major organizational framework for the chapter. Emphasis will be placed on methods—particularly the more recent ones—which have been most completely verified in the literature, at the expense of aiming for a complete review of all the work which has been forthcoming in the field of carbamate residue chemistry. Particularly useful and recent comprehensive reviews in the field include those of Dorough and Thorstenson (2) and Magallona (3); other reviews covering aspects of the subject include ones by Cochrane (4), Kanazama (5), Ruzicka (6), Thornburg (7), and Williams (8).

8.2. COMPOUND-SPECIFIC METHODS

8.2.1. General Considerations

The use of GLC has followed a somewhat circuitous route as a primary technique for carbamate residue determination, unlike its early, straightforward adoption and virtual monopoly since the early 1960s in the determination of organochlorine and organophosphorus pesticides. The lack of a halide, sulfur, or phosphorus "handle" in the majority of carbamates, and early misgivings about their thermal stability contributed to this situation. Reinforcing the latter, Zielinski and Fishbein (9) found only 1-naphthol, confirmed by a color reaction, in the GLC effluent upon injection of carbaryl on columns of Carbowax 20M and SE 30; this was in marked contrast to the behavior of simple carbamates, urethane and its derivatives, which chromatographed intact on a variety of columns (10). In studying 11 aryl N-methylcarbamates, many pesticidal in nature, Strother (11) and Wheeler and Strother (12) found extensive GLC-induced decomposition to phenols, while Peck and Harkiss (13) used mass spectrometry to confirm the break-down of carbaryl to 1-naphthol on a 2% SE 30 column.

On-column conversion of carbamates to phenols has been made quantita-tive and reproducible by inserting a phosphoric acid-treated plug of the solid support at the head of the column (14). This may form the basis for residue-level quantitation of those carbamates in which the phenol moiety contains elements allowing selective and sensitive detection; examples are Mobam and methiocarb, the phenols of which contain sulfur.

From the middle 1960s to the present considerable effort has been devoted to overcoming the decomposition tendency through the formation of stable derivatives. The propensity of carbamates to hydrolyze in mild alkali or under strongly acidic conditions led to derivatization of the hydrolysis products, that is, phenols and alkyl (usually methyl) amines. This approach has the added advantage of allowing for introduction of an electron-capturing or heteroatom-containing "handle," with a consequent gain in detection sensitivity, through manipulation of the derivatizing reac-tion. Somewhat later the advantages of more direct derivatization of the intact carbamate, usually at the carbamoyl nitrogen, were recognized; the unique structural features of the carbamates, particularly the presence of the carbamoyl nitrogen, were preserved in the derivative which in turn could be formed under conditions minimizing the derivatization of potential interferences.

Major efforts in recent years have aimed at overcoming the thermal decomposition tendency through the use of relatively inactive or specially deactivated GLC columns. It now appears that most carbamates can be reproducibly chromatographed as intact molecules with appropriate precau-tions; detection is based on nitrogen-selective electrolytic conductivity or alkali-flame ionization (thermionic) detectors, or the electron-capturing or heteroatom-containing parts of particular carbamates. These general considerations, embodied in the summary of Figure 8.1, will form the basis for subsequent discussion of analytical methods based on GLC.

8.2.2. Direct Gas Chromatography (Table 8.2)

Riva and Carisano (15) described a set of conditions under which carbaryl could be gas chromatographed essentially intact. Using a 1-m glass column packed with 0.5% SE 30 on silanized Gas Chrom P maintained at 168° and a relatively low temperature (185°) glass-lined inlet, with periodic on-column silanization, less than ca. 20% breakdown to 1-naphthol was observed for carbaryl injections of 0.3–10μg. Decomposition, which increased when smaller quantities were injected and when on-column deacti-vation was omitted, was measured by the response of the KCl thermionic

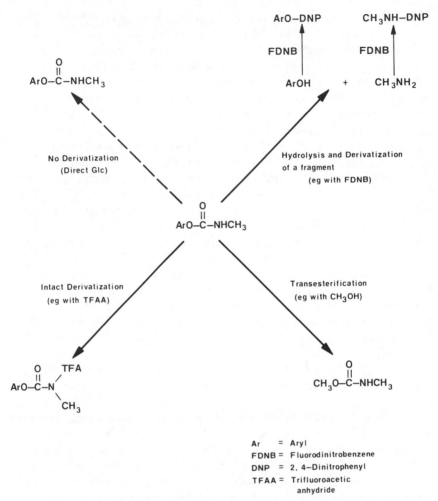

Figure 8.1 Summary of methods of sample preparation for gas chromatography of aryl *N*-methylcarbamate insecticides.

detector, or with the FID response relative to a methyl pentadecanoate internal standard. Similarly, Sans (16) chromatographed carbaryl, carbofuran, and methomyl intact on a 10% QF-1 column, and noted that aqueous methanol solutions could be injected directly without adversely affecting column or detector (AFID) performance.

Using columns of 5 and 10% DC 200 on Chromosorb W HP deactivated by on-column silanization with Silyl-8 during conditioning, Laski and Watts (17) obtained a 50% FSD N-response with the Coulson electrolytic conduc-

TABLE 8.2. Examples of Direct GLC Determination of Carbamate Residues

Carbamates	GLC Column Conditions	Detector[a]	Applications	Ref.
Carbaryl	1 m 0.5% SE 30 on 100/120 mesh silanized Gas Chrom P, 168°C	AFID	None	15
Aldicarb, aminocarb, carbaryl, carbofuran, Landrin (2 isomers), metalkamate, methomyl, mexacarbate, Mobam, propoxur	6 ft 10% DC 200 on 80/100 mesh Chromosorb W HP on-column silanized, 180°C	CECD	Carbaryl in apples, cherries, peaches	17
Carbaryl	0.3 m 3% SE 30 on 80/100 mesh Gas Chrom Q, 145–150°C	EC	Contaminated water	19
Methomyl	1.8 m 6% OV 210/4% OV 101 on 60/80 mesh Gas Chrom Q, 160°C	MC	None	22
Carbofuran and metabolites	2 ft 20% SE 30 on 60/80 mesh Gas Chrom Z on-column silanized, 165°C	MC	42 commodities	24, 26–8
Carbofuran and metabolite	6 ft 6% OV 210/4% OV 101 on 60/80 mesh Gas Chrom Q, 160°C	CECD	Blueberries, cranberries, raspberries, strawberries	32
Carbofuran and metabolites	4 ft 3% Apiezon L 80/100 mesh Chromosorb W HP 165°C	NP or CECD	Paddy water, rice foliage, rice grain	29–31
Metalkamate	2 ft 5% SE 52 + Reoplex 400 (4:1) on 60/80 mesh AW Chromosorb W, 160°C	MC	Corn, rice, soil, water	33
Carbaryl, methiocarb, mexacarbate, promecarb	1.74 m 60/80 mesh AW Chromosorb W surface-modified with Carbowax 20M, 138–183°C	AFID	Lettuce, tomatoes	34
Pirimicarb	0.5 m 3% phenyldiethanolamine succinate on 100/200 mesh Gas Chrom Q, 180°C	AFID	Variety of crops	37

[a] AFID = Alkali-flame ionization detector, CECD = Coulson electrolytic conductivity detector, EC = electron capture detector, MC = microcoulometric detector, NP = nitrogen-phosphorus (thermionic) detector.

tivity detector for 11 carbamates and some N-containing metabolites. Injection levels, 20–100 ng, were compatible with quantities available in residue extracts. The isolation scheme of Porter et al. (18) was coupled with GLC for carbaryl residue determination on cherries, apples, and peaches without noteworthy interference.

Carbaryl and presumably some other aryl N-methylcarbamates are sufficiently responsive to electron-capture for residue sensitivity. Using a short (0.3 m) 3% SE-30 on Gas Chrom Q column, and low column (145–150°) and inlet (170°) temperatures, Lewis and Paris (19) chromatographed carbaryl intact, with levels as low as 0.2 ng detectable by electron capture. Breakdown was minimized by long (48 hr) column conditioning at 225°, and by periodic conditioning with a concentrated carbaryl standard. Departure of column temperature from the 145–150° optimum decreased sensitivity. No interferences were noted in uncleaned isooctane extracts of water containing high microbial populations and organic nutrients. A similar column was used with on-column injection for carbaryl formulation analysis (20). More extreme inlet (270°) and column (190°) temperatures were successfully employed, perhaps because of the large amounts of carbaryl analyzed and the on-column injection techniques followed. Ragab (21) chromatographed 3-ketocarbofuran intact on a 36 cm 3% OV-210 column maintained at 155°C, with a relatively low (200°C) inlet temperature. As little as 1 ppb was determined in extracts of fortified water samples using electron-capture detection.

Williams (22) inferred that methomyl may also be gas chromatographed apparently intact, in contrast with the findings of others (23) working with this oxime carbamate. Conditions included a 1.8-m column of 6% OV 210/4% OV 101 mixed phase on Gas Chrom Q at 160° and use of a microcoulometric detector. Methomyl response was not linear throughout the concentration range tested, but a measurable and reproducible response was obtained with 20 ng. Unfortunately no data were presented to verify the applicability of these conditions to environmental samples. Furthermore, since the nature of the eluted component(s) was not confirmed, the possibility of fragmentation in the heated inlet or chromatographic column can not be dismissed.

The first direct GLC method with an apparently wide degree of applicability to residues of the aryl N-methylcarbamates was reported by Cook et al. (24). Carbofuran and its 3-hydroxy metabolite were determined by GLC on a 2-foot aluminum column packed with freshly precipitated SE 30 (20%) on Gas Chrom Z. The column was conditioned by on-column silanization and daily pretreatment with sample extracts to maintain a low state of surface activity. Carbofuran and its 3-hydroxy derivative eluted at 3 and 6 min, respectively, at a column temperature of 165° (Figure 8.2). A microcoulometric detector operated in the N mode gave a good response for

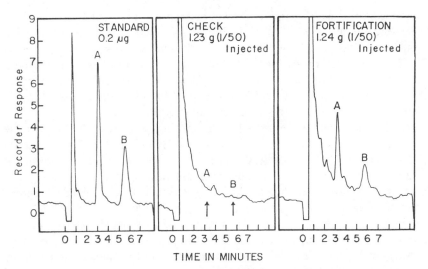

Figure 8.2 Gas chromatograms of underivatized carbofuran (*A*) and 3-hydroxycarbofuran (*B*) standards, and corn kernel extracts from fortification studies. A 2-ft 20% SE 30 on 60/80 mesh Gas Chrom Z column deactivated by on-column silanizaton, column temperature of 165°C, and microcoulometric detector were used (24).

injections of 100–200 ng. Samples (corn kernels, stover, and cobs) were extracted by refluxing with 0.25 *N* HCl which also served to free the aglycone of the 3-hydroxy metabolite from its glucoside conjugate. Residue from the methylene chloride extract of the acid phase was cleaned by hexane–acetonitrile partitioning, followed by chromatography on a Nuchar-Attaclay and silica gel mixed bed column eluted with methylene chloride and 30% hexane in ethyl acetate. Average recoveries of the two compounds from kernels and stover fortified at 0.05–0.9 ppm were 63–72%. Efficiency of the extraction and cleanup steps was checked with radiolabeled carbofuran. The acid extraction was further verified by comparison with exhaustive solvent extraction procedures (25).

This basic method, reported in detail in several references (26–28), has been extended to 42 sample commodities and may be applied without modification to 3-keto carbofuran metabolite (28). Acid alumina deactivated with 3% water may be substituted for silica gel, and a 3% Apiezon L on 80/100 mesh Chromosorb W HP column and Coulson electrolytic conductivity detector may be used in GLC (29). Direct GLC with the Apiezon column was coupled with a recent introduced NP (thermionic) detector for analysis of carbofuran extracted from rice paddy water (30) and, after a further modification of the column cleanup, to carbofuran and its carbamate metabolities from rice foliage and grain (31). Williams and Brown (32) determined carbofuran residues in strawberries, raspberries, blueberries, and cranberries with a similar extraction, substituting a silica gel–alumina cleanup column and a 6% OV 210/4% OV 101 GLC column coupled with a Coulson conductivity detector. Tucker (33) applied the method of Cook et al. (24) to the determination of metalkamate (Bux) residues in corn plant, corn ears, rice straw, rice grain, soil, and water at levels as low as 0.02 ppm. A 2 ft column containing 5% SE 52-Reoplex 400 (4:1) on acid-washed and silanized Chromosorb G was a notable substitution.

A second direct GLC procedure promising general utility to residue analysis of carbamates is that of Lorah and Hemphill (34). The extraction and cleanup of Holden (35) (see Section 8.2.4) was followed for determination of carbaryl, methiocarb, promecarb, and mexacarbate in vegetable samples, but without the methylene chloride–KOH partition designed to remove phenolic degradation products. A column of Chromosorb W surface-modified with carbowax 20M (36) and operated isothermally at 160–183° or temperature programmed in the same temperature range gave well-resolved, symmetrical peaks for the four carbamates, with response detected by alkali-flame ionization (Figure 8.3). From the linear calibration curves and consistent recoveries—90–100% for tomato and lettuce samples fortified at 0.25–25 ppm—it appeared that no degradation occurred on the column. Furthermore, no deterioration in column performance was noted over a 3-month period.

Bullock (37) described a direct GLC method for determining residues of the N,N-dimethylcarbamate insecticide pirimicarb in a variety of crops. A 0.5–m column containing 3% phenyldiethanolamine succinate on Gas Chrom Q at 180° and CsBr thermionic detector were employed, following chloroform extraction and acid partition cleanup of the basic pirimicarb and its desmethyl metabolite.

From the foregoing examples it appears feasible to analyze many of the common carbamates at the residue level by direct GLC. Close attention to column preparation and deactivation (as well as injector port design and temperature, from this author's observations) is undoubtedly more important to success here than in residue analyses for the more stable pesticides. New developments to enhance the sensitivity of the alkali-flame ionization (38–40) and electrolytic conductivity (41–43) detectors may encourage further use of these techniques. The potentially large increase in resolution offered by capillary GLC columns might, when coupled with high sensitivity N-detection, give an indication of things to come in direct GLC analysis of the carbamates (44, 45).

8.2.3.   Intact Derivatization (Table 8.3)

The carbamoyl N—H represents a reactive position common to most of the N-methylcarbamates. For example, stable N-acyl (46–48) and N-tri-

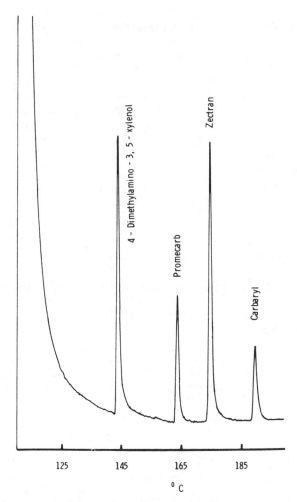

Figure 8.3 Temperature-programmed gas chromatogram of three underivatized carbamates and the phenol corresponding to mexacarbate (Zectran): 3 min. hold at 125°C, programmed at 5°/min. A 1.75-m column packed with 60/80 mesh AW Chromosorb W, surface modified with carbowax 20 M, and alkali-flame ionization detector were used (34).

methylsilyl derivatives (49) may be prepared by relatively straightforward techniques. Lau and Marxmiller (50) were the first to utilize N-derivatization of an intact carbamate for practical residue determination; Landrin was converted to its *N*-trifluoroacetyl (*N*-TFA) derivative by overnight treatment with trifluoroacetic anhydride (TFAA) in ethyl acetate at room temperature. A symmetrical and sensitive response of this derivative resulted using GLC on a 1.7-m column of 2% Reoplex 400 on 60/100 mesh

TABLE 8.3. Examples of GLC Determination of Carbamate Residues Based on Intact Derivatization at the Carbamoyl Nitrogen

Carbamates	Derivatization Conditions[a]	Applications	Ref.
Landrin, both isomers, and metabolites	TFAA in ethyl acetate at room temperature overnight	Corn ears, ensilage, stover, oats, soybeans	50
Aminocarb, carbaryl, carbofuran and metabolites, Landrin 3,4,5-isomer, metalkamate, methiocarb, Mobam, propoxur	TFAA in benzene at 100°C for 2 hr	None	51
Carbanolate, carbaryl, 11 other carbamate pesticides	TFAA in ethyl acetate at room temperature for 48 hr	None	52
Aminocarb, carbaryl, carbofuran, Landrin 2,3,5-isomer, methiocarb, mexacarbate, propoxur	PFPA and pyridine in isooctane at room temperature for 1 hr	Ethylene glycol air samples	55
Carbofuran and metabolites	TFAA in ethyl acetate at 45°C for 16 hr	Bird, fish, oyster, shrimp, skate	58
Carbofuran and metabolites	TFAA in benzene at 100°C for 2 hr	Alfalfa	59
Carbaryl and 1-napththol	HFBA and pyridine in benzene at 30°C for 1 hr	Water	64
Carbofuran and metabolites	HFBA and pyridine in isooctane at RT for 1 hr	Carrots, celery, tomatoes, corn	65
Methiocarb and metabolites	TFAA in ethyl acetate at 100° for 2 hr	Blueberries	66
Carbaryl	CH_3SCl and pyridine at 60–65°C for 2 hr	Beans, carrots, lettuce, tomatoes	70

[a] TFAA = Trifluoroacetic anhydride; PFPA = pentafluoropropionic anhydride; HFBA = heptafluorobutyric anhydride.

344

Gas Chrom Q at 170°, or a 1.7 m column of 3% OV-17 on 100/120 mesh Gas Chrom Q at 185° (Figure 8.4), both with an electron-capture detector. The two major isomers of Landrin and two hydroxylated metabolities formed similar derivatives, all separable by GLC. The determination was applied to Landrin fortified to corn (ears, stover, and ensilage), oats and soybeans at 0.2 ppm; samples were extracted with acetonitrile, and cleaned by hexane partitioning and column chromatography on alumina-Darco-Solka-Floc eluted with 3:1 hexane:ether. Recoveries were 70–110%.

N-TFA derivatization was applied to eight aryl N-methylcarbamate insecticides, and the 3-keto and 3-hydroxy metabolites of carbofuran (51).

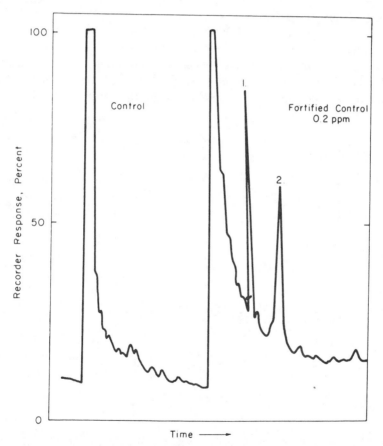

Figure 8.4 Gas chromatograms of corn ear extracts showing recovery of the 2,3,5-isomer (1) and 3,4,5-isomer (2) of Landrin as their N-trifluoroacetyl derivatives. A 6-ft 3% OV-17 on 100/120 mesh Gas Chrom Q column at 185°C and electron capture detector were used (50).

From examination of reaction variables quantitative derivatization was obtained with TFAA in either ethyl acetate or benzene solvent at 100° for 2 hr, but reagent background was lower with the benzene solvent. Better chromatographic efficiency was obtained with the derivatives than with the parent compounds (Figure 8.5); GLC columns of SE-30 or AN-600 gave symmetrical peaks. Good sensitivity resulted with either electron-capture or alkali-flame ionization detectors. Retention times of the TFA, pentafluoropropionyl (PFP), and heptafluorobutyryl (HFB) derivatives were similar. The HFB derivative, however, gave a slight increase in electron-capture response when compared with the TFA and PFP derivatives. The improvement in chromatographic efficiency offered by perfluoroacyl derivatization allowed for fair resolution of the six major peaks of the multicomponent carbamate metalkamate.

Suzuki et al. (52) employed thin layer chromatography (TLC) to separate a mixture of 13 aryl N-methylcarbamates, including carbanolate and carbaryl, into four fractions, each of which was resolved by GLC as their N-TFA derivatives on an OV-17 column. Without TLC fractionation, the 13 compounds (as their derivatives) were not completely resolved by GLC.

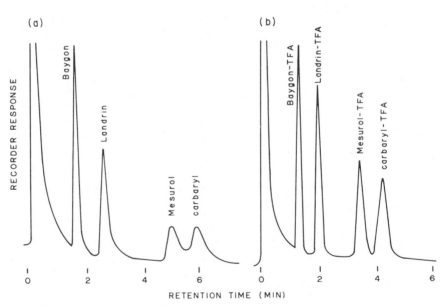

Figure 8.5 Gas chromatograms of a mixture of four N-methylcarbamates (*a*) before derivatization and (*b*) after derivatization with trifluoroacetic anhydride. 100 ng of each compound was injected on a 4-ft 3% SE 30 on 60/80 mesh AW DMCS-treated Chromosorb G column at 180°C, using an alkali-flame ionization detector (51).

Pyridine catalyzes the *N*-perfluoracyl derivatization, as well as the formation of N-chloracetyl derivatives. *N*- and *O*-TFA derivatives of carbaryl and its ring hydroxylated metabolites were formed with TFAA or *N*-trifluoroacetylimidazole in benzene containing pyridine in 15 min at 25° (53). The products were characterized by mass spectrometry (54). *N*-Hydroxymethyl carbaryl did not derivatize intact, but gave rise to the TFA derivative of 1-naphthol. A similar procedure (PFPA in isooctane containing pyridine) was used to analyze 7 carbamates extracted from ethylene glycol air sampling medium (55) (see Section 8.3). Pyridine serves both as a catalyst and as a competitive base in the perfluoroacylation reaction, preventing the formation of the unreactive perfluoroacid conjugate of the carbamate. The minimum pyridine/carbamate ratio (2.86:1) for high yield derivatization was determined by NMR for TFAA with carbofuran (56). Trimethylamine may be substituted for pyridine; this base in benzene solvent promoted high yields for TFA and HFB derivatives of several carbamates after 15–20 min at room temperature (57).

Uncatalyzed TFA derivatization, followed by electron-capture GLC, formed the basis for determination of carbofuran and its 3-hydroxy and 3-keto metabolites in animal tissue (58). Samples were extracted by refluxing with 0.25N HCl for 1 hr, followed by methylene chloride partitioning of the acid extract. Cleanup was carried out by successive (1) column chromatography on Florisil, (2) petroleum ether partitioning of a solution of residue in 9:1 acetonitrile:water, and (3) a second Florisil chromatography. Average recoveries from fortified oyster, shrimp, fish, and avian tissues were *ca* 84% for carbofuran. Limits of detection for the three compounds were *ca* 0.5, 0.05, and 0.07 ppm, respectively.

Archer analyzed alfalfa for residues of carbofuran and its 3-hydroxy and 3-keto conversion products by a somewhat similar cleanup and derivatization procedure (59). Samples were extracted with refluxing 0.25N hydrochloric acid, and residues were recovered from the extract by partitioning with dichloromethane and ether. Extracts were cleaned by a modification of the Florisil column chromatography of Knaak et al. (60) and by preparative TLC, such that the carbamates and their corresponding phenols could be recovered in separate fractions. TFAA derivatization was performed by the method described above (51). Recoveries of carbofuran and the related compounds carried through the entire procedure of extraction, cleanup, TLC and GLC determination were from 77 to 100% at fortification levels of 1 and 10 ppm. A similar method was extended to the analysis of carbofuran and its metabolites on strawberries, except that milder derivatization conditions (55) were employed for the carbamates (61).

Ryan and Lawrence compared the response of several N-perfluoroacyl carbamate derivatives to the electron capture and electrolytic conductivity detectors (62). The electron capture response of carbofuran derivatives increased with fluorine content in the order TFA < PFP < HFB < pentadecafluorooctanoyl. Minimum detectability was generally 10–100 times better by electron capture than by electrolytic conductivity (halide mode). The detector comparison was extended to crop samples; HFB derivatives of 3-ketocarbofuran were detectable at 0.1–1.0 ppm fortification levels using the Coulson electrolytic conductivity detector (halide mode), while underivatized 3-ketocarbofuran was not detectable at these levels using the same detector in the nitrogen mode (63).

In other applications of perfluoroacylation, trace amounts (5 ppb) of carbaryl and 1-naphthol were determined as their HFB derivatives by electron capture GLC using a hexachlorobenzene internal standard (64). A notable innovation in this procedure was the use of an Amberlite XAD-8 column to resolve the carbamate and its phenol from organochlorine pesticide interferences. Chapman and Robinson determined residues of carbofuran and its carbamate and phenol metabolites as their HFB derivatives (65). In this case, detection by mass fragmentography monitoring at the 228 amu peak allowed for a sensitive (0.02–1.0 ppm) residue procedure with minimal sample preparation.

Trifluoroacetylation provided a method for separate analysis of methiocarb, its sulfoxide, sulfone, and phenol analogs (66). Methiocarb sulfoxide and its phenol formed anomalous di-TFA derivatives, from initial reaction at the sulfoxide moiety, in addition to the usually reactive free NH or OH groups. All of the trifluoroacetyl derivatives of methiocarb and its metabolites could be determined by gas chromatography on a 2-ft 5% DC 200 column using the S mode of the flame photometric detector. A field-treated sample of blueberries was analyzed with very little cleanup by this method, peaks for residues of methiocarb, and its sulfoxide and sulfone derivatives being visible with no noteworthy interference. Further clarification of the mechanism of the sulfoxide trifluoroacetylation, in which trifluoroacetoxymethyl sulfide analogs result, has recently been provided (67).

N-acetyl-Cl$_x$ derivatives of carbaryl were prepared by pyridine-catalyzed reaction of the carbamate with the appropriate acid chloride (3, 68, 69). Derivative stability decreased with increasing Cl content, and a relatively good electron-capture response (about ⅕ that of DDT) was obtained with the monochloroacetyl derivative. The N-nitroso- and N-trichloroacetyl derivatives were unsuitable for analytical use due to extensive decomposition on the GLC column. Applications of these principles to residue determination has not yet been forthcoming.

The N-thiomethyl derivative of carbaryl is suitable for GLC analysis using the flame photometric detector (70). The derivatization, carried out with methylsulfenyl chloride catalyzed by pyridine, was applied to extracts of beans, carrots, tomatoes, and lettuce fortified at the 5 and 10 ppm levels. Recoveries averaged ca. 90%.

Reactivity of the N—H functionality has also been used to convert N-methylcarbamates to N,N-dimethylcarbamates for confirmatory purposes (71). Methyl iodide was employed as the alkylating agent in DMSO using NaH as base. While a 50° reaction temperature was satisfactory for derivatization of organophosphorus, triazine, and urea compounds with reactive N—H moieties, little or no product was obtained from the carbamates examined (methiocarb and terbutol) at this temperature. Reaction at room temperature for 5 min was preferable for the carbamates, each of which gave single products under these conditions. Less successful was an attempt to prepare the N,N-dimethyl derivatives by reaction in the injector port with coinjected trimethylanilinium hydroxide (Methelute®) (72). Carbaryl and carbofuran give the aryl methyl ether under these conditions, while the more stable N-aryl carbamates gave the desired N-methyl derivative. A treatment of derivatization procedures as confirmation techniques may be found in the review of Cochrane (4). The use of N-acetylation of carbamates for metabolite purification and characterization has also been described (73, 74), and its use for residue determinations in conjunction with electrolytic conductivity or electron-capture detectors was proposed (68).

Perfluoroacylation of carbamate insecticides for residue determination has been reported in several examples, many of which are cited above. Given these examples and applications to related compounds, dithiocarbamates (75), urea (76, 77), urethane (78), diflubenzuron (79, 80), anilines (81), and urea herbicides (82), it does appear to offer a promising route for further exploitation.

8.2.4. Derivatization of the Phenol Fragment (Table 8.4)

Phenol fragments, hydrolysis products of the aryl N-methylcarbamates, may be readily converted to derivatives for GLC analysis. A desirable derivatizing reagent is one which confers improved chromatographic properties to the relatively polar phenols and introduces the elements of sensitivity and selectivity for detection purposes. The more suitable derivatives are ethers and esters, but they include products from bromination of the aromatic ring. The derivatization procedure may be carried out in two steps, that is, hydrolysis to the phenol followed by coupling with an active reagent or halogenation, but one-step procedures are common as well. A complication arises from the likely presence in environmental samples of phenols derived from the carbamates during weathering and metabolism; these must be

TABLE 8.4. Examples of GLC Determination of Carbamates Based on
Derivatization of the Phenol Fragment

Carbamates	Derivatives[a]	Applications	Ref.
Ethers			
Butacarb, carbaryl, methiocarb, propoxur	DNP	Apples, lettuce, peas, river water	85
Bay 32651, Bay 78537, carbanolate, carbaryl, carbofuran, Ciba C-9643, decarbo- furan, Hercules 5727 and 9007, Landrin, metalkamate, methiocarb, mexacarbate, Mobam, promecarb, propoxur	DNP	Asparatus, carrots, cucumbers, eggplant, green beans, kale, lettuce, spinach, squash, tomatoes	34
Carbanolate, carbaryl, carbofuran, pro- poxur	DNP	Apples, corn, green beans, leafy vegetables	86
Carbofuran metab- olites	DNP	Potatoes, milk, animal tissue, eggs	94
Carbaryl, methiocarb, propoxur	DNP and DNT	Apples, lettuce	95
Carbaryl, carbofuran	DNT	River water	96
Metalkamate, carbaryl	PFB	—	97
Carbofuran, 3-keto- carbofuran, carbaryl, metmercapturon, Mobam, propoxur	PFB	Water and soil	100
Methiocarb and metabolites	TMS	Apples, pears, corn, sugar beets, fodder beets, bovine tissue and milk	101
Esters			
Carbanolate, carbaryl	MCA	Apples, Bermuda grass, cucumbers, milk, tomatoes	103
Carbaryl	MCA	Alfalfa, bees, honey, pollen	104
Propoxur	MCA	Air	106

350

TABLE 8.4. (Continued)

Carbamates	Derivatives[a]	Applications	Ref.
Propoxur, carbaryl, carbofuran, amino-carb, Mobam, mexa-carbate, Hercules 5727	DCBS	Cabbage, lettuce	95
Carbofuran	DMTP	Corn ensilage, milk	120
Carbaryl, Mobam, carbofuran	TCA	Apples, potatoes, range grass, sugar beets	107
Carbofuran and metabolites	TCA	Cucumbers, lettuce, potatoes, soil, toma-toes, tomato foliage	108
Propoxur and metabolites	TCA	Alfalfa, corn, oats, pasture grass	109
Propoxur and metabolites	TCA	Animal tissue, milk	110
Brominated Phenol			
Carbaryl	Brominated naphthol	Green beans	113
Carbaryl	Brominated naphthol acetate	Apples, beans, bees, broccoli, corn, chicken, trout, soil	114
Formetanate	Brominated 3-amino-phenol	Animal tissue, citrus, grapes, green crops, milk, pears, raisins, soil	117

[a] DNP = 2,4-dinitrophenyl-, DNT = 3,5-dinitro-4-(trifluoromethyl)phenyl-, MCA = monochloroacetate, DCBS = 2,5-dichlorobenzenesulfonate, DMTP = dimethyl-thiophosphate, TCA = trichloroacetate, TMS = trimethylsilyl-.

removed from extracts prior to hydrolysis and coupling (and determined separately if they are of analytical interest) lest they interfere with the analysis of the carbamates.

The preparation (83), GLC retentions, and electron-capture response of 2,4-dinitrophenyl (DNP) ethers of phenols was described in detail by Cohen et al. (84). The successful procedure for derivatization of carbamates involved simultaneous hydrolysis and coupling (85); a solution of 1-fluoro-2,4-dinitrobenzene (FDNB) in acetone was added to the cleaned carbamate residue along with pH 11.0 aqueous phosphate buffer and the mixture allowed to react at 50° for 30 min. The DNP ether was recovered in hexane and determined by electron-capture GLC on a 1% XE 60 column. At the pH of the reaction medium amines do not react and aryl organophosphorus compounds are not readily hydrolyzed. For analysis of water, samples were treated with ceric sulfate to oxidize phenols prior to extraction of the carbamates for derivatization. For crop material an acetone extraction was used, followed by cleanup with ammonium chloride–phosphoric acid coagulation, then ceric sulfate oxidation of phenols in the aqueous-acetone coagulation filtrate. The method allowed for determination of propoxur, Butacarb, and methiocarb at levels down to 0.005 ppm in river water, and 0.1 ppm in crops. It was unsuccessful when applied to soil, for which a more stringent cleanup would have been required.

A general procedure for determining carbamates in plant materials based on similar methodology was described by Holden (35). Residues were extracted by blending with acetonitrile, and the extract cleaned by petroleum ether partition and ammonium chloride–phosphoric acid coagulation. Carbamates were recovered from the coagulation filtrate in dichloromethane, from which free phenols were removed by mild (0.1N KOH) alkali extraction. Hydrolysis-derivatization was carried out in a highly agitated aqueous KOH solution containing FDNB, to which borax was added during the reaction to lower the pH and presumably retard hydrolysis of the derivatizing reagent and the DNP ether product, competing side reactions. Conversions were quantitative and detection limits in crops were as low as 0.05 ppm with generally excellent (90–110%) recoveries. The method was subsequently verified by interlaboratory collaboration for determining propoxur, carbofuran, carbanolate and carbaryl in apples, corn, green beans, and leafy vegetables fortified at 0.1, 0.2, and 0.5 ppm (86). No interferences from crop-derived impurities were experienced by any of the 8 participating laboratories at levels in excess of 0.01 ppm apparent residue. Slightly lower recoveries were reported for carbaryl, apparently from some instability of this compound in polar solvents used in the procedure.

DNP derivatization has been applied to the determination of carbofuran and its metabolites in soil, water, corn, Chinese cabbage, and other substrates (29, 87–90); to carbaryl in soil (91); and to several carbamate insecticides in rice (92). Useful modifications have been incorporated in some procedures. For example, Asai et al. (93) found 0.02M aqueous NaOH was an optimum medium for the reaction, as little as 5 min contact being suitable for hydrolysis of the major 3,4,5-isomer of Landrin to its phenol, and 2 min for the coupling. No attempt was made to remove Landrin-derived phenols from treated soil samples prior to derivatization, since the phenols were only poorly (<13%) recovered from soil even with prolonged acetone extraction. Separation of 3-hydroxycarbofuran and its phenol during dichloromethane/0.25N NaOH partitioning is improved after conversion of both products to their 3-ethoxy derivatives with HCl in absolute ethanol; 3-ethoxycarbofuran may be determined directly by GLC, and the 3-ethoxyphenol after conversion to its DNP derivative (29). The three major phenolic metabolites of carbofuran were determined in this procedure by GLC with a Coulson conductivity detector (N mode) as their DNP derivatives (94). Method sensitivities for each phenol component were: potatoes, 0.10 ppm; milk, 0.02 ppm; tissue, 0.05 ppm; and eggs, 0.05 ppm. Recoveries averaged 69% for all phenols in all matrices.

Ernst et al. (95) described in detail a procedure in which carbaryl, methiocarb, and propoxur were extracted from vegetables and fruits with methylene chloride, and detected qualitatively by TLC cholinesterase inhibition techniques or quantitatively by GLC of their DNP or 3,5-dinitro-4-trifluoromethylphenyl (DNT) derivatives. After hydrolysis (0.5N NaOH in ethanol), phenols were cleaned by steam distillation from acidified magnesium sulfate solution. The distillate was made basic with KOH and heated for 20 min with FDNB or DNT-Cl. The derivatives were recovered in isooctane for electron-capture GLC on OV 210/OV 17 mixed phase (Figure 8.6). Detection limits were 0.05 ppm, with excellent recoveries at 0.5–2.0 ppm fortification levels in lettuce and apples. The DNT ethers were superior to the DNP derivatives in electron-capture response and GLC resolution characteristics.

A more complete comparison of the two types of diaryl ethers, and an aralkyl ether formed with pentafluorobenzyl bromide (PFB-Br), was carried out by Seiber et al. (96). The physical properties of DNT derivatives of 29 phenols, many derived from carbamate insecticides, were reported. Microscale derivatization of carbamates proceeded as follows: to a solution of microgram quantities of the carbamate in 0.2 ml of acetone in a 10-ml volumetric flask was added 20 μl of 5% aqueous potassium carbonate. The stoppered flask was then heated (80°) in an oil bath for 1 hr to effect hydrolysis, diluted with 9 ml of acetone, and treated with 0.25 ml of a 5 mg/ml solution of derivatizing reagent in acetone. The volume was adjusted to 10 ml with acetone, and the flask shaken and then allowed to stand in the dark for 2 hr prior to electron-capture GLC analysis. Use of KOH as base led to unacceptable reagent background, and to the hydrolysis of aryl organophosphates to potentially interfering phenols. By omitting the heat-promoted hydrolysis step, only phenols originally present in the residue were determined, and were thus differentiated from the carbamates. Among the three derivatives retention times varied regularly for individual phenols in the order PFB < DNT < DNP; relative electron-capture response, ease of preparation, and derivative stability were quite similar for the three types of ethers. The retention time variation could be useful for confirmatory

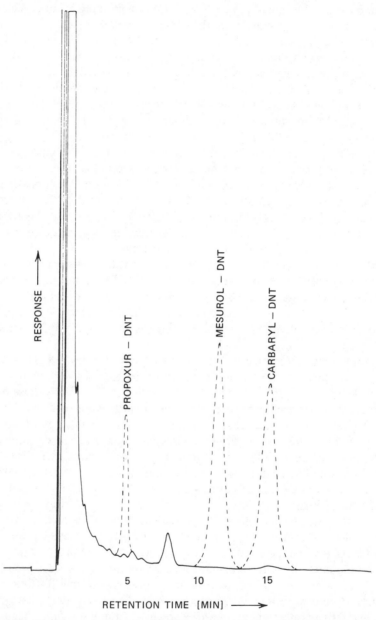

Figure 8.6 Gas chromatogram of unfortified apple sample (−) with peaks corresponding to carbamate DNT derivatives at the 0.5 ppm fortification level superimposed (---). A 1.6-m column packed with a 4:1 mixture of 10% OV 210 and 10% OV 17 on Chromosorb W HP, operated at 210°C, and electron-capture detector were used (95).

354

analysis of suspect carbamates and phenols. Carbaryl and carbofuran were recovered by the DNT procedure following dichloromethane extraction of water fortified at 25 ppb.

PFB ether formation was used in the electron-capture GLC determination of several phenols, including 1-naphthol (carbaryl) and the sec-amylphenol mixture resulting from metalkamate (97). The derivatization, using K_2CO_3 and PFB-Br in acetone at 50°, was essentially that of an earlier procedure (98, 99). A novel aspect was the use of a microsilica gel column to remove excess PFB-Br from the reaction mixture. Working with standards, 0.02–20 μg quantities of phenols were derivatized efficiently. Derivatized sec-amylphenols were resolved by GLC on OV 1 into component isomers, an advantage for confirmatory work. A similar derivatization procedure was used to determine phenolic residues in carbofuran-treated alfalfa (59) and strawberries (61). Several N-methylcarbamate insecticides were determined as residues in water and soil by PFB derivatization followed by electron capture GLC (100). The detection limits in water were as follows: carbofuran and 3-ketocarbofuran, 0.1 ppb; carbaryl, metmercapturon, and Mobam, 0.5 ppb; propoxur, >1.0 ppb.

Thornton and Dräger described a method for determination of residues of methiocarb and its carbamate metabolites which apparently involves formation of a phenol trimethylsilyl ether (101). After initial extraction and precipitation cleanup steps, the extract was oxidized with potassium permanganate to convert methiocarb and its sulfoxide to the sulfone. The sulfone residue in acetone was then treated with Regisil [bis(trimethylsilyl)trifluoroacetamide] at room temperature overnight. It was postulated that a trimethylsilyl group replaced the methylcarbamoyl group during derivatization, but this was not proved. At any rate the derivative could be determined by gas chromatography (5% DC 200) using a flame photometric detector, providing a residue method suitable for methiocarb and the two metabolites at levels down to 0.05 ppm in apple, pear, corn, sugar beet, and fodder beet samples, and in bovine tissue and milk.

Among carbamate-derived phenol ester derivatives, monochloroacetylation (MCA) and trichloroacetylation (TCA) have been most extensively studied for residue analysis. Argauer (102) described the preparation and GLC properties of 32 phenol monochloroacetates; the procedure was adapted to residue analysis of carbanolate in a subsequent paper (103). Crop samples were extracted with dichloromethane and subjected to cleanup on deactivated Florisil; phenols were removed by careful extraction of the dichloromethane column effluent with 0.25N NaOH; the carbamate was hydrolyzed in 0.25N NaOH; and the phenol from hydrolysis was derivatized with chloroacetic anhydride in benzene-aqueous NaOH for electron-capture GLC analysis. Recoveries of carbanolate from Bermuda grass fortified at 3 and 0.1 ppm were 86% and 95%, respectively, with a limit of detection of ca. 0.04 ppm. The method was satisfactory for carbanolate residues in apples, cucumbers, milk, and tomatoes, as well as Bermuda grass, but carbaryl, when taken through the same procedure, had a reagent-derived GLC interference at levels below 0.1 ppm (Figure 8.7). While the method can be used to quantitate carbamate derived phenols from field degradation, levels of free carbanolate phenol encountered in a disappearance study of Bermuda grass were so low as to be practically insignificant. An acetone extraction was preferred for analysis of milk samples. Electron-capture responses for MCA derivatives varied from carbofuran ($\frac{1}{200}$th that of heptachlor epoxide) to carbaryl ($\frac{1}{10}$th) among several standards examined, but it has been pointed out (3) that the procedure was not optimized and poor response could have been due to inefficient conversion to the derivative or its hydrolysis.

The derivatization procedure was subsequently modified by addition of Na_2SO_4 to the hydrolysis mixture just prior to ester formation (104); this increased the rate of ester formation, as did Na_2CO_3, but without the foaming which was encountered with the latter. The modified procedure was used to confirm carbaryl and 1-naphthol in honeybees, pollen, alfalfa, and honey that had been quantitated by fluorometry.

To overcome hydrolysis of derivatives in the strongly alkaline reaction medium, Shafik et al. (105) substituted a homogeneous reaction medium (chloroacetic anhydride in benzene containing pyridine as a catalyst) for the 2-phase system. The modified method was used to analyze 1-naphthol in urine, and propoxur entrained from air in an aqueous NaOH impinger trap (106).

Trichloroacetylation was accomplished by Butler and McDonough as follows (107, 108): The carbamate residue was mixed with a small amount of 85:15 methanol–water containing NaOH; the methanol was evaporated and the solution combined with dichloromethane-pyridine; the solution was heated for ca. 1 min at 95–100° to effect hydrolysis, and, after cooling, TCA-Cl was added and the heating continued for ca. 2 min longer. The TCA ester was recovered in hexane, freed of excess reagent and pyridine by washing with $NaHCO_3$ solution, and analyzed by electron-capture GLC. In the first report (107) food crops and grass fortified with carbaryl, Mobam, and carbofuran were extracted by blending with chloroform, and cleaned by coagulation as well as chromatography on an activated Florisil column. Alternatively, cleanup was effected only by column chromatography, on a combination adsorbent of alumina, Florisil, and Norit A. In the second report (108) acid extraction (24) was used, to convert conjugated metabolites to their aglycones; carbamates were cleaned up on an alumina column, since recoveries from Florisil varied greatly from one batch of

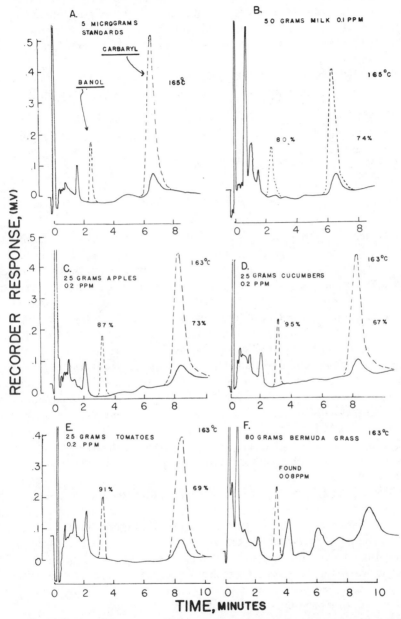

Figure 8.7 Gas chromatograms showing carbanolate (Banol) and carbaryl recoveries as MCA derivatives. A 5-ft column packed with 2% XE 60 on 80/100 mesh AW DMCS-treated chromosorb W, and electron-capture detector were used (103).

357

adsorbent to another. Phenols did not interfere, since they were not eluted from the alumina. Residues of carbofuran and its toxic metabolites were recovered efficiently from crops and soil fortified at levels as low as 0.1 ppm. Method sensitivity was ca. 0.04 ppm for 3-hydroxycarbofuran and ca. 0.01 ppm for carbofuran.

The TCA procedure was extended to residues of propoxur (**1**) and its metabolites (**2, 3**) in plant tissue (109), and in animal tissue and milk (110).

| PROPOXUR | $\underline{o}$-HYDROXY METABOLITE | $\underline{N}$-HYDROXYMETHYL METABOLITE |

The rather extensive procedure for plant tissue, designed to determine both the parent and conjugated metabolites, is summarized in Figure 8.8. Somewhat similar methodology was used for animal tissue, with the exception that samples were extracted with hexane–acetonitrile, and the acetonitrile layer evaporated to give a residue separable into parent and conjugate fractions by dilute aqueous sulfuric acid–chloroform partition. Some points to note in the procedure are (1) cleanup by use of deactivated Florisil for propoxur, and deactivated silica gel for metabolites, (2) hydrolysis-derivatization in an aqueous medium, the exact procedure varying somewhat between the parent and metabolite fractions, and (3) a silica gel microcolumn cleanup of derivatized propoxur to reduce interferences. For crop samples control levels were <0.01–0.10 for propoxur and the two metabolites, and for animal tissue and milk <0.02–0.08 for the parent and O-hydroxy metabolite. The procedure, reported in detail elsewhere (111), has been applied with minor modifications to carbaryl and propoxur residues in tobacco (112).

Some of the first procedures for derivatization to enhance electron-capture response were based on bromination of carbamate-derived phenols (113–5). For example, carbaryl was determined in crops, animal tissue, and soil by a single step hydrolysis–bromination–acetylation giving 2,4-dibromo-1-naphthyl acetate, following extract purification by coagulation (114). The method warrants only passing mention here, since subsequent attempts to repeat it have had poor success (116), and at any rate it is limited to those carbamates that brominate in a specific way.

Bromination does offer an acceptable basis for determination of formetanate by derivative formation. Formetanate is a relatively new insec-

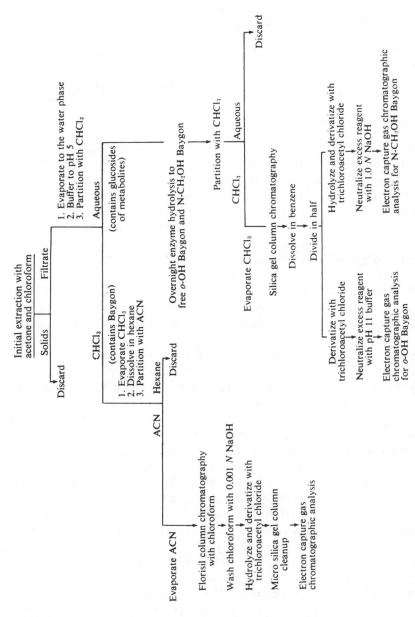

Figure 8.8 Outline of procedure for determination of propoxur (Baygon) and its metabolites in plant tissue (109).

ticide–acaricide having a hydrolyzable formamidine group in addition to the carbamoyl moiety (117; equation 1).

FORMETANATE HYDROCHLORIDE

Formetanate hydrochloride was hydrolyzed in refluxing $2N$ acetic acid, then refluxing $5N$ NaOH, to yield m-aminophenol. The product was brominated with $KBr/KBrO_3$ solution and the derivative recovered in toluene for electron-capture GLC determination on a 10% SE 30 column. Formetanate was adequately recovered from several crop samples fortified at 0.1–2.0 ppm by extraction with acidified methanol and dichloromethane, and cleanup by acid partition. We have used essentially the same method for determination of formetanate residues in fruit crops, substituting a Coulson conductivity detector for the electron-capture detector (118).

Bowman and Beroza (119, 120) prepared carbamate derivatives tailored to the use of the highly selective flame photometric detector. For example, residues of carbofuran and its phenol, separated by column chromatography on alumina, were subjected to hydrolysis (for carbofuran), cleanup by steam distillation, and coupling with dimethylchlorothiophosphate to form the thiophosphoryl ester (119). With the P mode of the flame photometric detector 0.5–5.0 ppm levels of the carbamate and its phenol were determined in corn silage and milk. Analogous derivatives were prepared from 8 other carbamate standards. A similar extraction–hydrolysis–cleanup procedure was used for methiocarb and 5 of its metabolites but, since all the corresponding phenols contain S, no coupling was required and the S mode of the flame photometric detector provided adequate sensitivity to the phenols (120). Thiophosphorylation has been reported more recently for analysis of phenols (121) but the technique has not achieved widespread use for carbamate residue analysis, perhaps because of the difficulty in removing excess reagent or because low yields were encountered with microgram quantities of the less acidic phenols (84).

Esters of sulfonic acids have recently been prepared as derivatives for phenols, including those derived from carbamate pesticides; either the electron-capture or S mode of the flame photometric detector may be employed for GLC determination (122). The p-bromo-, 2,5-dichloro-, 3,4-dichloro-, and pentafluorobenzenesulfonyl chlorides were examined as reagents. The corresponding derivatives had similar electron-capture response, but the pentafluorobenzenesulfonate had by far the shortest retention time on a 5% UC W-98 GLC column. Leafy vegetables were extracted by the method of Cook et al. (24), and phenols were removed by an alkaline wash of recovered residue in benzene. With no further treatment the residue was derivatized at pH 8 with the sulfonyl chloride in aqueous acetone at 80° for 15 min. Carbofuran and carbaryl were detected as the 2,5-dichloroben-zene sulfonates at the 0.05 ppm level in lettuce, and carbofuran at the 0.05 ppm level in soil. Recoveries decreased with increasing amounts of residue, apparently due to some hydrolysis of the carbamate during the basic parti-tion prior to derivatization. GLC determination was facilitated by an all-glass detector transfer line, to minimize metal-catalyzed degradation of the sulfonates. The method was also applied to analysis of carbamate pesticides encountered at the low ppb level in water (123).

It is difficult to compare the relative merits of these derivatization procedures since not all have been verified to the same extent and, sadly, there have been no systematic side-by-side comparisons. It bears repeating that, among the ester and ether derivatization procedures, only DNP ethers have been subjected to interlaboratory comparison for carbamate multi-residue analysis. One can only speculate that, with the technology available with these derivatives, a single procedure could be devised which, with minor modification, could be applied to all the aryl N-methylcarbamates, and their metabolites and conjugates, in most types of environmental samples.

8.2.5. Derivatization of an Amine or Carbamoyl Fragment (Table 8.5)

The carbamates, along with several other pesticide classes, can be converted to amines by simple hydrolysis (124). Formation of the diagnostic amine (methylamine for most of the carbamate insecticides) should be the most broadly applicable conversion among the alternatives, yet at the same time the least compound-selective. As in the case of phenols, further treat-ment of the amine fragment is necessary for conventional GLC with sensi-tive and selective detectors and, once again, several alternatives are possible. Major effort has focused on only one, however, the formation of secondary anilines for electron-capture GLC.

A procedure for determination of methyl- and dimethylcarbamate insec-ticides as the 2,4-dinitrophenyl derivative of the amine fragments was reported by Holden et al. (125). Plant material was extracted by blending with dichloromethane; the residue was cleaned by ammonium chloride–phosphoric acid coagulation; recovered residues were hydrolyzed by warm-ing (40–42°) in 0.5N KOH containing a little pentane to retard loss of the volatile amine; the acidified product was cleaned by dichloromethane parti-tion to remove non-basic materials, and the acid-soluble amine was treated with a benzene solution of FDNB at pH 7 for 30 min at ca. 100°. The deriv-

TABLE 8.5.   Examples of GLC Determination of Carbamates Based on
Derivatization of Amine or Carbomoyl Fragment

Carbamates	Derivative	Application	Ref.
Aldicarb, aminocarb, carbanolate, carbaryl, carbofuran, dimetilan, mexacarbate, Mobam, propoxur, pyrolan	2,4-Dinitroaniline	Apples, spinach, broccoli, string beans, cucumbers, tomatoes, lettuce	125
Carbaryl, 2-isopropyl-phenyl N-methyl-carbamate, cosban, hopcide, meobal, propoxur, Tsumacide	2,4-Dinitroaniline	—	126
Carbaryl, methomyl, mexacarbate	2,4-Dinitroaniline	Methomyl-rapeseed oil	127, 128
Carbaryl	4-Bromobenzamide	Spinach, chicory	129
Aldicarb, aminocarb, carbaryl, carbofuran, Hercules 5727, methiocarb, Mobam, mexacarbate, Tranid	Methyl N-methyl-carbamate	Mobam-lettuce, Carbo-furan-lettuce	130, 131
Methomyl	Methyl N-methyl-carbamate	Tobacco	132

ative was recovered in benzene and analyzed by electron-capture GLC on 2% XE 60. Despite a somewhat high reagent blank carbaryl and dimetilan were recovered and determined efficiently at the 0.2 ppm level from fortified vegetables, and 10 carbamates were recovered from spinach at 0.05–2.0 ppm fortification levels with a 92.4% average efficiency.

Sumida et al. (126) modified the procedure to allow for hydrolysis and derivative formation in a single reaction mixture; the carbamate was heated in 5% borax solution for 30 min in the presence of FDNB. Dioxane was added to enhance solubility of the carbamate and FDNB, and to retard loss of methylamine. Excess derivatizing reagent was removed from the final mixture by reaction with glycine, forming a product which could be separated from the desired derivative by aqueous base extraction. Without this precaution a "giant" interfering peak from unreacted FDNB was presented in the gas chromatogram. Conversions for seven carbamate standards in the procedure were 91–106%.

Mendoza and Shields (127) obtained variable and generally low recoveries of methomyl with the derivatization procedure of Sumida et al. (126), and modified it further. The carbamate was hydrolyzed by heating (82°) in 5% borax containing NaOH, and FDNB in dioxane was added to form the derivative during a second 82° heating period. Glycine was added to remove excess reagent and the DNP derivative was recovered in benzene, cleaned on a microcolumn of silica gel G-HR, and analyzed by electron capture GLC on a SE 30/QF 1 column. Recoveries were 95–105% for 0.05, 0.5, and 1.0 ppm of methomyl added to rapeseed oils; recoveries from 0.1 ppm fortifications were slightly lower but acceptable. The procedure applies also to methomyl *anti* isomer, carbaryl, and mexacarbate, though the latter gives both methylamine and dimethylamine derivatives. Further details of the silica gel microcolumn cleanup of the derivative, which removes a broad GLC peak eluting just after the derivative, were reported in a subsequent paper (128).

The *p*-bromobenzamide of methylamine was considered as a derivative for residue analysis of carbaryl by Tilden and van Middelem (129). Sulfuric acid hydrolysis converted carbaryl to methylamine salt and crop extractives to water-soluble products. The derivative was formed with *p*-bromobenzoyl chloride in alkaline medium. A major disadvantage of the procedure was the occurrence of a reagent-derived GLC peak coeluting with the derivative which interfered in crop samples containing less than 1 ppm carbaryl.

Moye (130) described a novel on-column transesterification of N-methylcarbamates to form methyl *N*-methylcarbamate. Reaction occurred on 6 in of NaOH-treated glass microbeads preceding the GLC column (Porapak P) when an injection of carbamate in methanol solution $5 \times 10^{-3}M$ in NaOH was made. Yields were high and reproducible. Substitution of ethanol, 1-propanol, and 1-butanol gave corresponding alkyl *N*-methylcarbamates of different retention times (Figure 8.9). An optimum reaction temperature was 215°, but good conversion was obtained at 150° with detection of markedly less crop contaminant interference. Linear response of the alkali flame ionization detector was from 1 to 100 ng Mobam injected on the glass beads.

The procedure was used for determination of carbofuran and its 3-hydroxy metabolite in lettuce (131). Samples were extracted and cleaned up by the method of Cook et al. (24). Quantitative recoveries were obtained at 0.05–1.0 ppm fortifications of 3-hydroxycarbofuran. The advantage of the procedure is that total carbofuran residues, on which tolerances are based, were determined as a single GLC peak. An obvious shortcoming is that no differentiation would result when several carbamates were present in the

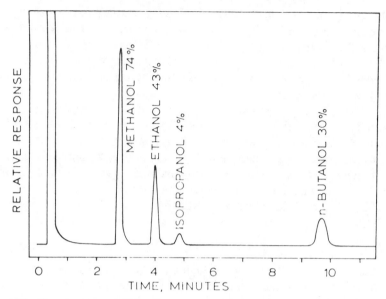

Figure 8.9 Gas chromatographic peaks obtained upon individual injections of Mobam in various alcohols. Sodium hydroxide-treated glass beads in the injection port liner (215°C), a 5-ft column containing 80/100 mesh Porapak Q at 180°C, and alkali-flame ionization detector were used (130).

same sample, unless separation preceded derivatization. These same comments apply to derivatization procedures based on methylamine. The Moye procedure was also applied to methomyl residues on cigar wrappers and flue-cured tobaccos (132), and to carbofuran residues in soil (133).

8.2.6. Derivative Methods for Methomyl, Aldicarb, and Other Oxime Carbamates

The "oxime carbamate" insecticides such as methomyl and aldicarb present analytical peculiarities which distinguish them somewhat from the aryl carbamates. Like the latter they may apparently be gas chromatographed intact, but only with adequate precautions taken to prevent decomposition (22). Several residue methods are based on derivatization of the amine or carbamoyl fragment and were referred to in Section 8.2.5. Since both methomyl (4) and aldicarb (5) contain sulfur, the most common methods take advantage in some way of this heteroatom for selective determination, methomyl as its S-containing oxime (6), and aldicarb as its carbamoylated sulfone (7).

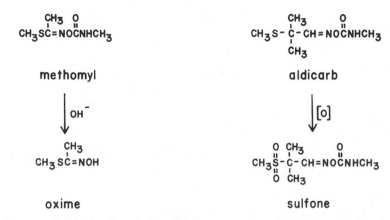

Pease and Kirkland (23) described a procedure for the determination of methomyl residues in animal and plant tissue and soil which has found widespread acceptance. Samples were extracted with ethyl acetate and the extract concentrated at room temperature in a beaker containing water. The aqueous concentrate after acidification was cleaned up by partitioning with hexane; methomyl was then recovered by chloroform extraction and hydrolyzed by heating on a steam bath in $0.1N$ NaOH. The oxime was recovered from the acidified hydrolysate in ethyl acetate and the solution concentrated by careful evaporation with triethylamine added as a keeper. The oxime was analyzed by programmed temperature (100–200°) GLC on 10% FFAP using the S mode of the microcoulometric detector (Figure 8.10). The oxime eluted essentially unchanged, with 75% of the theoretical sulfur being coulometrically titrated. Recovery of methomyl fortified at 0.02–4.0 ppm in 20 commodities averaged 93%. For high sulfur containing substrates, such as cabbage, careful conditioning of the GLC column at 200° must follow every 5 or 6 analyses.

This procedure has been reported in detail elsewhere (134), a notable substitution being the use of the S mode of the flame photometric detector in place of microcoulometry. It has been applied with some modification to a study of methomyl decline in soil (135) and to methomyl residues in mint hay and oil (136), greenhouse tomatoes (137), and rapeseed and rape plant (138). Reeves and Woodham (139) modified the extraction and cleanup; methomyl was extracted from soil, sediment, and water with dichloromethane, and from tobacco with dichloromethane:benzene (97.5:2.5) mixed solvent. Soil, sediment, and water extracts were purified by Florisil column chromatography, tobacco extracts by coagulation. Unhydrolyzed methomyl was injected on a 10% DC 200 GLC column operated isothermally (200°), and eluting peaks were detected by flame photometry. Fung (140) included GC/MS and TLC confirmation in a modification of the method of Pease and Kirkland (23) applied to methomyl residues in tobacco. A 4.5%

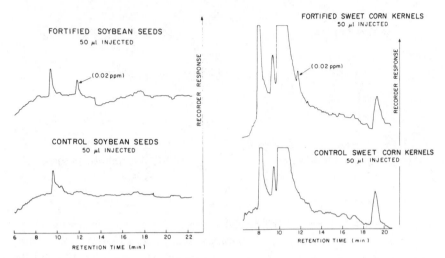

Figure 8.10 Gas chromatograms of soybean and sweet corn kernel extracts containing methomyl as its oxime derivative. A 4-ft column packed with 10% FFAP on 80/100 mesh Chrom W HP and microcoulometric detector were used. Samples were injected at 100°C, temperature programmed at 7.5°/min to 200°C, and held at 200°C (23).

carbowax 20M TPA column programmed from 100–200° was used for oxime GLC. Another application was to soil and water samples, in which the oxime was again confirmed as the GLC species by coupled MS (141).

Holt and Pease determined oxamyl in 26 substrates including plant tissue, animal tissue and soil following alkaline hydrolysis to the more volatile oxime fragment and flame photometric GLC (142). The method provided an average recovery of 90% for fortifications in the 0.02–10 ppm range. Bromilow and Lord formed the methylated oxime ("methoxime") by coinjection of oxamyl and trimethylanilinium hydroxide on GLC equipped with a flame photometric detector (143). Yields for several S-containing carbamate insecticides (oxamyl, methomyl, aldicarb, and methiocarb) were 90% or greater, whereas Tirpate, aldicarb sulfoxide, and aldicarb sulfone gave low or zero yields, apparently due to competing formation of the nitrile.

While aldicarb may be gas chromatographed directly, its retention time is too short to allow adequate separation from the solvent front. Thus a method was developed by Maitlen et al. (144) in which aldicarb was determined as its sulfone following oxidation. Samples were extracted by blending with chloroform, the residue being partially purified by filtration in acetonitrile through Celite. Column chromatography on Nuchar-Florisil resulted in the successive elution of aldicarb oxime (fraction I), aldicarb and its sulfone oxime (fraction II), aldicarb sulfone and sulfoxide oxime (fraction III), and aldicarb sulfoxide (fraction IV) with acetone–petroleum ether as eluting solvent. The sulfone oxime, sulfone, and sulfoxide were determined by GLC without derivatization; aldicarb in fraction II was oxidized (75° for 20 min in hydrogen peroxide–acetic acid) and the sulfone analyzed by GLC. A column 1:1 in 5% carbowax 20M and 10% DC 200, both on 60/80 mesh Gas Chrom Q, and flame photometric detection were used for GLC (Figure 8.11). As little as 0.01 ppm in oranges, apples, sugar beets, and potatoes was detected. The method requires correction of aldicarb residues determined as the sulfone for the contribution from sulfone oxime since these two products had the same GLC retention time.

Maitlen et al. (145) introduced a more rapid and direct method for determination of total toxic residues of aldicarb; oxidation preceded Florisil cleanup such that aldicarb and its sulfoxide and sulfone were simultaneously determined as the sulfone derivative. Oximes did not interfere, since their terminal oxidation product, the sulfone oxime, was resolved from the sul-

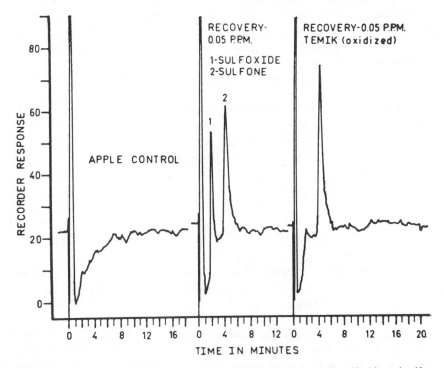

Figure 8.11 Gas chromatograms of oxidized aldicarb (Temik) and its sulfoxide and sulfone metabolites in apple extracts. A 1.22-m mixed column containing equal parts of 5% Carbowax 20 M on 60/80 mesh Gas Chrom Q and 10% DC 200 on the same support, column temperature of 115°C, and 394 nm flame photometric detector were used (144).

fone on the Florisil column. The sensitivity of the method was 0.007–0.010 ppm in apples, potatoes, cucumbers, alfalfa, and cottonseed, with recoveries from 72 to 114%.

These two procedures for aldicarb have been reported in detail elsewhere (146). Andrawes et al. (147) utilized the basic methodology to characterize aldicarb residues in cotton foliage and cottonseed. Smelt et al. determined aldicarb residues in soils during studies of the persistence of the parent carbamate and its metabolites (148). Iwata et al. applied a similar method to oranges, citrus byproducts, orange leaves, and soil (149). Carey and Helrich (150) modified the GLC determination by employing a high temperature (300°) glass wool-packed injection port in which aldicarb sulfone was converted to 2-methyl-2-(methylsulfonyl)propionitrile. Use of a water-moistened carrier gas blocked the thermal conversion of aldicarb sulfone oxime to the nitrile, eliminating this potential interference without the need for prior liquid chromatographic separation. The detection limit of the method was ca. 0.005 ppm for carrot, corn, green bean, orange, potato, and silage samples. An alternative removal of oximes was reported by Beckman et al. (151); acid hydrolysis effected oxime removal, carbamates were cleaned and fractionated on a silica gel column, and determined after alkaline hydrolysis as the oximes on a GLC equipped with a carbowax 20M/SE 30 column and microcoulometric detector. The detection limits were ca. 0.01 ppm in sugar beets.

A related method has been reported for thiofanox in which the parent oxime carbamate and its sulfoxide metabolite were converted to the sulfone for flame photometric GLC (152). Soil, water, cottonseeds, cotton gin trash, foliage, and sugar beets were the commodities studied, providing 95% average recoveries at 50–500 ppb fortifications of the parent or metabolites. Aldicarb and its sulfone did not interfere.

8.3. MULTIRESIDUE METHODS FOR CARBAMATES

The advantages of a single multistep procedure for assaying most of the pesticides in common use were recognized at a relatively early date by those engaged in enforcement analysis (1). The most common multiresidue procedure, described in AOAC (153) and FDA (154) manuals, is based on acetonitrile extraction, petroleum ether–acetonitrile partition and Florisil column cleanup steps, and GLC analysis of resulting fractions with electron-capture and phosphorus-selective detectors. More polar compounds, including the carbamate insecticides, are not accounted for by this methodology since they are not transferred from diluted acetonitrile to petroleum ether and/or are not eluted from Florisil with the most polar eluting solvent employed, 50% ethyl ether in petroleum ether (18, 155, 156). In one modification designed to accommodate carbaryl the initial acetonitrile extract of plant tissue was subjected to ammonium chloride–phosphoric acid coagulation, carbaryl being recovered from diluted acetonitrile by dichloromethane extraction (18). A Florisil cleanup of the dichloromethane phase yielded a sample suitable for determination by TLC, spectrophotometry, or polarography.

In a more recent method (156) carbaryl was included among 31 pesticides recovered from fruits and vegetables. Samples were extracted with acetone and the extracts partitioned with dichloromethane–petroleum ether to remove water. Organophosphorus and organonitrogen pesticides (including carbaryl) were determined without cleanup by GLC using a KCl-thermionic detector. Response for carbaryl increased with the accumulation of nonvolatile materials in the GLC injection area but, even then, a relatively large concentration of carbaryl (ca. 10 ppm in the sample) was needed to produce a half-scale deflection. Further cleanup and perhaps derivatization to enhance sensitivity would apparently be needed to render this method generally useful for carbaryl and related carbamates. A more comprehensive discussion of extraction and cleanup procedures applicable to carbamates in the presence of other pesticides and in several substrates may be found in the review of Magallona (3).

A multiresidue method of potentially more general applicability was reported recently by Johansson (157). Using apples as the substrate, organochlorine, organophosphorus, dinitrophenyl, and carbamate pesticides were extracted in toluene–hexane solution, and fractionated by Florisil column chromatography. Carbamates were determined by electron capture GLC as their TFA derivatives, and confirmed by GC/MS. Generally good recoveries were obtained at 0.5 ppm fortification levels and above.

The first comprehensive multiresidue procedure to appear in which carbamates figured as an integral part, rather than as an add-on in retrospect, was devised for determination of pesticides in air (55). The method, outlined in Figure 8.12, included the following steps: ethylene glycol, equivalent to an air sample of 20 m³ collected through a Greenburg–Smith impinger, was diluted with 2% sodium sulfate and extracted with dichloromethane. The extract was washed with sodium sulfate solution and water, dried, and evaporated. The residue was transferred in hexane to a chromatographic column containing 1 g of Woelm activity grade 1 silica gel activated at 175° for 48 hr and then deactivated with water (1 ml to 5 g silica gel) prior to use. Organochlorine insecticides eluted in fractions I and II were determined by electron-capture GLC, and organophosphorus insecticides in fractions II and III by flame photometric GLC. The carbamates, in fraction III, were derivatized with pentafluoropropionic anhydride (25 μg) in

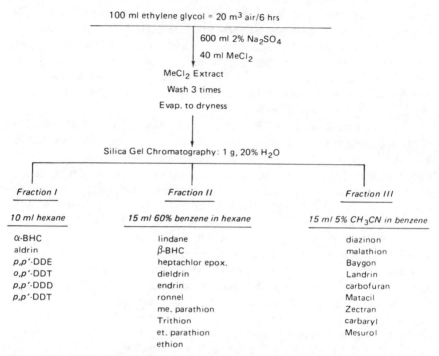

Figure 8.12 Multiresidue method for determining pesticides from air samples collected in ethylene glycol (55).

isooctane (2 ml) containing redistilled pyridine (0.01 ml) at room temperature for 1 hr. The reaction was stopped by addition of 3 ml of pH 7 phosphate buffer, and the PFP derivatives partitioned to 3 ml isooctane containing 0.05 ml acetonitrile. The isooctane layer was washed with water, dried, and analyzed by electron-capture GLC using a 5% SE 30 column at 165°. Since large amounts of diazinon in fraction III could interfere with carbaryl, diazinon was removed from this fraction, along with aminocarb (Matacil) and mexacarbate (Zectran), by extraction with dilute HCl. The seven carbamates were recovered from ethylene glycol with efficiencies of 78–101% when spiked at 1–4 μg, and with efficiencies of 52–126% when spiked at 20–80 ng. The latter, corresponding to 1–4 ng/m^3 in air, represent the approximate detection limits for air samples under ideal conditions. The multiresidue scheme was recently expanded to include a total of 45 pesticides, and applied to determination of airborne residues in and around agricultural operations (158).

A more realistic appraisal of the above sample-analysis scheme was reported in which pesticides were spiked to incoming air during the sampling period (159). In this way practical recoveries during sampling, and detection limits in the presence of coextracted matter from air, could be ascertained. For propoxur, the 2,3,5-isomer of Landrin, carbofuran, carbaryl, and methiocarb impinger recoveries were 51–76%, and detection limits were estimated to be 3–16 ng/m^3 for samples collected over a 12-hr period.

The deactivated silica gel cleanup/fractionation appears to offer a general approach for separation of carbamates in samples having a low content of organic matter. It has recently been used successfully for analysis of carbamates in soil (160) and in water (161). In the latter example DNP derivatization (35) was applied to the carbamate fraction due to high electron-capture background encountered with the PFP derivatives. The average recovery was 84% following cleanup on the silica gel column. For animal and plant tissues additional steps may be needed to reduce the quantity of organic coextractives such that the capacity of the silica gel is not exceeded.

8.4. CONCLUSIONS

Determination of residues of the carbamate insecticides represents an actively evolving area in pesticide microanalysis. Recent advances in GLC resolution and detection steps have not yet been matched with sufficient examples of practical application, but there is evidence that the technology as it now exists could encompass most of the analyses required in modern residue procedures. The development of somewhat uniform and fully verified procedures will require collaborative comparison of alternatives free of the bias often apparent in reports emanating from the laboratories in which specific methods originated. Such bias is, however, not without merit, since it provides the driving force for innovation which could eventually culminate in the "ideal" method. While this review has focused on GLC methods, equally exciting possibilities lie in the areas of high-speed liquid chromatography and, to a certain extent, TLC densitometry and automated photometric methods.

Regardless of the determination procedure used—and the availability of several suitable alternatives should be encouraged—a considerable effort is still needed in the development of extraction and cleanup procedures which will allow recovery of carbamate residues with a minimum of interfering coextractives from a variety of substrates. Recoveries have been generally based on simply spiking at the point of extraction, and tell little concerning the ability of a particular extraction technique to recover penetrated, bound, and conjugated residues formed during weathering. The requirements for generally applicable extraction and cleanup techniques are particularly

stringent given the physical and chemical limitations imposed by the carbamates and their environmental transformation products. It should be
apparent by now that a major departure from traditional methods aimed
primarily at organochlorine and organophosphorus pesticide is needed
before these goals can be achieved.

REFERENCES

1. J. A. Burke, *Residue Rev.*, **34,** 59 (1971).

2. H. W. Dorough and J. H. Thorstenson, *J. Chromatogr. Sci.*, **13,** 212 (1975).

3. E. D. Magallona, *Residue Rev.*, **56,** 1 (1975).

4. W. P. Cochrane, *J. Chromatogr. Sci.*, **13,** 246 (1975).

5. J. Kanazama, *Japan Anal.*, **22,** 155R (1973).

6. J. H. A. Ruzicka, *Pesticide Sci.*, **4,** 417 (1973).

7. W. Thornburg, *Anal. Chem.*, **49,** 98R (1977); **47,** 157R (1975); **45,** 151R (1973).

8. I. H. Williams, *Residue Rev.*, **38,** 1 (1971).

9. W. L. Zielinski, Jr. and L. Fishbein, *J. Gas Chromatogr.*, **3,** 333 (1965).

10. W. L. Zielinski, Jr. and L. Fishbein, *J. Gas Chromatogr.*, **3,** 142 (1965).

11. A. Strother, *J. Gas Chromatogr.*, **6,** 110 (1968).

12. L. Wheeler and A. Strother, *J. Chromatogr.*, **45,** 362 (1969).

13. J. M. Peck and K. J. Harkiss, *J. Chromatogr. Sci.*, **9,** 370 (1971).

14. M. C. Bowman and M. Beroza, *J. Agr. Food Chem.*, **15,** 894 (1967).

15. M. Riva and A. Carisano, *J. Chromatogr.*, **42,** 464 (1969).

16. W. R. Sans, *J. Assoc. Offic. Anal. Chem.*, **61,** 837 (1978).

17. R. R. Laski and R. R. Watts, *J. Assoc. Offic. Anal. Chem.*, **56,** 328 (1973).

18. M. Porter, R. J. Gajan, and J. A. Burke, *J. Assoc. Offic. Anal. Chem.*, **52,** 177 (1969).

19. D. L. Lewis and D. F. Paris, *J. Agr. Food Chem.*, **22,** 148 (1974).

20. L. G. Weyer, *J. Assoc. Offic. Anal. Chem.*, **57,** 778 (1974).

21. M. T. H. Ragab, *Anal. Lett.*, **10,** 551 (1977).

22. I. H. Williams, *Pestic. Sci,* **3,** 179 (1972).

23. H. L. Pease and J. J. Kirkland, *J. Agr. Food Chem.*, **16,** 554 (1968).

24. R. F. Cook, R. P. Stanovick, and C. C. Cassil, *J. Agr. Food Chem.*, **17,** 277 (1969).

25. C. H. Van Middelen and A. J. Peplow, *J. Agr. Food Chem.*, **21,** 100 (1973).

26. C. C. Cassil, R. P. Stanovick, and R. F. Cook, *Residue Rev.*, **26,** 63 (1969).

27. Pesticide Analytical Manual, U.S. Department of Health, Education, and Welfare, Food and Drug Administration, Rockville, Md. Vol. II. Section 180.254. 1973.

28. R. F. Cook, *Analytical Methods for Pesticides and Plant Growth Regulators*, Vol. 7 (G. Zweig, Ed.) Academic Press, New York-London, 1973, p. 187.

29. O. H. Fullmer, FMC Corporation, Richmond, Calif. Personal communication.

30. J. N. Seiber, M. P. Catahan, and C. R. Barril, *J. Environ. Sci. Health*, **B13(2)**, 131 (1978).

31. J. N. Seiber, E. A. Heinrichs, G. B. Aquino, S. L. Valencia, P. Andrade, and A. M. Argente, IRRI Research Paper Series No. 17 (May, 1978). The Int'l. Rice Research Institute, P.O. Box 933, Manila, Philippines.

32. I. H. Williams and M. J. Brown, *J. Agr. Food Chem.*, **21**, 399 (1973).

33. B. Tucker, *Analytical Methods for Pesticides and Plant Growth Regulators*, Vol. 7 (G. Zweig, Ed.), Academic Press, New York-London, 1973, p. 179.

34. E. J. Lorah and D. D. Hemphill, *J. Assoc. Offic. Anal. Chem.*, **57**, 570 (1974).

35. E. R. Holden, *J. Assoc. Offic. Anal. Chem.*, **56**, 713 (1973).

36. W. A. Aue, C. R. Hastings, and S. Kapila, *J. Chromatogr.*, **77**, 299 (1973).

37. D. J. W. Bullock *Analytical Methods for Pesticides and Plant Growth Regulators*, Vol. 7 (G. Zweig, Ed.), Academic Press, New York-London, 1973, p. 399.

38. B. Kolb and J. Bischoff, *J. Chromatogr. Sci.*, **12**, 625 (1974).

39. C. A. Burgett, D. H. Smith, J. C. Wirfel, and S. E. Goodhart, Application Note ANGC 2-76, Hewlett-Packard Corp., Avondale, Pa. (1976).

40. Verga, G. R., and F. Poy, *J. Chromatogr.*, **116**, 17 (1976).

41. R. C. Hall, *J. Chromatogr. Sci.*, **12**, 152 (1974).

42. B. P. Wilson and W. P. Cochrane, *J. Chromatogr.*, **106**, 174 (1975).

43. J.F. Lawrence and N. P. Sen, *Anal. Chem.*, **47**, 367 (1975).

44. M. J. Hartigan, J. E. Purcell, M. Novotný, M. L. McConnell, and M. L. Lee, *J. Chromatogr.*, **99**, 339 (1974).

45. H. Pyysalo, *J. Agr. Food Chem.*, **25**, 995 (1977).

46. J. Fraser, P. G. Clinch, and R. C. Reay, *J. Sci. Food Agric.*, **16**, 615 (1965).

47. L. J. Sullivan, J. M. Eldridge, and J. B. Knaak, *J. Agr. Food Chem.*, **15**, 927 (1967).

48. J. Fraser, I. R. Harrison, and S. B. Wakerley, *J. Sci. Food Agric. Suppl.*, **1968**, 8.

49. L. Fishbein and W. L. Zielinski, Jr., *J. Chromatogr.*, **20**, 9 (1965).

50. S. C. Lau and R. L. Marxmiller, *J. Agr. Food Chem.*, **18**, 413 (1970).

51. J. N. Seiber, *J. Agr. Food Chem.*, **20**, 443 (1972).

52. K. Suzuki, H. Nagayoshi, and T. Kashiwa, *Agr. Biol. Chem.*, **37**, 2181 (1973).

53. S. Khalifa and R. O. Mumma, *J. Agr. Food Chem.*, **20**, 632 (1972).

54. R. O. Mumma and S. Khalifa, *J. Agr. Food Chem.*, **20**, 1090 (1972).

55. J. Sherma and T. M. Shafik, *Arch. Environ. Contamin. Toxicol.*, **3,** 55 (1975).

56. R. J. Bose, *J. Agr. Food Chem.*, **25,** 1209 (1977).

57. J. F. Lawrence, *J. Chromatogr.*, **123,** 287 (1976).

58. L. Wong and F. M. Fisher, *J. Agr. Food Chem.*, **23,** 315 (1975).

59. T. E. Archer, *J. Agr. Food Chem.*, **24,** 1057 (1976).

60. J. B. Knaak, D. M. Munger, and J. F. McCarthy, *J. Agr. Food Chem.*, **18,** 827 (1970).

61. T. E. Archer, J. D. Stokes, and R. S. Bringhurst, *J. Agr. Food Chem.*, **25,** 536 (1977).

62. J. J. Ryan and J. F. Lawrence, *J. Chromatogr.*, **135,** 117 (1977).

63. J. F. Lawrence, D. A. Lewis, and H. A. McLeod, *J. Chromatogr.*, **138,** 143 (1977).

64. K. Nagasawa, H. Uchiyama, A. Ogamo, and T. Shinozuka, *J. Chromatogr.*, **144,** 77 (1977).

65. R. A. Chapman and J. R. Robinson, *J. Chromatogr.*, **140,** 209 (1977).

66. R. Greenhalgh, W. D. Marshall, and R. R. King, *J. Agr. Food Chem.*, **24,** 266 (1976).

67. R. Greenhalgh, R. R. King, and W. D. Marshall, *J. Agr. Food Chem.*, **26,** 475 (1978).

68. E. D. Magallona, F. A. Gunther, and Y. Iwata, *Arch. Environ. Contamin. Toxicol.*, **5,** 177 (1976).

69. E. D. Magallona and F. A. Gunther, *Arch. Environ. Contamin. Toxicol.*, **5,** 185 (1977).

70. S. B. Mathur, Y. Iwata, and F. A. Gunther, *J. Agr. Food Chem.*, **26,** 768 (1978).

71. R. Greenhalgh and J. Kovacicova, *J. Agr. Food Chem.*, **23,** 325 (1975).

72. R. G. Wien and F. S. Tanaka, *J. Chromatogr.*, **130,** 55 (1977).

73. R. P. Miskus, T. L. Andrews, and M. Look, *J. Agr. Food Chem.*, **17,** 842 (1969).

74. G. D. Paulson, R. G. Zaylskie, M. V. Zehr, C. E. Portnoy, and V. J. Feil, *J. Agr. Food Chem.*, **18,** 110 (1970).

75. W. H. Newsome, *J. Agr. Food Chem.*, **22,** 886 (1974).

76. P. W. Miller, *J. Agr. Food Chem.*, **19,** 941 (1971).

77. R. T. Evans, *J. Chromatogr.*, **88,** 398 (1974).

78. G. Walker, W. Winterlin, H. Fouda, and J. Seiber, *J. Agr. Food Chem.*, **22,** 944 (1974).

79. B. L. Worobey, and G. R. B. Webster, *J. Assoc. Offic. Anal. Chem.*, **60,** 213 (1977).

80. A. B. DeMilo, P. H. Terry, and D. M. Rains, *J. Assoc. Offic. Anal. Chem.*, **61,** 629 (1978).

81. D. E. Bradway and T. Shafik, *J. Chromatogr. Sci.*, **15**, 322 (1977).

82. D. G. Saunders and L. E. Vanatta, *Anal. Chem.*, **46**, 1319 (1974).

83. J. D. Reinheimer, J. P. Douglass, H. Leister, and M. B. Voelkel, *J. Org. Chem.*, **22**, 1743 (1957).

84. I. C. Cohen, J. Norcup, J. H. A. Ruzicka, and B. B. Wheals, *J. Chromatogr.*, **44**, 251 (1969).

85. I. C. Cohen, J. Norcup, J. H. A. Ruzicka, and B. B. Wheals, *J. Chromatogr.*, **49**, 215 (1970).

86. E. R. Holden, *J. Assoc. Offic. Anal. Chem.*, **58**, 562 (1975).

87. J. H. Caro, H. P. Freeman, D. E. Glotfelty, B. C. Turner, and W. M. Edwards, *J. Agr. Food Chem.*, **21**, 1010 (1973).

88. B. C. Turner and J. H. Caro, *J. Environ. Qual.*, **2**, 245 (1973).

89. J. H. Caro, D. E. Glotfelty, H. P. Freeman, and A. W. Taylor, *J. Assoc. Offic. Anal. Chem.*, **56**, 1319 (1973).

90. N. S. Talekar, L. T. Sun, E. M. Lee, J. S. Chen, T. M. Lee, and S. Lu, *J. Econ. Entomol.*, **70**, 689 (1977).

91. J. H. Caro, H. P. Freeman, and B. C. Turner, *J. Agr. Food Chem.*, **22**, 860 (1974).

92. I. Takase and S. Osuga, *Noyaku Seisan Gijutsu (Pesticide Tech)*, **30**, 22 (1973).

93. R. I. Asai, F. A. Gunther, and W. E. Westlake, *Bull. Environ. Contamin. Toxicol.*, **11**, 352 (1974).

94. R. F. Cook, J. E. Jackson, J. M. Shuttleworth, O. H. Fullmer, and G. H. Fujie, *J. Agr. Food Chem.*, **25**, 1013 (1977).

95. G. F. Ernst, S. J. Röder, G. H. Tjan, and J. T. A. Jansen, *J. Assoc. Offic. Anal. Chem.*, **58**, 1015 (1975).

96. J. N. Seiber, D. G. Crosby, H. Fouda, and C. J. Soderquist, *J. Chromatogr.*, **73**, 89 (1972).

97. L. G. Johnson, *J. Assoc. Offic. Anal. Chem.*, **56**, 1503 (1973).

98. F. K. Kawahara, *Anal. Chem.*, **40**, 1009 (1968).

99. F. K. Kawahara, *Environ. Sci. Technol.*, **5**, 235 (1971).

100. J. A. Coburn, B. D. Ripley, and A. S. Y. Chau, *J. Assoc. Off. Anal. Chem.*, **59**, 188 (1976).

101. J. S. Thornton and G. Dräger, *Int. J. Environ. Anal. Chem.*, **2**, 299 (1973).

102. R. J. Argauer, *Anal. Chem.*, **40**, 122 (1968).

103. R. J. Argauer, *J. Agr. Food Chem.*, **17**, 888 (1969).

104. R. J. Argauer, H. Shimanuki, and C. C. Alvarez, *J. Agr. Food Chem.*, **18**, 688 (1970).

105. M. T. Shafik, H. C. Sullivan, and H. F. Enos, *Bull. Environ. Contamin. Toxicol.*, **6**, 34 (1971).

106. C. W. Miller, M. T. Shafik, and F. J. Biros, *Bull. Environ. Contamin. Toxicol.*, **8**, 339 (1972).

107. L. I. Butler and L. M. McDonough, *J. Agr. Food Chem.*, **16**, 403 (1968).

108. L. I. Butler and L. M. McDonough, *J. Assoc. Offic. Anal. Chem.*, **54**, 1357 (1971).

109. C. W. Stanley, J. S. Thornton, and D. B. Katague, *J. Agr. Food Chem.*, **20**, 1265 (1972).

110. C. W. Stanley and J. S. Thornton, *J. Agr. Food Chem.*, **20**, 1269 (1972).

111. C. A. Anderson, *Analytical Methods for Pesticides and Plant Growth Regulators*, Vol. 7 (G. Zweig, Ed.), Academic Press, New York-London, 1973, p. 163.

112. E. Nesemann and F. Seehofer, *Beitr. Tabakforsch.*, **7**, 251 (1974).

113. J. W. Ralls and A. Cortes, *J. Gas Chromatogr.*, **2**, 132 (1964).

114. W. H. Gutenmann and D. J. Lisk, *J. Agr. Food Chem.*, **13**, 48 (1965).

115. C. H. van Middelem, T. L. Norwood, and R. W. Waites, *J. Gas Chromatogr.*, **3**, 310 (1965).

116. W. R. Benson, *J. Assoc. Offic. Anal. Chem.*, **53**, 351 (1970).

117. N. A. Jenny and K. Kossmann, *Analytical Methods for Pesticides and Plant Growth Regulators*, Vol. 7 (G. Zweig, Ed.), Academic Press, New York-London, 1973, p. 279.

118. W. O. Gauer, J. N. Seiber, and J. E. Woodrow, paper presented at the 9th Pacific Regional Meeting of the American Chemical Society, San Diego, Calif. Nov. 3, 1973.

119. M. C. Bowman and M. Beroza, *J. Assoc. Offic. Anal. Chem.*, **50**, 926 (1967).

120. M. C. Bowman and M. Beroza, *J. Assoc. Offic. Anal. Chem.*, **52**, 1054 (1969).

121. M. P. Heenan and N. K. McCallum, *J. Chromatogr. Sci.*, **12**, 89 (1974).

122. H. A. Moye, *J. Agr. Food Chem.*, **23**, 415 (1975).

123. A. H. Blagg and J. L. Rawls, *Amer. Lab.*, **1972**, 17.

124. D. G. Crosby and J. B. Bowers, *J. Agr. Food Chem.*, **16**, 839 (1968).

125. E. R. Holden, W. M. Jones, and M. Beroza, *J. Agr. Food Chem.*, **17**, 56 (1969).

126. S. Sumida, M. Takaki, and J. Miyamoto, *Agr. Biol. Chem.*, **34**, 1576 (1970).

127. C. E. Mendoza and J. B. Shields, *J. Agr. Food Chem.*, **22**, 255 (1974).

128. C. E. Mendoza and J. B. Shields, *J. Agr. Food Chem.*, **22**, 528 (1974).

129. R. L. Tilden and C. H. van Middelem, *J. Agr. Food Chem.*, **18**, 154 (1970).

130. H. A. Moye, *J. Agr. Food Chem.*, **19**, 452 (1971).

131. C. H. van Middelem, H. A. Moye, and M. J. Janes, *J. Agr. Food Chem.*, **19**, 459 (1971).

132. W. B. Tappan, W. B. Wheeler, and H. W. Lundy, *J. Econ. Entomol.*, **66**, 197 (1973).

133. D. H. Hubbell, D. F. Rothwell, W. B. Wheeler, W. B. Tappan, and F. M. Rhoads, *J. Environ. Qual.*, **2**, 96 (1973).

134. R. E. Leitch and H. L. Pease, *Analytical Methods for Pesticides and Plant Growth Regulations*, Vol. 7 (G. Zweig, Ed.), Academic Press, New York-London, 1973, p. 331.

135. J. Harvey, Jr. and H. L. Pease, *J. Agr. Food Chem.*, **21**, 784 (1973).

136. U. Kiigemagi, D. Wellman, E. J. Cooley, and L. C. Terriere, *Pestic. Sci.*, **4**, 89 (1973).

137. H. R. Krueger, R. K. Lindquist, J. F. Mason, and R. R. Spadafora, *J. Econ. Entomol.*, **66**, 1223 (1973).

138. Y. W. Lee, R. J. Ford, M. McDonald, K. S. McKinlay, L. G. Putnam, and J. G. Saha, *Can. Entomol.*, **104**, 1745 (1972).

139. R. G. Reeves and D. W. Woodham, *J. Agr. Food Chem.*, **22**, 76 (1974).

140. K. K. H. Fung, *J. Agr. Food Chem.*, **23**, 695 (1975).

141. K. K. H. Fung, *Pestic. Sci.*, **7**, 571 (1976).

142. R. F. Holt and H. L. Pease, *J. Agr. Food Chem.*, **24**, 263 (1976).

143. R. H. Bromilow and K. A. Lord, *J. Chromatogr.*, **125**, 495 (1976).

144. J. C. Maitlen, L. M. McDonough, and M. Beroza, *J. Agr. Food Chem.*, **16**, 549 (1968).

145. J. C. Maitlen, L. M. McDonough, and M. Beroza, *J. Assoc. Offic. Anal. Chem.*, **52**, 786 (1969).

146. R. R. Romine, *Analytical Methods for Pesticides and Plant Growth Regulators*, Vol. 7 (G. Zweig, Ed.), Academic Press, New York-London, 1973, p. 147.

147. N. R. Andrawes, R. R. Romine, and W. P. Bagley, *J. Agr. Food Chem.*, **21**, 379 (1973).

148. J. H. Smelt, M. Leistra, N. W. H. Houx, and A. Dekker, *Pestic. Sci.*, **9**, 279 (1978); (and earlier references).

149. Y. Iwata, W. E. Westlake, J. H. Barkley, G. F. Carman, and F. A. Gunther. *J. Agr. Food Chem.*, **25**, 933 (1974).

150. W. F. Carey and K. Helrich, *J. Assoc. Offic. Anal. Chem.*, **53**, 1296 (1970).

151. H. Beckman, B. Y. Giang, and J. Qualia, *J. Agr. Food Chem.*, **17**, 70 (1969).

152. Chin, W-T., W. C. Duane, M. B. Szalkowski, and D. E. Stallard, *J. Agr. Food Chem.*, **23**, 963 (1975).

153. W. Horwitz (ed.), *Official Methods of Analysis of the Association of Official Analytical Chemists*, 12th ed., Association of Official Analytical Chemists, Washington, 1975, p. 518.

154. Pesticide Analytical Manual, U.S. Department of Health, Education and Welfare, Food and Drug Administration, Rockville, Md., Vol. I.

155. R. W. Storherr, P. Ott, and R. R. Watts, *J. Assoc. Offic. Anal. Chem.*, **54**, 513 (1971).

156. M. A. Luke, J. F. Froberg, and H. T. Masumoto, *J. Assoc. Offic. Anal. Chem.*, **58**, 1020 (1975).

157. C. E. Johansson, *Pestic. Sci.*, **9**, 313 (1978).

158. A. Barquet, C. Morgade, T. M. Shafik, J. E. Davies, and J. X. Danauskas, Paper Presented to the Division of Pesticide Chemistry (Pest. 89), 170th National American Chemical Society Meeting, Chicago, Ill. August, 1975.

159. J. N. Seiber, J. E. Woodrow, M. T. Shafik, and H. F. Enos, *Environmental Dynamics of Pesticides*, (R. Haque and V. H. Freed, Eds.), Plenum, New York, 1975, p. 17.

160. H. A. Moye, University of Florida, Gainesville, Fla. Personal Communication.

161. J. F. Thompson, S. J. Reid, and E. J. Kantor, *Arch. Environ. Contamin. Toxicol.*, **6**, 143 (1977).

RESIDUE ANALYSIS AND ANALYTICAL CHEMISTRY OF INSECT PHEROMONES AND IGRs

DAVID A. CARLSON

Insects Affecting Man and Animals Research Laboratory,
Agricultural Research, Science and Education Administration,
Department of Agriculture,
Gainesville, Florida

CONTENTS

9.1. INTRODUCTION

9.1.1. Justification

If a chemist were faced with the analytical problem of determining trace amounts of pheromones, hormones, insect growth regulators (IGRs) and so on, the following approaches might be investigated, since complete residue procedures for these chemicals have not necessarily been reported in the literature or made available by chemical manufacturers.

As only three insect sex attractant pheromones have been registered by the U.S. Environmental Protection Agency and only one insect growth regulator, it would seem worthwhile to include other materials which may be considered for registration in the future. Pheromones, hormones, and insect growth regulators are considered by EPA at this time to be pesticides, as they are considered to have "an effect on insects" and therefore must undergo the same requirements for registration as any new pesticide, regardless that they are natural products. Analytical methods for several as yet unregistered chemicals that show promise for survey or control have therefore been included; residue methods can be derived from this information as a basis for analysis of those or similar materials, and several are included. The procedures described here are not exotic, to the extent that their publication may not have been considered.

The analyst's concern for residues is more meaningful for hormones than for pheromones, as hormones are more likely to affect nontarget organisms. All controlled release systems depend on dose rate and longevity for efficacious performance, necessary to be documented by chemical means for registration. Test methods allow varied environmental treatments of controlled-release formulations, requiring subsequent analysis of captured or residual material by gas chromatography (GLC), high pressure liquid chromatography (HPLC) or combined gas chromatography-mass spectrometry (GCMS). All GLC methods mentioned in this chapter use the flame ionization detector unless specifically mentioned otherwise. The term HPLC is used throughout for the original high pressure (high performance or high resolution) liquid chromatography, and LC for gravity-fed liquid chromatography.

9.1.2. Survey of the Literature

Many reviews of insect sex pheromones and application techniques are available, including those of Beroza (1), Birch (2), Jacobson (3), Mayer and McLaughlin (4), MacConnell and Silverstein (5), Shorey (6), and Stallberg-Stenhagen et al. (7); Gilbert (8) has reviewed juvenile hormones. Several reviews of analytical techniques specifically for semiochemicals (chemicals affecting behavior) include those of Wood et al. (9), Jacobson (10), Inscoe and Beroza (11), and Young and Silverstein (12).

Journals in which such articles on pheromones have been published include *Ann. Ent. Soc. Amer., Bull. Ent. Soc., J. Austr. Ent. Soc., J. Chem. Ecol., J. Economic Entomol., Environ. Entomol., Experientia, J. Agric. Food Chem., J. Insect Physiol., Nature,* and *Science.*

Wolff (13) has written a chapter entitled "U.S. Government Regulations for Pesticide Uses With Special Emphasis on Analytical Chemical Aspects," which describes the registration process, tolerance-setting procedures in FDA and EPA as well as other EPA pesticide-related activities including FEPCA, the successor of FIFRA. Cardarelli (14) and

Zweig (15) have described registration criteria that are applicable to pheromones and IGRs.

Inscoe and Beroza (11) discussed approaches, methodology and appropriate microtechniques for pheromone systems, but specific details of equipment were not included. They briefly discussed analysis of synthesized compounds pointing out that specifications are critical for unformulated chemicals as purchased or as synthesized, and that cis–trans isomer ratios in unsaturated acetates are critical for some lepidopteran species. Responses of insects to particular cis–trans ratios may change with geographic location, and optical activity is a consideration for some synthesized pheromones. Each supplier of chemicals and each batch must be evaluated for purity or composition of critical ingredients. Formulations must eventually be analyzed by chemical methods, as cleanup of formulations may be necessary because of thickeners, emulsifying agents, antioxidants, uv-absorbers, sunscreens, keepers, stickers, etc. Emission rate studies may be done by weighing, collection of volatiles, use of radiolabeled or isotopically labeled materials.

9.2. RESIDUE ANALYSIS FOR REGISTERED PHEROMONES

A simple residue analysis of aged samples for an attractant of house flies, muscalure [(Z)-9-tricosene containing 5–15% E isomer], was done by extracting the grits with hexane. The crude extract was analyzed by GLC on a 3% SE-30 column on 100-120 mesh Gas Chrom Q® at 225° and helium carrier gas at 20 ml/min. Compared to a standard, it showed loss of about 20% of the material per day (16).

Persistence in soil and water, soil leaching and photo-degradation experiments using mixtures of (Z,E-) and (Z,Z-)-7,11-hexadecadien-1-ol acetate (1:1 gossyplure) have been reported. Ether extracts of soil and water were concentrated for analysis by GLC on a 2 m × 2 mm ID glass column packed with 10% QF-1 on Gas Chrom Q held at 150°. The two isomers eluted together in 7.2 min and the major soil degradation product, hexadecadienol, at 4.0 min (17).

Hollow fibers made of crystallizable polymers have been used for field applications of pheromones of the following insects; pink bollworm, grape berry moth, codling moth, spruce budworm, and tussock moth. Procedures were developed and submitted to EPA, Insecticide Registration Branch, for analysis of the pheromone of the pink bollworm, gossyplure, as residue in cotton seed (18). Analyses were performed with samples of cotton seed spiked with 1.0, 0.5 and 0.1 ppm of mature seed. Pheromone residue, if present, was below 300 ppb in concentrated cotton seed extracts from pheromone treated fields using the same procedure, summarized as follows.

Cotton seeds were ground in a rotating knife mill. Unspiked or spiked 500 g samples were then extracted at ambient temperature for 20 hr with 1.5 liters of carbon disulfide in separate three liter resin kettles fitted with mechanical stirrers and $CaCl_2$ drying tubes. The extract solution was decanted, the seed washed with 200 ml of carbon disulfide and the combined extract/washing was dried over 100 g anhydrous Na_2SO_4. The Na_2SO_4 dessicant was removed by vacuum filtration and the extract concentrated by fractional distillation at 70° through a six-ball Snyder column.

The concentrated extracts were analyzed using a Varian® 2860 GLC with 5 ft × ⅛ in columns packed with 5% neopentylglycol succinate on 100–120 mesh Chromosorb® W-A at 170° or 180°, the injector at 210°, the detector at 285°, and 30 ml/min of helium carrier gas. The injectors had Pyrex® inserts which were replaced before injection of a new sample. Retention time for gossyplure was 6.6 min at 180° and 14.2 min at 170°. The immature extracts contained a material which interfered with the analysis, in addition to peaks which eluted before gossyplure in the mature extracts, as well as after it. Extracts of the mature seed showed fewer minor peaks in the GLC analysis than the extracts of immature seed and virtually no background interference at the retention time of gossyplure. Losses of gossyplure during sample preparation were regarded as negligible.

Column substrate and instrument parameters which give optimum gossyplure resolution from interferences are variable. Extractables from mature cotton seed may vary from sample to sample due to differences in growing conditions, seed maturity, etc. which require that operating parameters for each analysis be determined by comparison of chromatograms for pure gossyplure and seed extract at several conditions. Conditions which show no interfering extract components at the retention time of gossyplure are used for the analysis.

In a residue method for pheromone-filled hollow fibers, a 3.0 to 4.5 g sample was allowed to stand in 30 ml of hexane overnight. The sample was then swirled under reduced pressure until the hexane began to boil, and the vacuum released. After a repetition, the solution was decanted, and the fibers rinsed. Vacuum was applied once more, the solution decanted, the fibers rinsed and the sample made up to volume with addition of benzyl benzoate as internal standard. Quantitation was done by a Varian 3700 GLC equipped with a Varian CDS-111 data system and Model 8000 automatic sampler set to inject 1 µl. A 5 foot × ⅛ in S.S. column packed with 1.5% OV-101 on 100/120 mesh Chromosorb G was used with N_2 carrier gas at 30 ml/min. The temperatures were 270°, 210°, and 280° for the injector, column, and detector, respectively (18).

9.3. RESIDUE ANALYSIS FOR UNREGISTERED PHEROMONES

9.3.1. Pheromone Release and Analysis by GLC

Butler and McDonough (19) have determined first-order evaporation rates and half-lives of acetate sex pheromones and n-alkyl acetate standards as released from natural rubber septa. After impregnation of septa with dichloromethane solutions of pheromones and aging in the laboratory, individual septa were extracted with 50 ml of hexane: dichloromethane (1:1) by shaking for 1 hr. This solution was analyzed directly by GLC on a 2 m × 2 mm column packed with 3% SE-30 on 80–100 mesh Gas Chrom Q with an external standard.

McDonough (20) also gives a theoretical treatment for evaluation of formulations which should give quantities of pheromone sufficient for sexual disruption in the field.

Flint et al. (21) used natural red rubber septa substrates (no. 1F-66F, The West Co., Phoenixville, PA) and commercial gossyplure, (Z,Z- and Z,E)-7,11-hexadecadien-1-ol acetate (1:1), (Farchan Chemical Co. Willoughby, OH) to produce first-order loss curves with a half-life of 159 days in the laboratory and 120 days in the field. Treatments of gossyplure in 100 μl of dichloromethane were absorbed into the septa, followed by absorption of another 100 μl of solvent. Septa were aged, then shaken in a flask for 1 hr with 50 ml of a mixture of hexane:dichloromethane (1:1). The resulting solution was concentrated to an appropriate volume with a rotating evaporator and analyzed by GLC using a column packed with SE-30.

Plimmer et al. (22) briefly discussed approaches to mating disruption in the gypsy moth program. Plimmer et al. (23) described air sampling using a cylinder containing 7.5–8.0 g of 4A molecular sieve (14–30 mesh) through which air was drawn at 2–3 m^3/hr to adsorb disparlure released from a 2% microencapsulated formulation (NCR) applied with a sticker (1% of RA 16453, Monsanto) to promote adhesion to foliage. Sampling was conducted for 2-hr periods on the day of application with 4-hr and 12-hr periods later. The molecular sieves were transferred to a vial, extracted with hexane:ether (1:1) and allowed to stand one hr before concentration and bromination to form an electron-capturing derivative (24).

Evaluation of a microencapsulated formulation of disparlure applied to a grass field or a forest showed rapid loss of pheromone to a low but constant level, increased evolution with rising air temperatures but retention by the capsules of much of the original material (23, 25).

Roelofs et al. (26) found similar release characteristics of microcapsules in the laboratory, as the rate of pheromone emission dropped from

0.7%/day to 0.02%/day by the 16th day when only 3.7% of a mixture of 11-tetradecenyl acetates had been released.

Beroza et al. (27) used traps containing 10 ml of the solvent hexane for trapping of airborne (Z)-7,8-epoxy-2-methyloctadecane (disparlure) dispensed from a solution of hexadecane in stainless steel planchets. Loss of disparlure from the trapping solvent was negligable. Analysis was by GLC with a Hewlett-Packard 810 or 7620 equipped with a 90 cm × 6 mm o.d. copper column packed with 5% SE-30 on 70–80 mesh Anakrom® ABS at 165°, or 10% Carbowax 20-M on Anakrom® AB at 180°.

Release rates from controlled release formulations have been measured in an apparatus that used a vacuum source to pull air across the formulation and through a tube of Tenax GC polymeric absorbent, which removed the airborne material from the airstream. The trapped material was recovered by elution of the absorbent with hexane and analyzed on a Varian 3700 GLC, with a Varian CDS-11 integrator and a 1 m × 2 mm stainless steel column packed with 12% OV-101 on 120–140 mesh Chromosorb W, held at 180° for (Z)-9-tetradecen-1-ol formate (Tr = 3.2 min, KI = 1704). A 2 m × 2 mm i.d. glass column packed with 5.2% Carbowax 20-M on 120–140 mesh Chromosorb W was used at 140° for (Z)-7-dodecen-1-ol acetate (Tr = 2.6 min, KI = 1893) and at 170° for (Z,Z)-3,13-octadecadien-1-ol acetate (Tr = 7.0 min, KI = 2590). Release rates were measured under the climatic conditions similar to those encountered in the field. Known quantities of a relatively volatile pheromone, (Z)-7-dodecen-1-ol acetate, and the relatively less volatile, (Z,Z)-3,13-octadecadien-1-ol acetate, were collected with 97% and 77% efficiency, respectively. The sensitivity, precision and internal consistency of the method were determined (28).

A pheromone formulation used for one project was successfully adapted for use in another project. In the first project, (Z)-9-tetradecen-1-ol acetate was formulated in 4.8 mm o.d. polyethylene tubing. In the second project, rates of formate ester released from 4 sizes of tubing were measured periodically for 50 days. One tubing size was selected that combined longevity with an adequately high release rate. Next a mating disruption test was run in the field and release rates were made on the tubing as it aged. It proved possible to correlate these two types of results and in a second mating disruption test, excellent (87%) disruption was achieved for 66 days against *Heliothis zea* (24).

Porous polymers were used for several methods of trapping volatile compounds that could be considered models for pheromones (30).

9.3.2.   Pheromone Analysis by Capillary GLC

For separation of all four geometric isomers of 3,11-hexadecadien-1-ol acetate, 2 isomers of which constitute the pheromone of the pink bollworm,

Pectinophora gossypiella (Saunders), a capillary GLC column is necessary, as shown by Bierl et al. (31).

Natural and synthetic isomers were separated on a 1500×0.05 cm EGGS-X SCOT® column at 170°, with retention times in min of: (Z,Z)- 17.6; (Z,E)- 17.0; (E,Z)- 17.3; (E,E)- 16.6, and retention indices (relative to hexadecylacetate at 16.00) of 16.78, 16.70, 16.75, and 16.61, respectively. Retention times in min on a 1500×0.05 cm DEGS SCOT column at 160° were: (Z,Z)- 12.03; (Z,E)- 11.55; (E,Z)- 11.82; (E,E)-11.25, while a Carbowax 20-M SCOT column gave poorer separations. The natural material was confirmed as a mixture of (Z,Z)-, (Z,E)-isomers (1:1), while other isomers were not detected.

After initial separation of natural alcohol pheromone precursors by open column LC and acetylation steps, a 240×0.32 cm HPLC column of Porasil® A-60 (Waters) was used, and the active compounds were eluted with benzene at 1 ml/min in 30–38 min (31).

The four monoterpene compounds (I, $(+)$-(Z)-2-isopropenyl-1-methylcyclobutaneethanol; II, (Z)-3,3-dimethyl-1-cyclohexaneethanol; III, (Z)-3,3-dimethyl-1-cyclohexaneacetaldehyde; and IV, (E)-3,3-dimethyl-1-cyclohexaneacetaldehyde) which comprise the pheromone of the male boll weevil, *Anthonomus gradis*, were determined by GLC. Bull et al. (32) effected a total separation of the four components on a 183 cm $\times 4$ mm glass column packed with 10% QF-1 on 80 to 100 mesh Chromosorb W held at 110°. In later work by Hedin et al. (33), essential oils were obtained by extraction of frass or steam distillation of 100 insects in an extraction apparatus with water and ether. The extract was concentrated to 20 μl in a small conical glass tube, and standard paraffins were added. A 250 ft $\times$ 0.03 in stainless steel capillary column coated with OV-17 was used at 160° with 5 ml/min N_2 carrier gas to separate all four components. The Kovats' Indeces were: compound I, 1363; compound II, 1383; compound III, 1428; and compound IV, 1439. The ratio was approximately 6:6:2:1, which may or may not reflect the actual production.

The Kovats Index system is based on the adjusted retention time [tR' = tR (total retention time) $-v$ (dead volume)] of a compound relative to the tR' of n-alkane standards expressed as the carbon number of the standards multiplied by 100; e.g., pentane = 500, n-tetracosane = 1400. Under fixed conditions including isothermal oven temperature, a plot of carbon number of a homologous series (e.g., alkanes, olefins, alcohols, long-chain acetates, etc.) vs. log tR' gives a straight line (34). This system is especially valuable in pheromone systems where standards are not always available.

For confirmation the capillary column was interfaced with a Hewlett-Packard 5930 quadrupole mass spectrometer. The column temperature was 160° and the helium flow was 8 ml/min. Mass spectra were obtained at 70 ev (33).

9.3.3. Derivatives Formed from Pheromones

An electron-capturing derivative of epoxy-containing disparlure was prepared with samples trapped from air by addition of dibromotriphenyl-
phosphorane in dichloromethane to a concentrated extract. The dibromo
addition product was cleaned up by passage of the resulting solution
through a short column containing 10 g of Florisil® using hexane as the
eluant. An aliquot of the concentrated eluate was analyzed by GLC with a
^{63}Ni electron-capture detector using a 3% DC-200 column at 190°, with
injector and detector temperature levels maintained at 250° and 290°, respectively. Laboratory tests gave 80–95% (average 85%) recovery from spiked
samples, with a sensitivity of 0.2 g/m^3 of air (24).

McDonough and George (35) described stereospecific epoxidation of 8-
and 9-carbon olefins, (Z)- and (E)-7-dodecen-1-ol acetate, and methyl (Z)-
and (E)-9-hexadecenoate. The resulting Z oxiranes eluted later than E
isomers on several polar columns (1.9 m × 3.1 mm o.d.) including 5%
Carbowax 20-M (terephalate end capped). Synthetic (Z)-7,8-epoxydodecan-
1-ol acetate was nearly separated from the (E)-isomer on a Carbowax 20-M
column with 2400 theoretical plates.

Kuwahara and Casida (36) determined diunsaturated acetates and corresponding alcohols from phyticid moths via electron-capturing derivatives
for GLC. Tips of 10 female moths were immersed in benzene for 24 hr.
After removal of the extracted tips, Nujol was added and the solution was
taken to dryness under vacuum at 40°. The extract was treated in either of
two ways: (1) addition of 1 ml of trichloroacetyl chloride solution (1% in
benzene) to derivatize the alcohol originally present; (2) hydrolysis of
acetates present by treatment with 0.2 ml of methanolic sodium hydroxide
(0.1N) before acetylation as above to derivatize all alcohols present. After
18 hr, the solutions were washed with water, neutralized, extracted into
hexane, dried and the solvent removed. Hexane (50 μl) was added and a 2-μl
aliquot removed for injection onto a 1.5 m × 0.3 cm i.d. glass column
packed with 5% OV-17 on 60–80 mesh Chromosorb W held at 190° in a
Yanaco Model G-80 E GLC equipped with a ^{63}Ni detector. Sensitivity of
the method was 1–20 ng, while conversion and recovery was 98–131% from
acetate starting material using a conditioned column.

9.3.4. Analysis by HPLC

Analysis of the Douglas fir beetle anti-aggregation pheromone has been
developed by Look (38) using a Waters ALC-200 system including a Model
6000 A pump, a U6K sample injector, a Model 400 absorbance uv detector
set at 254 nm, and 4 ft × 0.125 in stainless steel columns packed with

Corasil-1 (Waters). The mobile phase was dioxane-trimethylpentane (3:7) at 2 ml/min. Samples to be analyzed had been trapped on Porapak Q®-S from formulations exposed to a stream of air in a 2-liter flask, then eluted with trimethylpentane. This very volatile pheromone could be distinguished by HPLC from solvent impurities which caused difficulties during GLC analysis.

Preliminary cleanup of samples of abdominal tips of 1-day-old European corn borer females was carried out by HPLC using a 1.5 m × 12 mm stainless steel column packed with a steric exclusion resin (Bio-Beads SX-2, Bio-Rad Laboratories), with benzene used as the eluent at 1.0 ml/min. High molecular weight materials started eluting with 30 ml of solvent, and the active compounds, 11-tetradecen-1-ol acetates, eluted with 79–90 ml of solvent. Ratios of Z—E isomers present in natural and synthetic 11-tetradecen-1-ol acetates were determined by GLC on a 3.7 m × 2 mm column of 3% phenyldiethanolamine succinate (PDEAS) on 100–120 mesh Chromosorb WAW-DMCS (37).

9.4. ANALYSIS OF INSECT GROWTH REGULATORS

A survey of chromatographic analysis methods up to 1975 has been assembled for insect juvenile hormones, their analogs, and Altosid (39).

A new class of insect growth regulators (IGRs) includes methoprene (isopropyl (2E,4E)-11-methoxy-3,7,11-trimethyl-2,4-dodecadienoate), known commercially as Altosid. This compound possesses juvenile hormone activity, is in commercial use as a mosquito larvicide when applied to aquatic areas, and has been fed to cattle as a systemic agent for control of horn fly and stable fly larvae in feces.

Preparative and analytical HPLC can be done by adsorption or reverse phase. Methoprene is efficiently separated on a 25 × 0.46 cm column of Zorbax-SIL (Dupont) with a solvent system such as hexane:diethyl ether (97:3) and the lower limit of sensitivity should be about 10 ng (40).

Radiolabelled (5^{14}C) (2E,4E)-methoprene was separated from a small amount of the (2Z,4E)-isomer on a 50 cm × 0.8 cm preparative column of μBondapak® C18/Porasil B using methanol:H_2O (65:35) at 9.0 ml/min in 160 min. Note that mineral acid catalyzed demethoxylation of methoprene, while radioisomerization and light-induced Z—E isomerization is possible. Analytical separation of the Z,E isomers was done on a μBondapak C18 30 × 0.4 cm column using a solvent system of methanol:water (70:30), while the solvent system for kinoprene and hydroprene was changed to methanol:water (85:15) (40).

Detailed methods for determination of low levels of residual methoprene used 20-g samples for extraction. Acetonitrile was used as the primary extraction solvent during high-speed blending and vacuum filtration. Further cleanup was accomplished by low temperature precipitation of fats, partitioning with petroleum ether, and chromatography on Florisil and neutral alumina. After concentration, GLC on columns of differing polarity showed the lower limit of detection to be: water, 0.0004 to 0.001 ppm; soils, blood, and urine, 0.001 ppm; forage grasses, legumes, and rice foliage, 0.005 ppm, and milk, eggs, fish, shellfish, poultry, and tissues or feces of cattle, 0.010 ppm (Miller et al. 41). Methods for mass fragmentography or suspected residues of methoprene describe use of a multiple ion detection system using a Hewlett-Packard 5930/5932A GCMS data system (42). While methoprene is relatively stable in GLC systems, allowing analysis of cold material, radio-labelled IGRs and metabolites can be separated with good resolution in HPLC systems for scintillation counting, especially using reverse-phase columns (43).

Frozen samples from dissected bluegill fish that had been treated with 5-^{14}C methoprene were homogenized in methanol and the extract developed on silica gel TLC plates using hexane:ethyl acetate (3:1), followed by benzene:ethyl acetate:acetic acid (100:30:3). Methoprene and its hydroxy ester from TLC-purified samples were determined on a 25 × 0.21 cm reverse phase column of Zorbax-ODS (DuPont) eluted with a solvent system of methanol:water (3:1). Residues of intact methoprene declined rapidly in living fish but they experienced an accumulation of radioactivity which was shown to reside in natural lipids including cholesterol, free fatty acids, glycerides, and protein. Simple radioassay methods can thus be highly misleading if simple, total radioactivity measurements are the only ones made (44).

Formulation analysis and residue analysis methods are thoroughly discussed for methoprene, with analysis by GLC using a 2m × 2mm i.d. glass column packed with 3% OV-101 on 100–120 mesh Chromosorb W-AW-DMCS operated at 165° with helium carrier gas. Retention time for methoprene is 12 min, with n-docosane eluting in 20 min. Cleanup methods using LC are included for plant foliage, water, soil, milk, eggs, fish and shellfish, poultry, cattle tissue and blood, cattle urine, poultry and cattle feces, and feed supplements (45).

Methoprene has been determined by GLC and HPLC using 2 g of bovine fat which was blended with acetonitrile, sodium sulfate, and Celite. The sample was decanted, concentrated, eluted from a column of acid alumina with hexane, and concentrated to 0.5 ml for preparative HPLC. A 120 cm × 0.9 cm preparative column of Porasil A was used for cleanup with a solvent system of hexane:tetrahydrofuran (20:1) at 7 ml/min. A Waters ALC 202/401 LC system was used with the uv detector set at 254 nm. The appropriate fraction was collected, evaporated to 0.1 ml and injected on a 6.35 mm × 30 cm μPorasil column, using hexane:tetrahydrofuran (99:1) solvent at 3 ml/min and a 254 nm uv detector. Residues of about 0.008 ppm could be quantitated, the lower limit of sensitivity being about 10 ng, with the peak observed at 8 min, and recoveries averaging 88% (range 84–96%) at levels of 0.008–7.000 ppm from fat (46).

Quantitation of methoprene was confirmed by GLC using a Micro-Tek Model 220 and a 6.35 mm × 1.8 m glass column packed with 3% OV-101 on Chromosorb W H/P, 100–120 mesh. The column oven was operated at 185°, the injector at 240° and the detector at 230°, using nitrogen carrier gas at 55 ml/min (46).

As natural JH compounds are slightly more polar than methoprene, interfering peaks can be shifted using dichloromethane, or a solvent system of hexane:dichloromethane (1:1) (40). Half-saturation of all solvents with water gave better results in adsorption HPLC than did addition of varying amounts of alcohol.

Other IGRs include hydroprene and kinoprene, commercially known as Altozar® and Enstar,® respectively.

Analytical procedures for diflubenzuron [TH6040, 1-(4-chlorophenyl)-3-(2,6-difluorobenzoyl)urea], which inhibits chitin synthesis in insects, have been summarized by Rabenort et al. (47).

Procedures were established for determination of diflubenzuron residues in a variety of agricultural and nonagricultural samples, in which methods for water, soil, sediment, fish, shellfish, agricultural crops, aquatic vegetation, forest products, cow and poultry tissues, milk, and eggs were summarized. They involve homogenizing 20 g solid samples with acetonitrile, Celite, liquid–liquid partitioning, Florisil–alumina–silica gel LC, and detection, using a Waters Model ALC/GPC/204 HPLC with a Model 6000 A pump, U6K injector and Model 440 uv detector set at 254 nm. A 30 × 0.4 cm i.d. μBondapak C18 reverse-phase column was used with solvent systems of acetonitrile:water (60:40) and methanol:water (75:25). Alternatively, a μPorasil (Waters) normal-phase column of the same size was used with solvent systems of isooctane:isopropyl alcohol (93:7) or dichloromethane:methanol (500:1). The methods are sensitive to 0.01 ppm in water and 0.05 ppm in solid materials. A set of standards ranging from 2 to 30 ng (at 1 ng/μl) was used to determine the response curve (48).

Diflubenzuron was determined in cattle feces by reverse phase HPLC after cleanup. Acetonitrile was used to extract a homogenized manure sample. After filtration, the extract was evaporated to dryness, the residue redissolved in acetonitrile, partitioned against hexane, concentrated and eluted from a 30 × 2.0 cm column of Florisil with a solvent system of hexane:dichloromethane (1:1). Analysis was performed on a Waters ALC-

100 system equipped with a Model 6000 pump, 254 nm uv detector, and a 30 × 0.6 cm reverse-phase column of μBondapak C18. A solvent system of acetonitrile:water (57:43) pumped at 1.5 ml/min gave a peak for TH6040 at 12 min. Sensitivity was about 0.5 ppm, or about 100 ng (49).

Lanzrein et al. (50) isolated and identified the three known natural juvenile hormones, JH I, II, and III, from hemolymph extracts of the cockroach *Nauphoeta cinerea* by the combined use of HPLC and chemical ionization GCMS. Extracts from 30 insects were first purified on preparative TLC plates (0.5 mm, silica gel G-Merck) developed with 6% ethyl acetate/benzene. The zone of Rf 0.3–0.5 was scraped and eluted with ether. HPLC was done with a Waters Associates Model 6000 pump, a variable wavelength uv detector (Schoeffel Instrument Corp.) adjusted to 215 nm and two 30 cm × 4 mm i.d. columns packed with μPorasil (Waters). Separations of 200 ng of JH I, II, and III were obtained using hexane:ether (95:5) at a flow rate of 2.5 ml/min, and compounds eluted at 26, 30, and 36 min, respectively. The same peaks were seen in purified haemolymph extract from 300 adult females, while a collected fraction contained JH III and dibutyl phthalate by GCMS. Immature insects carried through the same sequences showed more JH I and II, but little III.

Girard et al. (51) extracted 60 kg of houseflies with methanol, fractionated the extract on a 150 cm × 4 cm column of Sephadex LH-20 using a solvent system of benzene:methanol (1:1). The most active fraction was separated on a 130 cm × 1 cm HPLC column of Porasil A (Waters) using a solvent system of hexane:ether (90:10). A biologically active fraction was further chromatographed on a 25 cm × 3 mm column of 10-μm silica gel with a solvent system of hexane:ether (99:1). The most active fractions showed many peaks after diazomethane treatment by GC on 1% or 6% OV-1 glass columns (1.8 m), temperature programmed 80–200° at 8°/min. When subjected to GCMS on an LKB-9000 instrument with a column packed with OV-1, only methyl esters of aliphatic acids were found; associated biological activity suggests presence of a trace amount of an unidentified JH-active compound.

REFERENCES

1. Morton Beroza (Ed.), *ACS Sympos. Ser. No. 23.* **1976.**

2. M. C. Birch, *Pheromones*, American Elsevier, New York, 1974.

3. M. Jacobson, *Insect Sex Pheromones, in The Physiology of Insects*, Vol. 3 (M. Rockstein, Ed.), Academic Press, New York, 1974, pp. 229–76.

4. M. S. Mayer and J. R. McLaughlin, *An Annotated Compendium of Insect Sex Pheromones*, Florida Agric. Exp. Stations Monograph Series, Univ. of Florida, No. 6 (1975).

5.   J. G. MacConnell and R. M. Silverstein, *Angew. Chem. Int. Ed. Eng.*, **12,** 644 (1973).

6a. H. H. Shorey, *Animal Communication by Pheromones*, Academic Press, New York, 1976.

6b. H. H. Shorey and J. J. McKelvey, Jr. (Eds.), *Chemical Control of Insect Behavior: Theory and Application*, Wiley, New York, 1977.

7.   S. Ställberg-Stenhagen, E. Stenhagen, and G. Bergström. *ZOON, Suppl.*, **1,** 77 (1973).

8.   L. I. Gilbert, *Juvenile Hormones*, Plenum, New York, 1976.

9.   D. L. Wood, R. M. Silverstein, and M. Nakajima (Eds.), *Control of Insect Behavior with Natural Products*, Academic Press, New York, 1970.

10. M. Jacobson, *Insect Sex Pheromones*, Academic Press, New York, 1972.

11. M. Inscoe and M. Beroza, *in Analytical Methods for Insecticides and Plant Growth Regulators*, Vol. VIII, (G. Zweig and J. Sherma, Eds.), Academic Press, New York, 1976.

12. J. C. Young and R. M. Silverstein *in Methods in Olfactory Research*, (D. G. Moulton, A. Turk, and J. W. Johnston, Jr., Eds.), Academic Press, New York, 1975.

13. J. Wolff, *in Analytical Methods for Insecticides and Plant Growth Regulators*. Vol. III (G. Zweig and J. Sherma, Eds.), Academic Press, New York, 1976.

14a. N. F. Cardarelli, *Controlled Release Pesticide Formulation*, CRC Press, Cleveland, OH and West Palm Beach, FL, 1976.

14b. N. F. Cardarelli and K. E. Walker, *Development of Registration Criteria for Controlled Release Pesticide Formulation*. EPA 540/9-76-012. (To be published and distributed by NTIS.)

15a. G. Zweig and J. Sherma (Eds.), *Analytical Methods for Pesticides and Plant Growth Regulators*, Vol. VIII, Academic Press, New York, 1976.

15b. G. Zweig, Current EPA views on the Requirements for the Registration of Encapsulated Pesticide Formulations. *Proceedings 5th International Symposium in Controlled Release of Bioactive Materials, Aug. 1978.*

16. D. A. Carlson and M. Beroza, *Environ. Entomol.*, **2,** 555 (1973).

17. R. D. Henson, *Environ. Entomol.*, **6,** 821 (1977).

18. R. B. Davis, C. E. Kramer, W. F. Hill, P. W. Carlson (FRL, Dedham, MA), T. W. Brooks, and D. W. Swenson (Conrel, 110 A Street, Needham Heights, MA), unpublished data.

19. L. I. Butler and L. M. McDonough, *J. Chem. Ecol.*, **5,** 825 (1979).

20. L. M. McDonough, *Agricultural Research Report*, USDA-SEA, Berkeley, CA 94705 ARR-W-1/May 1978.

21. H. M. Flint, L. Butler, L. M. McDonough, R. L. Smith, and D. E. Forey, *Environ. Entomol.*, **7,** 57 (1978).

22. J. R. Plimmer, B. A. Bierl, R. E. Webb, and C. P. Schwalbe, *ACS Symp. Ser. No. 53*, 168–83, **1977.**

23. J. R. Plimmer, J. H. Caro, and H. P. Freeman, *J. Econ. Entomol.*, **71**, 155 (1978).

24. J. H. Caro, B. A. Bierl, H. P. Freeman, and P. E. Sonnet, *J. Agric. Food Chem.*, **26**, 461 (1978).

25. J. H. Caro, B. A. Bierl, H. P. Freeman, D. E. Glotfelty, and B. C. Turner, *Environ. Entomol.*, **6**, 877 (1977).

26. W. L. Roelofs, R. T. Carde', E. F. Taschenberg, and R. W. Wieres, Jr., *ACS Symp. Ser. No. 23*, 75 **1976.**

27. M. Beroza, B. A. Bierl, P. James, and D. DeVilbiss, *J. Econ. Entomol.*, **68**, 369 (1975).

28. J. H. Cross, J. H. Tumlinson, R. E. Heath, and D. E. Burnett, *J. Chem. Ecol.*, in press.

29. J. H. Cross, E. R. Mitchell, J. H. Tumlinson, and D. E. Burnett, *J. Chem. Ecol.*, in press.

30a. L. E. Browne, M. C. Birch, and D. L. Wood, *J. Insect Physiol.*, **20**, 183 (1974).

30b. K. J. Byrne, W. E. Gore, G. T. Pearce, and R. M. Silverstein, *J. Chem. Ecol.*, **1**, 1 (1975).

31. B. A. Bierl, M. Beroza, R. T. Staten, P. E. Sonnet, and V. E. Adler, *J. Econ. Entomol.*, **67**, 211 (1974).

32. D. L. Bull, R. A. Stokes, D. D. Hardee, and R. C. Gueldner, *J. Agric. Food Chem.*, **19**, 202 (1971).

33. P. A. Hedin, D. D. Hardee, A. C. Thompson, and R. C. Gueldner, *J. Insect Physiol.*, **20**, 1707 (1974).

34. E. Kovats, pp. 229–247 *in Advances in Chromatography*, Vol. I (J. C. Giddings and R. A. Keller, Eds.), Dekker, New York 1965.

35. L. M. McDonough and D. A. George, *J. Chromatogr. Sci.*, **8**, 158 (1970).

36. Y. Kuwahara and J. E. Casida, *Agr. Biol. Chem.*, **37**, 681 (1973).

37. J. Kochansky, R. T. Carde', J. Liebherr, and W. L. Roelofs, *J. Chem. Ecol.*, **1**, 225 (1975).

38. M. Look, *J. Chem. Ecol.*, **2**, 481 (1976).

39. L. L. Dunham, D. A. Schooley, and J. B. Siddall, *J. Chromatogr. Sci.*, **13**, 334 (1975).

40. D. A. Schooley, Zoecon Corp., Palo Alto, CA (personal communication).

41. W. W. Miller, J. S. Wilkins, and L. L. Dunham, *J. Assoc. Off. Anal. Chem.*, **58**, 10 (1975).

42. L. L. Dunham and R. J. Leibrand, *Adv. Mass Spectrom.*, **6**, 251 (1974).

43. D. A. Schooley, J. B. Bergot, L. L. Dunham, and J. B. Siddall, *J. Agric. Food Chem.*, **23**, 293 (1975).

44. G. B. Quistad, D. A. Schooley, L. A. Staiger, B. J. Bergot, B. H. Sleight, and K. J. Macek, *Pest. Biochem. Physiol.*, **6**, 523 (1976).

45. L. L. Dunham and W. W. Miller, *in Analytical Methods for Pesticides and*

Plant Growth Regulators, Vol. X (G. Zweig and J. Sherma, Eds.), Academic Press, New York, 1978.

46. L. M. Hunt and B. N. Gilbert, *J. Agric. Food Chem.*, **24,** 669 (1976).

47. B. Rabenort, et al., pp. 57–72 *in Analytical Methods for Pesticides and Plant Growth Regulators*, Vol. X (G. Zweig and J. Sherma, Eds.), *New and Updated Methods*, Academic Press, New York, 1978.

48. S. J. Diprima, R. D. Cannizzaro, J.-C. Roger, and C. D. Ferrell, *J. Agric. Food Chem.*, **26,** 968 (1978).

49. D. D. Oehler and G. M. Holman, *J. Agric. Food Chem.*, **23,** 590 (1975).

50. B. Lanzrein, M. Hashimoto, V. Parmakovich, K. Nakanishi, R. Wilhelm, and J. Luscher, *Life Sci.*, **16,** 1271 (1975).

51. J. E. Girard, K. Madhavan, T. C. McMorris, A. DeLoof, J. Chong, V. Arunachalam, H. A. Schneiderman, and J. Meinwald, *Insect Biochem.*, **6,** 347 (1976).

GOVERNMENT REQUIREMENTS FOR PESTICIDE RESIDUE ANALYSES AND MONITORING STUDIES

MARGUERITE L. LENG

Health and Environmental Sciences, The Dow Chemical Company,
Midland, Michigan

CONTENTS

10.1. INTRODUCTION

Requirements for analysis of pesticide residues have increased vastly since the 1950s when regulations were first implemented to require tolerances (maximum residue limits) for these useful chemicals in food crops intended for human and animal consumption. The evolution of changes in such requirements can be traced in reviews published between 1955 and 1971 by Gunther and Blinn (1), Fogelman (2), Harris (3), Vorhes (4), Frehse (5), Harris and Cummings (6), and Bevenue and Kawano (7).

In 1968 the U.S. Food and Drug Administration issued FDA Guidelines for Chemistry and Residue Data Requirements of Pesticide Petitions (8). These official guidelines described the studies needed at that time for obtaining analytical data suitable for establishment of tolerances for pesticides in food and feed crops and their byproducts, as well as in meat, milk, poultry, and eggs.

During the early 1970s bills were passed in the U.S. House and Senate proposing major revisions in the Federal Insecticide, Fungicide, and Rodenticide Act (FIFRA), resulting in passage of the Federal Environmental Pesticide Control Act (FEPCA) on October 8, 1972. Detailed regulations for implementation of Section 3 on Registration of Pesticides were finalized by publication in the Federal Register on July 3, 1975 (9). At that time certain sections were "Reserved" for later issuance. Registration Procedures were published as regulations on September 9, 1975 (10). A preliminary draft of sections on *Petitions for Tolerances* (11) was made available by EPA in 1975 for comment by outside experts, but a revised version had not been issued as of May 1980. Review of the draft indicated that requirements for residue data by the Environmental Protection Agency (EPA) under FIFRA as amended will be similar but more extensive than those outlined by FDA under the Food, Drug and Cosmetic Act in 1968 (8).

Proposed guidelines for studies to meet Section 3 Regulations were published in the Federal Register on June 25, 1975, along with extensive appendices outlining how the studies should be conducted (12). Revised proposed guidelines were published in 1978 incorporating some changes based on comments by other government agencies, industry, and any interested parties.

This chapter presents a brief outline of how certain government agencies control pesticides, of the laws under which they operate, and of regulations implemented to carry out this task. It also provides a detailed description of current requirements for residue data based on proposed EPA Guidelines and on personal experience gained during more than ten years of compiling data in petitions for tolerances, including many volumes of residue data for phenoxy herbicides, dinoseb and picloram. Finally, it discusses monitoring studies for pesticide residues in raw agricultural commodities, total diet samples and environmental samples, as well as pitfalls in interpretation of "positive" findings.

10.2. REGULATION OF PESTICIDES

10.2.1. Government Agencies

Prior to December 1970, the Pesticides Regulation Division (PRD) of the U.S. Department of Agriculture (USDA) was responsible for registration of pesticides. PRD reviewed data on efficacy, toxicology, and residues submitted by manufacturers in applications for registration of formulated products, and approved labels bearing directions for use in crops and noncropland areas. The labels also gave appropriate warnings and directions for handling to prevent toxic effects in man.

A companion function was handled by the Food and Drug Administration (FDA) of the U.S. Department of Health, Education and Welfare. FDA reviewed more extensive data on toxicology and residues as presented in Pesticide Petitions and Food Additive Petitions. These petitions proposed tolerances at levels adequate to cover maximum residues of the active pesticide ingredients that might be found in raw agricultural commodities and processed foods due to treatment of crops or animals with the formulated pesticide products being registered by PRD. Such tolerances generally provide a safety factor of at least 100-fold over the no-effect level found in long-term feeding studies in laboratory animals.

Following establishment of the Environmental Protection Agency (EPA) on December 3, 1970, these functions of PRD and FDA were transferred to the Pesticide Registration and Pesticide Tolerance Divisions, respectively, which were eventually combined as the Registration Division of EPA. The Product Manager system was instituted in 1975, through which one person in EPA is assigned the responsibility for all registration activities pertaining to a family of chemicals.

Several other agencies have also been actively involved in the regulation and transportation of pesticides. According to a 1964 Memorandum of Agreement signed by the Secretaries of Agriculture, Interior, and Health, Education and Welfare, scientists in those three departments were given the opportunity to review and comment on proposals relating to pesticide registrations, reregistrations and tolerances. The agreement was reinforced by passage of Public Law 94-140 on November 28, 1975. It includes a requirement for the EPA Administrator to formally submit proposed and final regulations to the Secretary of Agriculture and to a Scientific Advisory Panel for review and comment, and to furnish a copy to the Committee on Agriculture of the House of Representatives, and to the Committee on Agriculture and Forestry of the U.S. Senate (13). The Scientific Advisory Panel was described as consisting of 7 members appointed by the EPA

Administrator from a list of 12 nominees, 6 nominated by the National Institute of Health and 6 by the National Science Foundation.

In the past, regulation of pesticides and pesticide residues by the Federal Government encompassed only those pesticides or residues in foods shipped in interstate commerce, while pesticides and food supplies produced and used within a single State were under the jurisdiction of that State. As of October 1975, Federal registration must be obtained for all products, including those manufactured and used only within individual States. In addition, each product must also be registered or licensed for sale by each State in which it is to be used.

Several agencies conduct monitoring studies for pesticide residues in foods and environmental samples. FDA handles surveillance programs on raw agricultural commodities and total diet studies based on market basket samples collected in various geographic regions of the country. USDA analyzes samples of meat and poultry and of crops collected through its Animal and Plant Health Inspection Service (APHIS). The U.S. Department of the Interior, EPA, and USDA all have programs for monitoring pesticide residues in soil, water, and/or air.

Other government agencies such as the Interstate Commerce Commission and the Department of Transportation have regulations pertaining to shipment of pesticides, including specifications for containers to be used for various products, precautionary labeling required, and many more.

10.2.2.   Laws Regulating Pesticides

Several basic laws govern the sale and application of pesticides and the residues resulting from approved uses.

10.2.2.1.   The Federal Insecticide, Fungicide, and Rodenticide Act (FIFRA)

The FIFRA of 1947 replaced the original Insecticide Act of 1910, which regulated pesticide products such as Paris green, lead arsenate, etc. The 1947 Act included fungicides and rodenticides, while an amendment in 1959 added herbicides, nematicides, plant regulators, defoliants, and desiccants. Another amendment in 1962 declared certain forms of plant and animal life, and viruses, to be pests under certain conditions.

A recent major change was finalized by passage of the *Federal Environmental Pesticide Control Act* of 1972, originally called FEPCA but now referred to as *FIFRA as amended.* Under this amendment, all uses of pesticides are to be classified as general or restricted, based on hazard to the user or the environment. Products labeled for restricted use are to be applied only by or under the supervision of applicators certified by each

State within which they operate. The Act also requires reregistration of all pesticide products according to revised requirements for toxicology data, and eventual renewal of registration with submission of all data needed to fulfill all current requirements for registration of new products under the Act.

A minor amendment enacted as Public Law 94-140 on November 28, 1975, effectively eliminated the mechanism for registration of similar products simply on the basis of "established use practice." It requires actual submission of all necessary data or specific reference to where such data can be found in EPA files. It also provides for compensation by the registrant to the owner of data submitted since January 1, 1970, and referenced by the registrant in his application for a new product or a new use (13). Finally, explicit provisions for protection of ownership of data were incorporated into FIFRA by passage of the Federal Pesticide Act (FPA), which was signed into law by President Carter on October 1, 1978.

10.2.2.2. The Federal Food, Drug, and Cosmetic Act (FD&C Act)

The FD&C Act of 1938, as amended, permitted establishment of residue tolerances through public hearings for harmful and deleterious chemicals, providing it could be demonstrated that the chemical was necessary and that residues could not be avoided by good agricultural practice. The *Pesticide Chemicals Amendment* of 1954, better known as the Miller Amendment, directed that FDA establish safe tolerances on raw agricultural products after USDA had approved the usefulness of such chemicals for controlling specific pests.

The Food Additives Amendment of 1958 extended the tolerance setting function to provide for establishment of safe limits for pesticide residues in processed foods. Section 409(c)(3)(A) of this amendment is the much publicized *Delaney Clause* which permits no residue in processed foods of chemicals considered to be carcinogenic. Under this clause, confirmation of finding traces of diethylstilbestrol (DES) in beef liver at levels as low as 1 ppb (1×10^{-9}) necessitated banning the use of this growth promoter in feed for livestock. It is also responsible for the controversy over banning saccharin as an artificial sweetener in foods for human consumption.

Sale of pesticide products may continue pending hearings on cancellation procedures but not during appeals on suspensions. Suspension of registrations for DDT, aldrin/dieldrin, and heptachlor/chlordane in 1972-1975 were based on decisions by the EPA Administrator that uses of these products presented an "imminent hazard" to man or his environment. His decision on DDT overruled conclusions drawn by the EPA Administrative Law Judge, who presided over hearings to review available scientific data on these pesticides. (Details of this controversy are given in a 1977 review by the National Academy of Sciences on *Decision Making in the Environmental Protection Agency* (69).

10.2.2.3. Implementation of Laws on Pesticide Residues

The implementation of laws on pesticide residues and a revolutionary change in analytical pesticide chemistry methodology were brought about as a result of a report issued in May 1963 by a committee of the National Research Council, National Academy of Sciences, on the use of pesticides in the United States. Improved techniques had been developed which were capable of detecting trace levels of pesticides in crops for which earlier registration had been granted on a no-residue basis. As a consequence, a notice was issued on April 13, 1966, implementing the conversion of all such no-residue or "zero tolerance" registrations to *negligible residue tolerances* (14). A deadline of December 31, 1967, was set for submission of petitions requesting tolerances, or of progress reports on studies either completed or underway toward obtaining the necessary additional data.

Rules for obtaining such negligible residue tolerances were proposed by FDA on April 15, 1967, including a listing of permitted *crop groupings* for tolerance purposes (15). A policy statement was proposed by FDA on February 29, 1968, for pesticide tolerances regarding *milk*, *eggs*, *meat*, and/or *poultry* produced by animals fed raw agricultural commodities bearing pesticide residues (16). Three categories were established: (1) that finite residues will actually be incurred in these animal products; (2) that it is not possible to establish with certainty whether finite residues will be incurred, but there is a reasonable expectation of finite residues; and (3) that it is not possible to establish with certainty whether finite residues will be incurred but there is no reasonble expectation of finite residues. The deadline set for establishment of tolerances was December 31, 1970. It was later extended one year due to the number of residue petitions filed with FDA during late 1970, all of which still had to be reviewed by the newly formed EPA.

Partly as a result of inflation in costs, and partly because of increased scope and complexity of pesticide petitions, *filing fees* were increased 333% in mid-1972 (17). The minimum fee now required for review of the initial petition for a tolerance (or tolerances) of a chemical is $10,000 and each amendment containing new data must be accompanied by an additional $3000. Subsequent petitions for additional tolerances are covered by a basic filing fee of $1000 plus $1000 per crop or crop grouping, provided the levels requested are the same or lower than those established previously. However, the fee is again $10,000 if a higher tolerance is requested. Filing fees for requesting temporary tolerances to cover residues resulting from large-scale tests under Experimental Use Permits are a minimum of $6000, and each annual renewal costs $3000. Such filing fees are not required for petitions submitted by Government Agencies, such as USDA's Interregional Research Project No. 4 (IR-4), or the U.S. Corps of Engineers.

Initially, requirements were less for negligible residue tolerances replacing former no-residue registrations than had been required for tolerances at permissible levels for pesticides in use for many years. However, EPA eventually requested as much residue and toxicology data for old chemicals as for new chemicals, regardless of residue level. Since more time was needed to conduct all the additional studies, *interim tolerances* were proposed and established by EPA during 1972 for most uses of those chemicals for which petitions were still active. A few interim tolerances are still in effect at this writing (May 1980), including some "old" nonproprietary pesticides for which protective patents expired long ago (18).

10.3. RESIDUE DATA REQUIREMENTS

10.3.1. Analytical Methodology

The methods chosen for analysis of pesticide residues must be of sufficient sensitivity and specificity to assess the magnitude and nature of all significant residues remaining in or on treated crops, byproducts, processed products, meat, milk, and eggs. The estimated sensitivity of each method should be stated when applied to each commodity, based on the lowest concentration of the residue which can be detected with a reasonable degree of assurance, taking into account the magnitude and variability of control values for untreated samples.

Appropriate cleanup procedures should reduce or eliminate interferences present in the samples themselves, or arising from the reagents or processes used in the analyses. In gas–liquid chromatographic (GLC) methods, the peak(s) used for quantitation of the component(s) should be reasonably discrete (in a "window" on the tracing) rather than responses which appear as shoulders on peaks for solvents or other interfering substances.

Each analytical method used to obtain residue data should be described in sufficient detail, in a stepwise fashion, so a competent analyst can apply it, even though he or she may be unfamiliar with the procedure. Published methods can be used if modifications for specific commodities are described fully.

One method may suffice for simultaneous analysis of several components of a pesticide residue, especially if all components contain a common chemical moiety. One method must also be suitable for regulatory purposes to assure that total residues do not exceed established tolerances for each pesticide and its metabolite(s) in or on raw agricultural commodities (racs) being shipped in interstate commerce.

10.3.1.1. Validation of the Method

Each analytical method must be validated by an adequate number of control and recovery values for the commodities being analyzed. Although Guidelines issued by FDA (8) and proposed by EPA (12) do not specify how many analyses represent an "adequate" number, the 10:10:10 rule of thumb would appear to be minimal (6). This means 10 determinations on untreated commodities (control values), 10 analyses of fortified samples (recoveries), and 10 analyses of samples treated with specific pesticide products according to label directions that are most likely to cause maximum residues.

Control values are those obtained in analyses of untreated samples carried through the entire procedure. These "blanks" should be reasonably low in relation to the level of the proposed tolerance. According to Jerome P. Alpert, former Chief of the Chemistry Branch in FDA, control values should be no more than one-tenth of the tolerance level for permissible residues or less than half the claimed sensitivity for negligible residues. For example, a blank value should be less than 0.05 ppm for a negligible residue of an insecticide in beef fat, but 1 to 5 ppm might be adequate for a tolerance of 100 ppm for an herbicide in or on grass. Values higher than 5 ppm for control samples indicate that the method should be improved, even for grass.

Recoveries should exceed 85% in analyses of fortified samples carried through the entire procedure, although lower values have been accepted by EPA in exceptional cases. A range of 75–85% might be acceptable to EPA if consistent values are obtained in replicate analyses of the same sample. The commodity or its macerate (not simply an extract) should be fortified at levels ranging from the claimed sensitivity of the method to the maximum residue anticipated in that commodity. The lowest level tested should be less than the tolerance to be proposed for that commodity. For example, samples should be fortified at 0.05 ppm to assure adequacy of a method for negligible residue tolerances of 0.1 ppm. Recoveries at 0.01 ppm (10 ppb) or lower may be required for residues of relatively toxic pesticides, metabolites, or impurities, particularly in milk which can constitute the entire diet of an infant. On the other hand, the method need not be so sensitive for residues in minor feed items such as almond hulls or sugarcane bagasse.

10.3.1.2. Extraction Efficiency

Recovery studies for a pesticide added to an untreated sample or to a macerate demonstrate the capability of procedural steps following extraction. However, this does not assure that the same extraction procedure would remove "ingrown" or weathered residues from field-treated samples. Similarly, data obtained by surface stripping of samples are not acceptable. One exception is in the case of hard fruits treated shortly before harvest, but only if other data on that crop have established that the total residue is in fact on the surface.

Conjugation of pesticides with carbohydrates or amino acids in plants and animals generally alters their relative solubilities in solvents used for extraction, even without actual degradation of the parent molecules. Some residues may also become "bound" in certain tissues so they are not readily extracted. Completeness of removal of ingrown residues can be demonstrated by repeated exhaustive extraction of field-treated samples using different solvents, coupled with hydrolysis of finely ground samples, extracts, and extracted fractions using acid, base, or enzymes, followed by the usual analysis for the parent chemical. Of course, the conditions used must not alter the parent chemical as shown by similar studies demonstrating stability of the chemical.

Studies by Dow analysts demonstrated that much higher residues of phenoxy herbicides were found in stem tissues of treated plants (such as corn stalks, rice straw, or sugarcane) by extraction with a polar solvent under alkaline conditions than with nonpolar solvents under acid conditions (see Chapter 6 in this volume). Hydrolysis was not necessary in this case, indicating that nonpolar solvents may simply not penetrate the aqueous matrix of the plant cells where the residues are stored. On the other hand, actual hydrolysis was needed to release residues of phenolic metabolites of phenoxy herbicides in animal tissues. Evidence was also obtained that the phenolic residues were in extractable conjugated form in kidney but were "bound" in liver of test animals fed high levels of the herbicides for several weeks immediately before slaughter.

Radiotracer field or laboratory studies can provide much useful information on "bound" or conjugated residues of pesticides or their metabolites in plants and animals. Studies in laboratory animals can sometimes be translated to large animals provided that degradation of the pesticide by rumen bacteria does not produce different metabolites than those obtained by usual digestive processes. In all cases, caution should be exercised in interpreting data from radiotracer studies where the nature of the residues is not fully elucidated, as discussed more fully below.

10.3.1.3. Total Residue Including Metabolites

A pesticide may be irreversibly altered within a plant or animal and stored in that form without further degradation. Thus, pesticide residue methods must include analyses for toxicologically significant metabolites that are likely to be present in food or feed crops, or their processed products. Radiotracer studies are essential for detection of such metabolites, and for their identification. However, chemical methods must also be developed which are not dependent on radiotracer counting, and which liberate bound or conjugated fragments found to be significant in the radiotracer studies. Where only traces (0.01–0.1 ppm) are present, characterization of metabolites may be sufficient without complete identification.

A problem inherent in radiotracer studies is equating radioactivity to parts per million of parent chemical or indeterminate metabolites. An example of this was the conclusion that relatively high levels of 2,4-D "metabolites" were present in fish maintained for several weeks in plastic pools containing water treated with ^{14}C-2,4-D as the dimethylamine salt (19). It was later shown conclusively that the ^{14}C-activity in the fish was due to degradation of 2,4-D by aquatic microorganisms and/or sunlight, followed by incorporation of labeled fragments into natural components of the fish (20). Similarly, microbial degradation of herbicides applied to soil can lead to uptake of labeled fragments by plants which could erroneously be attributed to significant residues from plant metabolism.

Thus, the routes of degradation or metabolism of a pesticide in plants, animals, and soil need to be delineated, preferably by portrayal in a flowsheet. The distribution of total residues in various parts within plants and animals should also be determined, as well as rates of dissipation of such residues with time. The methods used for analysis should be capable of detecting total residues; in some cases this can be accomplished by simultaneous determination of several components after conversion to a common chemical moiety; in others, completely different techniques may be required to account for all metabolites and degradation products.

10.3.1.4. Regulatory Method

A complete stepwise description of the method must be provided in sufficient detail that a competent analyst unfamiliar with the method can apply it. A uniform standard of style and format for presentation of methods is given by the Association of Analytical Chemists in a 1972 edition of the *AOAC Style Manual* (21). Adherence to this format would ensure that essential steps are not omitted.

As mentioned in Section 10.3.1.1, blank values should be below the tolerance level proposed for each commodity so that gross (uncorrected) values for treated samples do not exceed the tolerance. According to draft guidelines for petitions for tolerances (12), an enforcement method should:

1. Be applicable without the use of a sample of the untreated commodity as a control or as an internal standard since such samples are not available to enforcement analysts
2. Use equipment and reagents that are readily obtainable
3. Be reasonably rapid
4. Be sufficiently specific, accurate, and reproducible to measure and essentially identify the residue in the presence of other pesticides that might reasonably be encountered in or on the same commodity.

Nonspecific methods may be useful for accumulating residue data where the identity of the pesticide is known, but such methods are not applicable for regulatory purposes. For example, inhibition of cholinesterase can be adapted to quantitation of residues of specified organophosphates or carbamates, but this technique does not differentiate among residues of different inhibitors in samples from unknown sources. Analyses for total phosphorus may be acceptable when used in conjunction with identification techniques such as paper chromatography or thin layer chromatography (TLC). However, the latter visual comparison methods are not usually considered suitable enforcement procedures because of their semi-quantitative nature.

The specificity of gas–liquid chromatography (GLC) can be enhanced by use of extraction cleanup procedures and various detection systems such as microcoulometry, flame photometry, or electron capture. Nonetheless, confirmatory methods should usually be provided to avoid potential problems with residues that appear to exceed tolerance limits. Techniques that may be acceptable as a means of increasing specificity of GLC analyses include the use of "p values" or fraction of solute partitioning into the upper phase of an equivolume two-phase system (22), conversion to other products by bromination or dehydrohalogenation vs the usual formation of esters, use of parallel columns of different polarity, and/or others, depending on the nature of the residue being measured. Mass spectrometric analysis can be used for final confirmation provided the component of interest has been adequately separated from interfering substances.

Methods proposed for regulatory purposes may be, and often are, tested in EPA and FDA laboratories to evaluate specificity, repeatability, and reproducibility of independent analyses for various samples. Repeatability is a measure of how well an analyst can expect to agree with himself from day to day, whereas reproducibility is how well analyses in one laboratory are likely to agree with results from other laboratories. Techniques for statistical analysis of results of interlaboratory collaborative tests are described in a 1975 edition of the *AOAC Statistical Manual* (23) as discussed in Section 10.5.5 of this chapter.

If the methods are judged to be satisfactory, descriptions will be made available to interested parties by publication or reference in the *FDA Pesticide Analytical Manual* (*PAM*). Volume I of the Manual includes methods which are capable of analyzing a number of pesticides in the presence of one another. Volume II contains methods for residues of individual pesticides. Volume III presents methods which detect pesticide residues in human and environmental media. A companion Food Additive Analytical Manual includes methods for food additives rather than pesticides. Although such manuals are updated frequently, some methods may become obsolete with time. For instance, the method given for 2,4-D in plant and animal tissues is dated 7/1/67 and still recommends extraction with nonpolar solvents under acid conditions, although ample data have been presented to show that this method may not be adequate for residues in green plant tissues. (See Chapter 6 in this volume.)

10.3.2. Residue Studies in Plants

When tolerances were first established for pesticide residues in 1955, it was sufficient to analyze a few samples of treated food crops collected at normal harvest, using a method capable of detecting 1 to 5 parts per million (ppm) of the parent chemical. As outlined in Section 10.3.1, government agencies now require data to delineate the nature and amount of residue in all crops, crop byproducts, and processed products which could be used for food for feed purposes. The nature of the residue should be determined by appropriate metabolism studies in plants and animals, and by degradation studies in soil and water. The analytical methods used should be capable of detecting residues down to at least 0.1 ppm of the parent chemical, its metabolites and degradation products, whether free, conjugated or bound, in the wide variety of plant and animal tissues likely to be involved.

10.3.2.1. Field Experiments

Residue data are required for samples of raw agricultural commodities taken from appropriate field studies. These field experiments must reflect the proposed pesticide use with respect to formulated products, dilutions, and the rates, modes, number, and timing of applications. Since data are needed to demonstrate maximum residue levels, samples from efficacy trials may not be satisfactory. Special plots should be established where only maximum rates are applied, the maximum number of times, with the shortest preharvest interval to be recommended. Enough plots should be set out to permit sampling at intervals after the last application (such as 0, 2, and 4 days, or 1, 2, and 4 weeks) as needed to establish appropriate preharvest or no-grazing periods. Treatments at exaggerated rates or number of applications are also desirable to provide data for ensuring that proposed or existing tolerances are not exceeded. The size and distribution of the plots should be adequate to provide random selection of representative samples of the various commodities to be analyzed.

A complete description of the treatment and handling of the samples should be provided on the residue data sheet for each experiment. Treatments should be applied according to proposed label directions for products to be registered for that use, including addition of other agents such as surfactants, "stickers," etc. Residue data for samples from small plots treated by hand or with backpack applicators will generally not suffice. Large scale experiments must be conducted with commercial ground equipment, and also by air or in ultralow volume (ULV) if such uses are to be registered.

Treatment rates should be expressed in pounds of *active ingredient* (ai) per acre of area actually treated. For phenoxy herbicides or other acid compounds applied as various esters or salts, the rates should be expressed in pounds of *acid equivalent* (ae). When the treatment is applied to only part of the area, the actual amount of chemical used is less than if the entire area is treated, but the rate applied should be specified on a broadcast or total area basis. For example, treatment at a broadcast rate of 6 lb/A on 8-inch rows spaced 24 inches apart requires only 2 lb of active ingredient per acre of field. However, the soil and crop *in the row* was treated *at the rate* of 6 lb/A. Confusion may arise when the band rate is equated to *lower* rates on a broadcast basis. For example, the above treatment at 2 lb/A "on the row" is equivalent to 6 lb/A for overall treatment (i.e., $2 \times 24/8 = 6$). However, it was reported erroneously in one case to be equivalent to only 0.67 lb/A of treated field (i.e., $2 \times 8/24 = 0.67$). Thus, it is extremely important to state treatment rates very precisely with adequate description of amounts used so the reviewer knows exactly how much pesticide was applied to the crop or to the soil in which the crop was actually grown (i.e., the row itself). For full coverage sprays as in orchards, the dosage should also be expressed as the amount of active ingredient (or acid equivalent) per unit volume of spray. For example, a spray containing 2 lb/100 gal applied at 400 gal/A is equivalent to treatment at the rate of 8 lb/A.

The total number of field studies and samples required depends on the relative importance of the crop and on the toxicity and/or persistence of the pesticide, its metabolites, or even its impurities. It also depends on the total pounds of pesticide expected to be used for treatment of that crop in any one growing season. Field experiments should be conducted in each principal growing region for that crop, including irrigated arid regions where appropriate. Important crop varieties, such as winter varieties, should be included.

In general, more residue data are needed to establish the first tolerance for a new chemical, or for a new crop, than to demonstrate that existing tolerances are not exceeded by a change in formulation or a modification in directions for use.

10.3.2.2. Sampling

Residue data are needed for all plant parts, byproducts, or processed products which could be used for food or feed purposes. Samples should be collected at appropriate stages of growth until normal harvest. A protocol should be drawn up in advance which states precisely when, what, and how much should be collected. A set of general instructions is helpful such as that developed by M. E. Getzendaner as manager for residues/metabolism/environmental research in the Agricultural Products Department at Dow (24). Sample size is usually determined by the requirements for collecting a representative sample rather than on getting enough for analysis. Table 10.1 lists recommendations for size of samples of different types of crops to be analyzed for pesticide residues.

Samples to be analyzed should be prepared from the whole raw agricultural commodity as it moves in interstate commerce. The following exceptions are listed under definitions and interpretations in §180.1(j) of Title 40 of the Code of Federal Regulations as issued 2/2/76, for examination for pesticide residues:

1. *Bananas* shall not include any crown tissue or stalk
2. Shell shall be removed and discarded from *nuts*
3. Caps (hulls) shall be removed and discarded from *strawberries*
4. Stems shall be removed and discarded from *melons*
5. Roots, stems, and outer sheaths (or husks) shall be removed and discarded from *garlic* bulbs and only the cloves analyzed
6. Roots and tops of *root vegetables* marketed together shall be analyzed separately, except carrot tops which should be discarded
7. Leaves at the top of *pineapples* should be removed and discarded
8. *Lima beans* means beans and pod.

The commodity should not be brushed, stripped, trimmed or washed except to the extent that these are commercial practices before shipment. For example, data are required for cabbage, celery, and lettuce with and without stripping or trimming, since these crops may be shipped either way. In the

TABLE 10.1. Sizes of Samples Recommended for Analysis for Pesticide Residues[a]

Commodity to be Analyzed for Residues	Amount per Sample to be Analyzed		
	Preferred Amount	Preferred Minimum[b]	Absolute Minimum[b]
Apples, peaches, potatoes, citrus, and other similar sized fruits and vegetables	30 fruit	20 fruit	10 fruit
Berries, cherries, etc.	2 kg	1 kg	0.5 kg
Hay, forage (wet or dry)	1–2 kg	1 kg	0.5 kg
Sweet corn (kernels plus cob with husk removed)	12 ears	6 ears	6 ears
Grains (dry), nuts, cottonseed, etc.	1–2 kg	1 kg	0.5 kg
Small vegetables, peas, etc.	1–2 kg	1 kg	0.5 kg
Canned goods (No. 2 cans)	4 cans	2 cans	1 can
Cabbage, lettuce, melons, other similar sized fruits and vegetables	6–8 fruit[c]	4–5 fruit[c]	2–3 fruit[c]
Soil samples	1–2 kg	0.5 kg	0.5 kg
Water samples	1–4 liters	500 ml	100 ml

[a] Adapted from Getzendaner (24).

[b] If analytical method permits.

[c] Or cut sections from 6–8 individual fruits or vegetables so that each sample contains about 2 kg.

case of sweet corn, the sample should consist of kernels plus cob with husk removed. Contrarily, samples of field corn consist only of grain, while the stalks, cobs, etc., are included under the term fodder. Similarly, corn silage is covered by corn forage which is harvested while still green.

A valid statistical basis should be used for selection of samples, although deliberate selection of some samples likely to have high residues may be helpful in evaluating maximum levels to be covered by a tolerance.

Comparable untreated samples must also be collected from nearby plots on the same schedule as for treated samples. Great care must be used to avoid cross contamination, particularly in the case of low-level or negligible residues where the response for treated samples may not be detectably different from that for untreated samples. In some cases where no detectable residue can be demonstrated, a waiver may be obtained for some of the studies deemed necessary to determine permissible levels for measurable residues.

For example, although subacute 90-day feeding studies in laboratory animals were sufficient for establishment of negligible residue tolerances according to 1967 regulations (15), the toxicity data needed now is the same for all tolerances, regardless of level (9). Current requirements include chronic lifetime feeding studies in at least two species of rodents, reproduction studies through two generations in rats, teratology studies, and oncogenic (tumor incidence) studies by oral and/or dermal administration.

All samples should be stored in such a way as to maintain integrity of the samples and their residues until analysis. The distribution and nature of residues should be elucidated. Data should demonstrate whether the residue is *on* the surface or has been absorbed *into* the plant; whether it is translocated primarily to the growing tip, to the fruit, or to the roots from foliar or soil applications; whether it is in free, conjugated, or bound form; and whether the chemical structure of the pesticidal active ingredient has been altered in various species or varieties of plants.

According to 1975 regulations regarding worker reentry intervals, data are also needed on dissipation of residues of cholinesterase inhibitors on all treated foliage (9). Thus, leaves as well as fruit must be sampled and analyzed at intervals after application of organophosphates and carbamates in orchards. Draft guidelines for such studies were reviewed by EPA's Science Advisory Panel in February 1980. At that time EPA was proposing to require such data for most pesticides, although initially intended for only those deemed to present a potential hazard to farm workers.

10.3.2.3. Residues in Crop Byproducts

Crop byproducts must also be analyzed if they can be used for food purposes or as feed for livestock. Unprocessed items include straw and vines remaining after harvest of crops; hulls of almonds, peanuts and rice; and similar byproducts. Data may also be needed on residues in regrowth of perennial crops such as alfalfa, or in rotational crops if uptake of residues from soil is likely.

A major requirement is for data on distribution of residues among processed fractions, particularly if residues can become concentrated during processing. If the residue in the processed fraction exceeds that in the raw agricultural commodity, a food additive tolerance must be obtained in addition to the pesticide tolerance. Both types of tolerances are handled by EPA for pesticide residues, whereas other direct and indirect food and feed additive tolerances are handled by FDA.

For example, a letter dated January 2, 1979, from the Residue Chemistry Branch of EPA, listed the following process fractions as generally requiring analysis for clearance of uses in the indicated crops:

Barley	milled fractions (such as bran, germ, hull, and flour)
Citrus	dried pulp, juice, molasses, oil (cold press)
Field corn	milled fractions (such as bran, shorts, starch, middlings, germ, and flour), oil (crude and refined)
Grapes	pomace, juice, raisins, raisin waste
Peanuts	meal, oil (crude and refined), hulls, soapstock
Pineapple	bran (processing residue)
Rough rice	bran, hulls, polishings, polished rice
Sorghum	syrup (sweet sorghums), molasses (sour sorghums), milled fractions (such as germ, bran, and middlings)
Soybeans	hulls, meal, oil (crude and refined), soapstock
Sugarbeets	molasses, pulp, crude syrup fractions, refined sugar
Tomatoes	concentrated tomato products (puree, sauce), pomace
Wheat	milled fractions (such as bran, germ, flour, and shorts).

The distribution of residues within a commodity may also be of interest, such as in the peel *vs* the edible pulp of bananas, or in fatty *vs* aqueous phases of milk.

Studies on residues in crop byproducts should be conducted by processing crops containing ingrown residues at or near the proposed tolerance level in a pilot plant or in semicommercial scale experiments. In some cases it may be necessary to spike the crop or a fraction of it at the proposed tolerance level. For example, the distribution of residues in sugarcane byproducts can be followed by milling raw sugarcane into bagasse and juice, and spiking the juice before it is processed—first into juice, then into syrup, then into molasses, and then into unrefined sugar under conditions comparable to commercial practice. Representative samples of each byproduct should be collected and analyzed to determine residue levels in each item which could be used for food or feed purposes.

An alternative is to treat the growing crop with a large excess of chemical to assure that enough readily measurable residue can be followed through the processing steps. For example, if the residues found in treated citrus fruit range from nondetectable to 0.06 ppm, a tolerance of 0.1 ppm would be needed to assure that no harvested fruit would contain illegal residues in excess of the tolerance. The tolerance needed for dried citrus pulp can be calculated using the concentration factor derived by comparison of the residue found in dried pulp with that in the fruit used in the processing study. For example, if treated fruit contained 0.04 ppm and the dried pulp obtained from it contained 0.48 ppm (12 times as high), the *maximum* residue likely to be found in dried citrus pulp derived from fruit containing the tolerance limit of 0.1 ppm would be 0.1 × 12 = 1.2 ppm. Thus, the tolerance for dried citrus pulp would probably be set at 2 ppm instead of 0.5

ppm, although 0.5 ppm exceeds the residue actually found in the processed fractions.

The example above illustrates that processing studies must be conducted on samples containing a residue at or near the tolerance level. If the raw fruit had contained no detectable residue, or a residue near the limit of sensitivity for the analytical method, the value of the conversion factor would not have been reliable enough to predict the tolerance level needed for dried pulp. In that case, it would be preferable to process fruit treated at twice the recommended rate and/or closer to harvest to ensure that a measurable residue was present.

10.3.2.4.   Crop Groupings for Residue Data

According to 40CFR §180.34, tolerances may be established for *non-systemic* pesticides in or on groups of crops on the basis of residue data for only one or more crops within 27 narrow groups. According to an amendment to this regulation published in 1967, *negligible residue tolerances* could be established for somewhat broader groups of similar crops on the basis of data on a representative number of commodities listed in the designated groups (15). (Negligible residues are discussed more fully below.) The basis for grouping was more liberal for negligible residues than for residues at higher permissible levels, and pertained to systemic as well as non-systemic pesticides. A total of 17 groups were originally listed of which 15 have been retained as follows: citrus fruits, cucurbits, forage grasses, forage legumes, fruiting vegetables, grain crops, leafy vegetables, nuts, pome fruits, poultry, root crop vegetables, seed and pod vegetables, small fruits, stone fruits, and stored commodities.

Fees for filing petitions for negligible residue tolerances were also calculated on the crop grouping basis. Commodities not listed in the groups were not considered as included in the groupings for the purpose of setting negligible residue tolerances and calculating fees. As a result, petitions for tolerances filed by registrants of pesticides have not generally included minor crops like sesame seeds, mushrooms, or tea. In some cases USDA's IR-4 program has undertaken to obtain residue data for certain pesticides in minor crops, and to file petitions for necessary tolerances by relying on toxicology and other data provided by the registrants.

10.3.2.5.   Negligible or Low-Level Residues

A notice published in the Federal Register on April 13, 1966, reported the findings of a committee appointed by the National Academy of Sciences, National Research Council, to evaluate the former system of registering pesticides for use on food crops on a zero-tolerance or no-residue basis (14).

The report included eleven recommendations, including abandonment of the no-residue and zero-tolerance concepts and adoption of the negligible-residue concept based on negligible *vs* permissible fraction of the maximum acceptable daily intake (MADI) for a pesticide as determined by appropriate safety studies.

A followup notice published one year later defined the term "negligible residue" as meaning any amount of a pesticide chemical remaining in or on a raw agricultural commodity or group of raw agricultural commodities that would result in a daily intake regarded as toxicologically insignificant on the basis of scientific judgment of adequate safety data (15). Ordinarily this would add to the diet an amount which would provide a 2000-fold safety factor over the amount that had been demonstrated to have *no effect* in feeding studies with the most sensitive animal species tested. Such toxicity studies were specified to include at least 90-day feeding studies in two species of animals.

According to the 1968 FDA guidelines for residue data (8), negligible residue tolerances would usually be 0.1 ppm or less and would require development of methods with good specificity. In most cases, the proposed negligible residue tolerance would be on the order of the sensitivity of the method, i.e., the level at which the method yields good recovery in the presence of the substrate. Also the proposed tolerance level should take into account variations in crop blanks due to reagents, instrumentation, analytical techniques, and geographical and varietal differences.

Since that time, many negligible residue tolerances have been granted by FDA and more recently by EPA for pesticides formerly registered on a no-residue basis for use in food and feed crops. Included among these are some based on pesticide petitions compiled by the author, such as for dinoseb (2-*sec*-butyl-4,6-dinitrophenol) at 0.1 ppm in about 70 commodities as finalized on December 12, 1974 (25), and for MCPA (2-methyl-4-chloro-phenoxyacetic acid) at levels from 0.1 ppm in or on commodities used as human foods to 300 ppm in or on forage grasses, as finalized November 3, 1975 (26). Additional tolerances were also granted for 2,4-D on March 19, 1976 (27), including some at negligible residue levels such as 0.1 ppm in blueberries and rice, and some at permissible levels from 0.5 to 5 ppm in foods such as corn, cranberries, grapes, and sorghum.

The April 1975 EPA draft of guidelines for residue data (11) discussed special considerations for "low-level tolerances" where the proposed tolerance would be of the same order of magnitude or slightly above the sensitivity of the method. Such tolerances could be established on a group basis with residue data on representative major crops and could include crops that were not candidates for registration. However, if registrations were sought later for such crops, additional residue data might be needed.

For example, establishment of tolerances for the grouping "seed and pod vegetables" might be based on residue data for peas and soybeans treated with a pesticide product according to proposed label directions. A subsequent request for registration of a similar and/or different use or product in beans might require additional residue data for beans, although the tolerances for seed and pod vegetables included beans. Residue data might also be needed for the same use in a similar crop such as lentils although the Federal Pesticide Act of 1978 includes provisions for requiring less data for minor crops or minor uses.

Of major concern is phaseout of the negligible residue concept. New regulations published July 3, 1975 (9), and proposed guidelines published June 25, 1975 (12), made no distinction between requirements for tolerances at permissible and negligible or low residue levels. The toxicology data required are the same for all registrations involving uses in food crops, and include lifetime feeding studies in two species of rodents, oncogenic studies to evaluate potential for causing tumors (benign or malignant), reproduction studies through three generations in rats, teratology studies in pregnant animals to determine potential for causing birth defects, and mutagenic studies when appropriate methodology has been developed. Additional studies such as neurotoxicity in hens may also be required for certain types of chemicals. The full spectrum of studies costs millions of dollars and takes a minimum of three years to complete. However, this interval provides ample time for conducting suitable residue studies as outlined herein. Based on the high cost of all these studies, it is imperative to avoid further delays in getting clearance for new uses of pesticides in food crops due to inadequacy of residue data.

10.3.2.6. *Temporary Tolerances*

Requirements for registration of a new use of a pesticide include efficacy data and residue data for large-scale trials under commercial use conditions according to proposed label directions. However, sale of treated crops or animals is prohibited in the absence of tolerances for residues of the pesticide in each raw agricultural commodity. To obviate having to destroy valuable crops or animals bearing illegal residues, temporary tolerances can be established by EPA in conjunction with Experimental Use Permits to conduct the large-scale trials (28). A petition for a temporary tolerance should specify the amount of pesticide which will be used under the conditions of the experimental permit, the crops and crop acreage which will be treated, and the geographic areas where the treatments will be made. Quarterly progress reports are also required for any studies conducted under Experimental Use Permits.

Requirements for temporary tolerances are similar to those for regular petitions, except that analytical method trials need not be conducted and the metabolism and residue data may not be as extensive as those for permanent tolerances. The amount of data required depends on such considerations as toxicity of the pesticide or possible degradation products or metabolites, the acreage to be treated, and the importance of the food or feed commodity. Additional data obtained while the Experimental Use Permit is in effect must be made available to the Registration Division of EPA for assessment of potential hazard.

Temporary tolerances are generally granted for a period of one year at the time of issuance of the Experimental Use Permit. Extension of temporary tolerances can also be obtained if additional large-scale trials are needed in subsequent years. For example, adequate residue data may have been obtained from experimental treatments of sugarcane, apples or citrus fruits, but additional data may be needed from commercial scale experiments to determine the tolerance levels needed for molasses, dried apple pomace or dried citrus pulp. Experimental use permits are also useful for obtaining residue/efficacy data from aerial treatments of large plots or for other uses where small plots cannot be treated with commercial equipment. The permit also allows early sales of a new pesticide to help defray the high cost of the many studies needed prior to full registration.

10.3.3. Residue Studies in Animals

Pesticide residues can occur in animal tissues, not only from intentional treatment of the animals but also from feeding treated forage or feed to livestock. As in residue studies in plants, all experimental treatments should reflect as closely as possible the conditions under which animals will be exposed to the pesticides.

Separate studies may be needed in dairy cows and in meat animals (preferably young cattle rather than goats or sheep), and may also be needed in swine and in poultry. Groups of at least three large animals or ten birds are needed for each dosage level or slaughter interval. In feeding studies treatment should be continued for at least 3 weeks or until a plateau in storage level for the pesticide and/or its metabolites has been achieved.

Comparable control animals should be carried through the experiment along with treated animals, receiving the same basal ration and sampled or slaughtered on the same schedule. Data are needed on residues in milk, meat (muscle, fat, liver, and kidney), poultry, and eggs, and sometimes in other tissues as specified below for individual types of studies. Information should also be provided on whether milk or eggs samples were composited from different animals in one test group or, for example, whether sheep passing through a dip tank were freshly shorn or unshorn.

The chief objective of animal residue studies may be to establish tolerances at suitable levels in meat, milk, poultry, and eggs, or to demonstrate that there is no reasonable expectation of residues in these food items. They are designed so that residues will be found and so that a relationship can be established between residue levels in the diet and those to be expected in various tissues. Appropriate tolerances can then be established based on tolerances set for feedstuffs. Analytical methods should be validated down to detection limits of 0.01 to 0.05 ppm in milk, and no more than 0.1 ppm in other edible tissues. Greater sensitivity is required for more toxic pesticides, or if toxic metabolites or impurities are a problem. The latter situation is discussed more fully under Special Monitoring Programs (Section 10.4.5).

10.3.3.1. Residues from Treated Plants

If residues are found in or on feed or forage crops, or in fruit or vegetable crops where processed fractions are fed to animals, data must be submitted to show whether there is a reasonable expectation of secondary residues in milk, meat, poultry, or eggs. The material fed to the livestock should correspond to the ingrown residue found in feed items, whether parent chemical or conjugates or metabolites thereof.

Several dosage levels should be fed, corresponding to expected maximum residues in all feed items, plus exaggerated levels (generally threefold and tenfold the maximum). The actual dosage levels in the total diet should take into account the proportion of the diet made up by feed items likely to contain residues. For example, dried citrus pulp would comprise no more than one third of the diet of dairy cows, whereas range grass could be the sole diet of some beef animals. However, due to the high cost of residue studies in large animals, it is recommended that the levels chosen cover potential future uses of the pesticide as well as those uses being proposed at the time of the study.

The dosage levels should be expressed in parts per million (ppm) in the total ration of the animals, excluding drinking water. Records should also be kept on the amounts of feed ingested daily by each treated animal during a preliminary acclimation period, during the treatment period, and after withdrawal from treated feed. This not only permits calculation of total pesticide consumed, but affords a warning if the animal has gone off feed for some reason. As a general rule, a young beef animal can be expected to consume about 3% of its body weight in feed each day on a dry feed weight basis. Thus a level of 300 ppm pesticide in the diet amounts to a dose of about 9 milligrams per kilogram of body weight per day. (Some publications have switched to using mg/kg instead of ppm as units for residues, a practice which leads to erroneous equating of mg/kg in feed or environmental samples with toxic dosage levels expressed in mg/kg of body weight.)

In studies to provide data on residues in milk, progressively higher levels can be fed to the same group of dairy cows (generally 3 per group) for 2 to 3 weeks at each level, followed by untreated feed until residues have declined to nondetectable levels. Samples of milk should be taken frequently during pretreatment, treatment, and posttreatment periods, and from a group of comparable control cows maintained on the same basal ration under the same conditions throughout the entire experiment. Samples of cream should also be obtained to determine the tendency for residues to concentrate in dairy products such as butter.

In studies in meat animals, all animals should be maintained on untreated feed until acclimated, then fed treated feed in groups of 3 animals per dosage level for 3 to 4 weeks or more prior to slaughter. Groups of animals should also be slaughtered at intervals after withdrawal from treated feed to determine rate of dissipation of residues. The intervals chosen depend on the persistence of residues, particularly in fat, liver, or kidney. Results of earlier metabolism studies in small animals may be helpful in designing these studies. The relative distribution of residues between muscle and fat is generally similar to that between milk and cream.

Feeding studies may also be needed in hogs and poultry since pesticides may be excreted or metabolized somewhat differently in nonruminant mammals and birds than in ruminants. This is particularly true for residues in major feed items such as corn, small grains, soybean meal, or dried alfalfa. On the other hand, if the feed item is fresh grass bearing high residues shortly after application of herbicides in pasture and rangeland, data on ruminants should suffice since grass can constitute the entire diet of ruminants but not of other animals. As with cattle, groups of hogs and poultry should also be slaughtered at intervals after withdrawal from treated feed to determine dissipation of residues, particularly in fat.

For poultry feeding studies with nonradioactive pesticides, groups of 10 birds should be fed at each of at least 3 levels (the maximum expected residue, threefold, and tenfold) for at least 4 weeks. Treated and untreated birds should be maintained side by side on the same basic ration. Residue levels should be determined in eggs gathered at frequent intervals, and in muscle, fat, and liver at slaughter. The distribution of residues between egg whites and egg yolks will likely be similar to that found between muscle and fat.

10.3.3.2. Residues from Treatment of Animals

Studies are also needed on residues from direct administration of pesticides to animals, either as oral or topical treatments. Pesticides can be administered orally such as in feed, drinking water, mineral blocks, boluses, or drenches. Animal treatments to control insects in cattle manure generally pass through the digestive tract of animals with no drug effects and are considered pesticides. However, orally administered treatments to control horse bots, cattle grubs, fleas, etc., are deemed to be drugs and are subject to the jurisdiction of the Bureau of Veterinary Medicine in FDA (29, 30). On the other hand, EPA regulations govern residues from topical treatments such as by dips, sprays, pour-ons, backrubbers or other automatic devices. In each case, the amount applied per animal should be described, along with sufficient information on recharging, placement, and duration of treatment so the degree of exposure of animals can be estimated.

Data are required to show the extent to which such uses can reasonably be expected to cause residues in meat (muscle, fat, liver, and kidney of livestock), poultry, milk, or eggs. The test treatment should reflect as closely as possible the conditions under which the pesticides will be used commercially, and should be designed to provide maximum likely exposure of animals to devices such as backrubbers where unlimited access is permitted. Comparable control animals should be maintained along with treated animals throughout each experiment.

In some cases data from one type of treatment may suffice for another type. For example, data from dips or high pressure sprays on cattle would be expected to cause higher residues than from dust treatments but not vice versa. On the other hand, more data may be needed for certain uses such as for dermal treatments with dusts in poultry where samples of skin should be collected and analyzed separately from muscle and fat. Information may also be needed on rate of clearance of drugs from kidney of birds. However, specific data on residues in kidney should not be needed since this item constitutes only a small part of edible poultry and is not packaged and sold separately like chicken livers or beef kidneys.

10.3.3.3. *Residues from Treatment of Agricultural Premises*

Uses on agricultural premises or in animal quarters should be described in sufficient detail to permit an evaluation of potential residues in meat, milk, and eggs. This includes possible residues on milking equipment, exposed feeds, drinking troughs, and feed bins. Directions for use should include precautions to avoid residues, but data reflecting exaggerated uses are highly desirable.

An example of such a study is one conducted by Dow in the mid-1950s using ronnel [O,O-dimethyl-O-(2,4,5-trichlorophenyl)phosphorodithioate] to control flies in a dairy barn on a Texas prison farm. The test afforded maximum opportunity for contamination of milk since the cows were hand-milked by inmates of the prison. The method of analysis was based on manometric measurement of inhibition of flyhead cholinesterase, and was sensitive to 0.005 ppm (5 ppb) ronnel in milk (31). Such a method was useful as a research tool but could not be used for regulatory purposes due to non-specificity for that insecticide.

Other uses on agricultural premises include incorporation of pesticides in litter, or in dusts added to livestock pens, or under caged layers or broiling chickens.

10.3.4. Miscellaneous Residue Studies

As discussed previously for residue studies in plants and animals, the design of experiments for other uses of pesticides should follow as closely as possible the manner in which they are to be used commercially. Several examples of miscellaneous studies are discussed briefly below.

10.3.4.1. Aquatic Sites

The type of study needed depends in large part on the nature of the pesticide (herbicide, insecticide, molluskicide, etc.), the levels required to obtain the intended effect, and the time interval needed for residues to degrade or dissipate to acceptable levels. As with other studies, specific use directions must be given including a description of the types of sites and pests to be treated; the dosage expressed in terms of amount of active ingredient per unit volume of flow per unit of time (for moving bodies of water), or per unit of surface area (for relatively still water), or parts per million based on average depth of water; the method of application; and the number and times of application relative to season or stage of growth.

The study should provide data on the nature of the residue in water, the decline of the residue in still water and its dispersal in flowing water, the residue adsorbed on bottom sediment and its rate of dissipation, the nature and amounts of residue in fish and shellfish, the amount of residue in water used for irrigation, the amount of residue transferred to representative irrigated crops, and the possible transfer of residues from water or irrigated crops to meat, milk, and eggs.

Tolerances may be needed for potable water, fish and shellfish, and for irrigated crops as appropriate. The tolerance levels will depend on the practicality of limitations such as treatment of only a fraction of the site; the interval between last application and use of the treated water for drinking, watering livestock, irrigation, recreation, and fishing; or precautions to minimize contamination of water in ditch bank treatments.

Representative samples should be obtained from treated sites. However, adequate sampling of water may be difficult since no clear definition of potable water has been given to date. During discussions with EPA personnel, opinion ranged from any water which could potentially be consumed by livestock or man, to water impounded in reservoirs used for human water supplies, to processed water fit for drinking by humans such as when it emerges from the tap. Hopefully, this situation will be clarified when Guidelines for Tolerance Petitions are finalized (12).

In the case of fish, they should be beheaded, gutted and scaled prior to analysis.

10.3.4.2.   Stored Commodities

Residue data are needed for uses of pesticides in sites where foods or feeds are stored or shipped. Residues from direct application of pesticides to foods or feeds may require food additive tolerances and/or pesticide residue tolerances. For fumigant uses, such conditions as time of exposure, pressure, air tightness of container, and aeration procedure should be specified. A sliding table for dosage, temperature, and time is often useful. Requirements may vary depending on the use, but the data should provide the reviewer with an adequate basis for estimating *maximum* residues likely to result from the proposed use.

10.3.4.3.   Food-Handling Establishments

Exposure of food must be avoided or minimized when pesticides are used in food-handling establishments such as processing plants, bakeries, canneries, restaurants, institutions, etc. (32). Directions for use should include the type of establishments to be treated, the spray concentration, the types of equipment used, the modes of application (sprays directed to crevices, cracks, and baseboards, or space sprays or fogs), maximum dosages per cubic or square foot, frequency of treatment, and time of treatment (such as after-hours in restaurants). Other pertinent information includes precautions to remove or cover food, dishes, and utensils, and to thoroughly clean food-processing surfaces before food preparation, processing, or serving resumes.

Supporting residue data should include studies reflecting use according to explicit label precautions for representative fatty and non-fatty foods, finely powdered foods (such as flour), and types of packaging. Data are also needed reflecting exaggerated uses with less stringent precautions.

10.3.4.4.   Seed Crops and Seeds

Tolerances are not needed for uses of pesticides on crops grown solely for seed purposes, nor in some cases on seeds prior to planting. However, residue data may be needed to demonstrate that the pesticide or toxicologically significant alteration products are not absorbed by the growing plant. Data may also be needed to demonstrate whether residues of pesticides such as defoliants applied shortly before harvest of a seed crop can be taken up by regrowth of a perennial crop such as alfalfa or grass, or by succeeding crops planted soon after harvest of the seed crop.

10.3.4.5. Postharvest Treatments

Residue data are needed for pesticides applied postharvest, such as fungicides and other agents used to preserve fresh fruits and vegetables during shipping and marketing. Label directions are not generally provided on retail packages because highly trained operators are required for applying wax coatings to fruit or for manufacturing pesticide-impregnated fruit wraps. However, each specific use must be described completely.

10.3.4.6. Minor Changes in Use or Formulation

Additional studies may be needed to determine whether minor changes in use pattern or formulation can cause residues in excess of established tolerances for a pesticide. Examples of changes in use patterns which are likely to require studies and possibly another petition for a higher tolerance are listed in draft EPA guidelines (12) as (1) significant changes in preharvest interval of a crop or in postharvest treatment of a commodity, (2) extension of use patterns to include aerial as well as ground application, or alteration of use patterns to include new types of usage (such as over-the-top *vs* directed sprays, etc.), (3) addition of a sticker or extender to the formulation, (4) conversion to a slow-release formulation, and (5) use in additional climatic regions.

Examples of minor changes which may *not* require additional residue studies are (1) a change in surfactant concentration, (2) substitution of a new but similar surfactant, and (3) substitution of one clay diluent for another. Exceptions must be made individually, based on a thorough knowledge of the chemistry of the ingredients involved.

10.3.4.7. Imported Foods

Although foreign uses of pesticides are not subject to requirements for registration in the United States, tolerances are needed for residues in imported foods. Petitions for such tolerances are submitted under Section 408(e) of the Food, Drug, and Cosmetic Act whereas tolerances proposed in conjunction with applications for domestic registration are submitted by pesticide manufacturers under Section 408(d) of the Act. Nevertheless, petitions for tolerances in imported foods should contain the same types of information, including descriptions of how the products are used (labeling), and complete residue data. In addition, the petition should briefly discuss any controls regulating the use of pesticides in the country of origin and some evidence that those requirements have been met.

10.3.4.8. Environmental Studies

According to FDA guidelines (8), environmental studies should preferably show the half-life of the pesticide or toxic conversion products in a variety of soil types reflecting conditions of moisture, temperatures, pH, and bacterial population likely to occur under the proposed conditions of use. Information on the extent of leaching or binding in the soil is useful. Some of the experiments should reflect exaggerated dosages.

Requirements for data on pesticide residues in soil and water were first described in *PR Notice 70-15* issued by USDA's Pesticide Registration Division on June 23, 1970 (33). Current requirements are outlined in Section 3 Regulations for implementation of FIFRA as amended (9) along with revised proposed guidelines as to how environmental studies should be conducted (34). Aquatic studies were discussed in Section 10.3.4.1.

Current objectives of field dissipation studies on pesticides in soil are listed as (1) to determine the rate of dissipation of the pesticide and its significant degradation products under fallow and cropped field conditions, (2) to determine the nature of significant degradation products of the pesticide under field conditions and to compare them with results of laboratory metabolism studies, (3) to determine the potential carryover of the pesticide and significant degradation products under field conditions for single applications and, where applicable, 3-year repeat applications, and (4) to evaluate leaching in the soil under field conditions.

Sites for the tests should be located in the regions of usage, on a major agricultural soil of the region, and available over the 3-year period needed to complete repeat applications. Descriptions of the soil profiles and heights of water tables are required. Additional information should be provided on topography (slope, drainage, land form), climate, soil analysis, pesticide formulation applied, sampling technique employed, and the analytical method used including evidence as to its adequacy for removing "bound" residues of the pesticide and/or its degradation products.

All soil samples should be taken at sufficient depth and at sufficient intervals to be all-inclusive of parent chemical and significant metabolites in order to show degradation and extent of downward movement. Identification of residues comprising more than 10% of initial application or 0.01 ppm, whichever is greater, is needed to construct decline curves of residues in foliage, litter, soil, and water (34).

Samples should be collected before application (checks), immediately after application, four or five times between application and fall freeze, and once before planting the following spring. Sampling depths should be in 6-inch (15-cm) increments in the 0-12-inch zone and deeper as needed. Great care must be exercised in collecting soil samples to avoid contamination of lower levels with soil from upper levels which generally contain much higher residues. (Also see Section 10.4.3.2). Replicate samples should be collected at zero time and at the last sampling before winter for separate analysis to estimate variability for each experiment. It should be recognized that inconsistent results are more the rule than the exception due to the inherent variability of soil profiles even within a single plot.

10.3.5. Reporting of Residue Data

No study is complete without a report, and the quality of the report often serves as a measure of the value of the data. Recommended reading for those who have difficulty writing such reports is a humorous paperback book by Rathbone on current uses and abuses in scientific writing (35). It includes many helpful suggestions in entertaining chapters entitled "The Peaceful Coexistence," "The Tenuous Title," "The Inadequate Abstract," "The Wayward Thesis," "The Improper Introduction," "The Artful Dodge," "The Ubiquitous Noise," "The Neglected Pace," "The Random Order," and "The Arbitrary Editor." A more conventional handbook for authors of chemistry reports and papers is available from the American Chemical Society (36).

Instructions for reporting residue data were included in guidelines for pesticide petitions issued by FDA in 1968 (8) and in the draft proposed by EPA in 1975 (12). Basically, the report should include an informative summary, a description of the study, an outline of the analytical method used, some raw data, and results of all analyses. Conclusions should be based on the data but should not include speculative interpretations. As learned through years of personal experience in compiling data from many sources for submission to government agencies, special attention should be given to providing the following information in reports of residue studies.

10.3.5.1. Identification of Report

The report should be identified in such a way that it can be cited as a reference and retrieved by modern searching techniques. The title should be descriptive of the contents of the report and should contain key words such as the name of the pesticide product, type of test, method, crop, pest, and so on. The author should be identified by name, affiliation (university, government group, contract laboratory, or company), and location, including city and state. Dating the report by month and year provides a relationship between that study and reports of other studies submitted in support of a pesticide registration. Such reports may be included in petitions requesting tolerances for a pesticide chemical, in an application for new registration of a new pesticide product, or in a series of applications for amendments to an initial registration stretching over many years.

10.3.5.2. Residue Summary

The summary is the most important part of the report and should be written with great care to avoid ambiguity. It should be informative and clearly state the purpose of the study, how and where it was done, and what conclusive data were obtained. It might also include a brief table of average values or range of residues found in the study.

A more extensive summary table should be included in the body of the report, keyed to where raw data can be found for each experiment within the overall study. For example, the report may include data from studies in several locations, comparisons of various rates, numbers, and timing of applications, in one or more crops, animal species, or sites. A suggested format for a summary table is presented in Table 10.2.

10.3.5.3. Body of Report

The purpose of an introduction is to inform readers of potentially diverse training about why the work was undertaken, what types of information it is to provide, where the study was done, and by whom. The main steps in the analytical procedure should be outlined, with adequate reference to where a complete description can be found (e.g., in an appendix).

The pesticide used must be clearly identified, not only by common or chemical name but by product name or formulation code number, so the data can be correlated with other studies to support a registration. The product or formulation should also be described as to type (granular, wettable powder, emulsifiable liquid, water-soluble solution, etc.), and by amount of active ingredient or ingredients, in percent for solids or in pounds per gallon for liquids.

Dosage rates should be specified accurately in precise terms that cannot be misinterpreted (as discussed in Section 10.3.2.1 of this chapter). For example, Dow recently conducted a study on residues of 2,4-D in sorghum using FORMULA 40® herbicide, a water-soluble liquid formulation containing 59.7% 2,4-D alkanolamine salts (of the ethanol and isopropanol

TABLE 10.2. Summary of Residues of 2,4-D in Sorghum Treated Postemergence with FORMULA 40® ^{a} **Herbicide at 1 lb/A** ^{b}

Sample Type	Days After Application	2,4-D, ppm	
		Range	Average
Over the top (from Table 7)^c			
Green Plants	0	48–73	56
	7	5.8–10.7	7.7
	14	2.0–2.7	2.4
	28	0.6–0.9	0.8
Stover	116	—	ND^d
Grain	116	ND–0.26	<0.1
Directed (from Table 10)^c			
Green Plants	0	1.6–2.8	2.1
	7	ND–1.2	0.8
	14	<0.5–0.7	0.6
	28	—	ND
Stover	109	ND–<0.5	ND
Grain	109	—	ND

^{a} Trademark of The Dow Chemical Company for a formulation containing 59.7% 2,4-D alkanolamine salts (of the ethanol and isopropanol series), or 4 pounds 2,4-D acid equivalent per gallon.
^{b} Unpublished data of The Dow Chemical Company.
^{c} Tables 7 and 10 in the study report summarized here.
^{d} ND = None detected with a detection limit of 0.2 ppm for green plants and stover and 0.06 ppm for grain when analyzed by gas chromatography using Dow residue determination method ACR 75.3.

series) or 4 pounds of 2,4-D *acid equivalent* per gallon, applied postemergence at 1 qt/A (1 lb a.e./A), as reported in Tables 10.2 and 10.3.

Brand names that are trademarks registered by individual companies should be recognized as such, and should not be considered as synonymous with chemical names or common names for an active ingredient in the product. On the other hand, residues are expressed in terms of the pesticide chemical, in some cases not the same as the active ingredient on the label. For example, residues of 2,4-D should be given in terms of the acid itself rather than of the salts or esters found in products such as FORMULA 40 used in the studies.

Each experiment should be described in sufficient detail for government

TABLE 10.3. **Residues of 2,4-D in Sorghum Treated Over-the-Top with FORMULA 40® [a] Herbicide at 1 lb/A[b]**

Sample Source:	Dow Field Station, Davis, California
Experiment by:	Dr. J. Geronimo, The Dow Chemical Company, Davis, CA
Soil Type:	Yolo Loam
Variety:	NB-501
Plot Size:	12 rows 400 feet long on 30-inch centers
Application:	Sprayed in a 12-inch band over-the-top with 8002E nozzles at 43 psi from a tractor–mounted four–nozzle boom and pump with a flow of 25 gal/A
Plant Size at Application:	8 inches
Application Dates:	Planted 7/3/73
	Emerged 7/10/73
	Treated 8/10/73
	Harvested 12/4/73
Other Chemicals:	None

Sample Number	Treatment Rate lb (a.e.)/A	Days after Application	2,4-D, ppm Gross	2,4-D, ppm Corrected[c]
Green Plants				
121913	0	7	0.00	—
121914	0	7	0.00	—
121919	0	35	0.00	—
121920	0	35	0.00	—
			Av = 0.00	
121881	1	0	42	48
121882	1	0	43	49
121883	1	0	64	73
121884	1	0	48	55
				Av = 56
121885	1	7	6.2	7.0
121886	1	7	5.1	5.8
121887	1	7	9.4	10.7
121888	1	7	6.6	7.5
				Av = 7.7
121889	1	14	2.4	2.7
121890	1	14	1.9	2.2
121891	1	14	1.8	2.0
121892	1	14	2.3	2.6
				Av = 2.4

TABLE 10.3. (**Continued**)

Sample Number	Treatment Rate lb a.e./A	Days after Application	2,4-D, ppm	
			Gross	Corrected[c]
Green Plants (*continued*)				
121893	1	28	0.48	0.6
121894	1	28	0.76	0.9
121895	1	28	0.61	0.7
121896	1	28	0.70	0.8
				Av = 0.8
Stover				
122466	0	109	0.14	—
122467	0	109	0.10	—
122468	0	109	0.06	—
122469	0	109	0.01	
			Av = 0.08	
122450	1	116	0.05	ND[d]
122451	1	116	0.11	ND
122452	1	116	0.05	ND
122453	1	116	0.03	ND
				Av = ND
Grain				
122462	0	109	0.02	—
122463	0	109	0.02	—
122464	0	109	0.02	—
122465	0	109	0.02	—
			Av = 0.02	
122446	1	116	0.04	ND
122447	1	116	0.02	ND
122448	1	116	0.23, 0.20 0.22	0.26[e]
122449	1	116	0.02	ND
				Av = <0.1

[a] Trademark of The Dow Chemical Company for a formulation containing 59.7% 2,4-D alkanolamine salts (of the ethanol and isopropanol series), or 4 pounds 2,4-D acid equivalent per gallon.

[b] Unpublished data of The Dow Chemical Company.

[c] Corrected for control value and for recovery of 2,4-D of 89% for plants, 98% for stover, and 88% for grain.

[d] ND = None detected with a detection limit of 0.2 ppm for green plants and stover and 0.06 ppm for grain.

[e] Based on repeat analysis of first extract and reanalysis of sample through complete procedure.

reviewers to evaluate its pertinence in relation to other studies with that pesticide and with similar products. For example, if the study involved spray applications of an herbicide, the description should include where it was done, under whose direction, in what crop and variety, at what stage of growth, at what rate, in what volume of solution, in what diluent, with what type of equipment, whether as a band treatment or broadcast, whether applied over-the-top or directed at the base of the crop plants or trees, whether as ter-restrial or aerial applications, etc. Additional information should include type of soil, identities and rates of other treatments applied to the crop, cli-matic conditions, amount and timing of rainfall before and after treatment, and all pertinent dates such as when the crop was planted or emerged, when treatment(s) were applied, and when samples were harvested. A suggested format is shown in Table 10.3.

The data reported should include both individual and average residue values, as well as some control and uncorrected residue values. Corrections made for control and recovery values should be indicated. When GLC is the method of analysis, chromatograms should *not* be corrected for background variations or discrete peaks at or near the retention time of the component sought.

Few residue studies result in clear-cut uniform data. Where extraneous variables may have contributed to the spread in values, the author(s) should include an objective discussion of factors which could have contributed to the findings. In some cases, explanations may suffice for reviewers to discard obvious "outliers" in their evaluation of residue data for establishment of tolerances for pesticides. Those reporting residue data might also consult the 1975 revision of the Statistical Manual of the AOAC (23) which presents simple and flexible statistical techniques using examples related to familiar questions concerning the significance of apparent differences among results.

10.3.5.4. Tables and Figures

Tables and figures should be designed to present data in a clear, concise fashion so they can be interpreted with a minimum of effort by a reader scanning the report. Each table and figure should include enough detail to stand alone and may repeat pertinent information described more fully elsewhere in the body of the report. Of major importance is the wording chosen for titles, for column headings, and for labeling of axes. The columns in a table should be organized so as to group pertinent items together, with emphasis on the net residue found. An adequate margin should be provided on all sides, particularly at the left so the report can be included in a bound volume without hiding pertinent information such as the title, the first column of the table, or the units on an axis of the figure.

Column headings should include the units for all values cited, such as quarts per acre, days after last treatment, or parts per million pesticide chemical in the sample. Distinction must be made between dosages in terms of formulated product *vs* active ingredient, and both should be cited if necessary to avoid confusion. Columns of numbers should be aligned according to decimal value, and a zero should precede the decimal point for numbers less than 1 to prevent error in interpretation from faint copies of a report. Long rows of numbers should be broken up into logical groups of data, such as by location, crop, treatment rate, interval between last treatment and sampling, etc.

Numerical data should be presented in tabular form, in the body of the report or attached as appendices. The data should include all values obtained in validating the method on fortified samples, and all values for analyses of untreated and treated samples in each experiment within the total study as in Table 10.3. Uncorrected gross values should be shown, along with net values after application of appropriate corrections for blanks and recoveries. Individual values should be shown for analyses of replicate samples and for replicate analyses on the same sample to distinguish between repeatability among analyses and variability among samples. Average values may also be shown where appropriate, to provide easy correlation with values listed in summary tables. Data in summary tables should be cross-referenced to tables for individual studies particularly if numerous studies are being reported. Table 10.2 represents data from only two studies, cross-referenced to Tables 7 and 10 in the study report.

The report should include representative raw data, such as standard curves, optical absorbance readings, and/or copies of appropriately labeled gas chromatograms. Some of the chromatograms should show the analyst's sample dilution factors, sample equivalent injected, and the way in which GLC peaks were quantitated (maximum peak height or total area under peak). Photographs of thin layer chromatography (TLC) plates, paper chromatograms, or radioautographs of plants treated with labeled pesticides should be included when such evidence would contribute significantly to the evaluation of the data.

As outlined in the FDA Guidelines issued in 1968 (8)—"A statistical treatment of data may be used to express the precision and accuracy of the analytical results. The null hypothesis technique is useful in calculating the confidence level at which a set of values from control samples do not differ significantly from a set of values from treated samples. A graphical representation of residues versus time (decline or dissipation rate curves) is also desirable. This is usually plotted on semi-log coordinates with time (days after treatment) as the linear function." (Statistical evaluation of results is discussed in Section 10.5.5.)

10.4.   MONITORING STUDIES

During the 1960s, concern grew about potential long-term effects of persistent pesticides in the environment. This concern was fed in part by reports of finding residues of DDT in samples such as pelicans, walruses, and even soil far removed from where the insecticide had been used. The impact of such reports has continued long after many of the results of early studies have been refuted by more specific analytical data. Positive findings originally attributed to DDT have often been shown to be due to the presence of PCBs (polychlorinated biphenyls) (37, 38). More recently, environmental and laboratory contamination by phthalate esters has been shown to cause responses frequently mistaken for DDT residues (39, 40). Thus, it is imperative that analytical methods used in monitoring for pesticides are specific for the chemicals being sought and that positive findings are confirmed by secondary methods based on different principles of analysis.

10.4.1.   National Pesticides Monitoring Program

The National Pesticides Monitoring Program (NPMP) was initially designed on the basis of the minimum monitoring needed to establish baseline levels of pesticides in substrates of food and feed, humans, soil, water, air, wildlife, fish, and estuaries, and to assess changes in these levels. A review of the components of this program was initiated in 1968 by the original Subcommittee on Monitoring and was completed in 1971 by the Monitoring Panel of the Working Group on Pesticides responsible to the Council on Environmental Quality (41). The focal point of Federal policies and activities for pesticides is within EPA but many of the monitoring activities have remained in other agencies.

The chief purpose of the NPMP is to establish mean levels of specific pesticides as baselines for assessing whether pesticide residues are increasing or decreasing. However, it is also important to consider the levels of concern for each pesticide in certain environmental components. Five bases for concern are: (1) any concentration of a pesticide known to be potentially harmful, (2) increasing trends, (3) exceeding standards, (4) recognition of adverse effects in humans, and (5) erratic variability. These five may be understood by scientists, administrators, and the public, but agencies should develop specific possible problem-area definitions for each medium being monitored.

The Federal Working Group on Pest Management is responsible to the Council on Environmental Quality and is comprised of representatives from EPA and from the U.S. Departments of Agriculture, Commerce, Defense, the Interior, Health, Education, and Welfare, State, Transportation, and Labor. Its Pesticide Monitoring Panel consists of representatives of FDA, EPA, the Agricultural Research Service (ARS), Animal and Plant Health Inspection Service (APHIS), Extension Service, Forest Service, Department of Defense (including the Corps of Engineers), Fish and Wildlife Service, Geological Survey, National Marine Fisheries Service, National Science Foundation, and Tennessee Valley Authority (TVA).

The *Pesticides Monitoring Journal* is published quarterly under the auspices of the Working Group and its Monitoring Panel as a source of information on pesticide levels relative to man and his environment. It was first issued in 1967 and its publication is now carried out by the Technical Services Division, Office of Pesticide Programs (OPP) of EPA.

The Working Group also issued two useful manuals in 1975, entitled *Guidelines on Sampling and Statistical Methodologies for Ambient Pesticide Monitoring* (42) and *Guidelines on Analytical Methodology for Pesticide Residue Monitoring* (43). The list of chemicals originally chosen for the National Pesticide Monitoring Program contained those believed to be of most interest because of their (1) extent and/or volume of usage, and/or (2) degree of hazard to man, fish, and wildlife, and/or (3) degree of persistence. The list has been revised periodically to take into account new pesticides which have been developed and changes in use patterns of older pesticides. Industrial pollutants such as PCBs (polychlorinated biphenyls) were added as they became identifiable. Also of interest are metabolic and/or breakdown products which are pesticidal or toxic, and toxic impurities such as certain dioxins, chiefly 2,3,7,8-tetrachlorodibenzo-p-dioxin (TCDD).

10.4.2. National Food and Feed Monitoring Program

Under the laws governing pesticides, EPA is responsible for registration of each product and for establishment of tolerances (maximum residue limits) in raw agricultural commodities and processed products derived from food and feed crops. A third element is enforcement of the regulations, a responsibility shared by EPA and FDA.

A main objective of FDA enforcement actions is to insure the safety of the U.S. food supply. According to Sam D. Fine, Associate Commissioner for Compliance in FDA (44), a food is unlawful and subject to regulatory action if it contains a pesticide residue in the absence of an established EPA tolerance or an applicable FDA action level for unavoidable residues (such as DDT in fish), or if it contains a pesticide residue in excess of a tolerance or action level. FDA also has the authority to initiate seizure of interstate shipments of foods adulterated with pesticide residues, and can deny entry of imported foods found to be adulterated.

The Federal program for monitoring pesticide residues in food and feed is comprised of *surveillance programs* maintained by FDA and by the Consumer Protection Program in the Consumer and Marketing Service of USDA. Sampling and analyses of raw agricultural commodities and market basket studies are conducted by FDA while sampling of meat and poultry is handled by the Animal and Plant Health Inspection Service (APHIS) of USDA.

The information gathered through the residue monitoring program is shared by USDA, FDA, EPA and appropriate State agencies. The three Federal agencies work in unison with the States to determine the steps necessary to correct any situation that results in above-tolerance residues. APHIS can condemn animals for human food uses. FDA can take action against individual producers and feed companies and may condemn foods for animal use. The Enforcement Division in EPA's Office of Pesticide Programs has the responsibility to be sure that pesticides are used according to label directions. Since violating FDA's residue regulations is a felony, those convicted of breaking the law are subject to stiff fines and face the possibility of prison sentences.

The purpose of the FDA pesticide surveillance program is to locate and remove from the marketplace foods containing pesticide residues greater than approved tolerance levels. Using a statistically based sampling plan, domestic and imported food and feed shipments are randomly sampled and tested for a wide variety of pesticide residues on an objective basis. This means that, prior to sampling, there is no suspicion of the presence of excessive residues or the misuse of pesticide chemicals.

Monitoring for residues in animal tissues is done by USDA's Animal and Plant Health Inspection Service using methods tested by FDA and the APHIS Meat and Poultry Inspection Program. An objective statistical method of selection provides a valid random sample of all animals slaughtered at Federally inspected plants (45). FDA also conducts another program involving examination of food ready to be eaten. This investigation measures the amount of pesticide chemicals found in a high-consumption varied diet, as purchased in retail markets and prepared for consumption before analysis. Market basket samples for the total diet studies are purchased from retail stores, bimonthly, in five regions of the United States.

The March 1976 issue of *Pesticides Monitoring Journal* reported results for total diet studies in the ninth year of the program (46). Samples were collected in 30 different grocery markets in 30 different cities between August 1972 and July 1973. Pesticide residues remained at the relatively low levels reported for the previous eight years as referenced in the report. A total of 988 residues of 37 different pesticides were found compared to 1003 residues of 35 compounds in 35 market baskets collected the previous year.

Of the 770 composites reported as containing pesticide residues, 302 or 39% contained only "trace" residues *below* the sensitivity of the analytical method.

Total residues as determined by diet studies indicate that consumption of individual pesticides is much lower than the estimated acceptable daily intake (ADI) for man. This is defined to be the daily exposure level of a pesticide residue, which, during the entire lifetime of man, appears to be without appreciable risk on the basis of all facts known at the time (47).* "Without appreciable risk" means the practical certainty that no adverse effect will result even after a lifetime of exposure. The ADI may be changed as new information becomes available.

10.4.3. Pesticide Residues in Environmental Samples

The National Academy of Sciences issued a book in 1975 entitled *Principles of Evaluating Chemicals in the Environment* (48). Chapter XX on Analysis and Monitoring discusses sampling considerations, extraction and concentration, cleanup, determination, and interpretation of data, as follows.

Validation of the method is as important or more important for environmental samples than for food crops. Quantitative data for the accuracy, precision, and lower limit of detection of an analytical method should be available when its use in a monitoring program is contemplated. The sensitivity and reliability required should depend primarily on the hazard associated with the substance in question, but may be mediated by analytical interferences in each substrate to be investigated. In monitoring programs involving significantly toxic chemicals, adequate interlaboratory programs should be initiated to assure reliability of analytical results obtained.

Confirmatory methods are of particular importance in monitoring studies for pesticide residues in environmental samples. Unfortunately, reports of finding undesirable contaminants are often released prematurely on the basis of unconfirmed results. Since such reports may cause serious sanctions to be applied, it is imperative that indicting analyses be confirmed by an independent procedure. The degree of assurance required will depend on the seriousness of the occurrence of the toxicant in the system. Confirmatory

---

* In general, the ADI provides a safety factor of at least 100-fold over the no-effect level in long-term chronic feeding studies in several species of animals. It is expressed in milligrams of the chemical per kilogram of body weight per day (mg/kg/day). Acceptable daily intakes for pesticides are recommended jointly by the Committee on Pesticides in Agriculture of the Food and Agricultural Organization (FAO, Rome), and the Expert Committee on Pesticide Residues in the World Health Organization (WHO, Geneva).

techniques are discussed by Ruzicka (37, pp. 36–39) and in guidelines issued recently by the Working Group (43).

Interpretation of analytical data is also of utmost importance. The analyst must report his interpretation as to reliability of his data (i.e., his certainty that the sample he analyzed actually contained the toxicant he reported and in the concentration he is reporting), with an indication of the variability encountered. These data must contain only positive concentration values; lower values must be reported in terms of "less than the lower limit of detection," including values often reported as "traces" which might be interpreted from very small responses of the detector. Only the analyst can know how much reliance can be placed on his results, and he has the responsibility for making this known since the ultimate recipient of the data cannot be expected to know their reliability and might therefore misinterpret their biological significance. (The analyst must also be very careful that he can back up his interpretation with reasonable data and not be influenced by a desire to find a residue for every "blip" of a recorder pen.)

The Bureau of Sport Fisheries and Wildlife has monitored for residues of organochlorine insecticides and their metabolites in *fish and wildlife* since the spring of 1967. Programs for sampling birds, fish, and shellfish are described in a series of articles in the June 1971 issue of *Pesticides Monitoring Journal* (41). Analyses are generally by the latest gas chromatographic method such as in the FDA *Pesticide Analytical Manual*, with a randomized system of cross-checking by thin layer chromatography. Interference values due to residues of polychlorinated biphenyls (PCBs) are estimated using Aroclor 1254 as a standard.

Programs for monitoring agricultural pesticides were initiated in 1964 for *soil*, in 1967 for *water*, and more recently for *air*. Analyses are primarily for chlorinated hydrocarbons, organophosphates, and phenoxy herbicides, although others are often included. The original list promulgated in 1967 contained chemicals believed to be of most interest because of (1) their extent and/or volume of usage, and/or (2) degree of hazard to man, fish, and wildlife, and/or (3) degree of persistence. The list has been revised periodically to include new pesticides or changes in use patterns of old pesticides. Metabolic and/or breakdown products are also included.

10.4.4. Monitoring in Humans

Concern about potential long-term effects of persistent chemicals led to the National Human Monitoring Program for Pesticides. In 1967, the Division of Pesticide Community Studies initiated a program in which adipose tissue (fat) was to be collected from postmortem and surgical specimens (41). The design provided for sampling which would yield valid data concerning pesticide incidence variation with geographic distribution, and with age, sex, and race variation of the donors.

Analyses of human adipose tissue were initially limited to the measurement of certain chlorinated hydrocarbons due to sensitivity limitations of available analytical methodology. The program originally included only DDT and its metabolites, dieldrin, heptachlor epoxide, and the isomers of BHC. As refinements of methodology progressed, additional pesticide chemicals or metabolites were added such as oxychlordane and mirex. Other pollutants such as PCBs were added as they became identifiable.

10.4.5. Special Monitoring Programs

Special programs are instituted as problems arise or are suspected. Many of these are a result of industrial pollution or accidental contamination rather than from direct use of pesticides. One such example was a monitoring program for mercury, or more specifically methyl mercury, in fish and bottom sediments in the Great Lakes attributed to dumping of wastes from giant electrolytic processes. A more localized problem was contamination of cattle and poultry in Michigan beginning in 1973 from ingestion of feed containing polybrominated biphenyls (PBBs).

Intensified programs are also put into operation when emergency use of cancelled pesticides is authorized. During recent years, EPA has taken stepwise action against DDT, mirex, aldrin/dieldrin, heptachlor/chlordane, and other chlorinated hydrocarbons are expected to follow. However, in the absence of adequate substitutes, EPA occasionally approves limited use of DDT to prevent spread of a specific pest. Similarly, EPA issued a permit in 1976 for aerial application of mirex to control fire ants on six million acres in Georgia, Mississippi, Louisiana, and Arkansas. Scientists from APHIS monitored water, soil, sediment, invertebrates (crabs and crayfish), and vertebrates (mostly birds) to determine the fate of residues in the nontarget environment. Such monitoring programs using specific methodology should provide more reliable information on the persistence of these useful pesticides in the environment.

Expansion of special programs is anticipated as EPA continues to review data for other pesticides on their list of candidates for Rebuttable Presumption Against Registration (RPAR). Several chlorinated phenols and phenoxy herbicides are included in the list, due in part to the presence of dioxin contaminants. The most widely publicized of these is the herbicide 2,4,5-trichlorophenoxyacetic acid (2,4,5-T), originally implicated as a teratogen in screening studies at high doses in susceptible strains of mice. The 2,4,5-T used in the first teratogenic studies by Bionetics (49) contained about 30 ppm of the highly toxic 2,3,7,8-tetrachlorodibenzo-p-dioxin

(TCDD). This dioxin isomer is formed during the manufacture of 2,4,5-trichlorophenol from tetrachlorobenzene at high temperatures under alkaline conditions. Current specifications are for less than 0.1 ppm TCDD in 2,4,5-T and other pesticides derived from 2,4,5-trichlorophenol, such as the herbicide silvex (2,4,5-trichlorophenoxypropionic acid) and the fungicide hexachlorophene [2,2'-methylene bis(3,4,6-trichlorophenol)].

Cancellation proceedings against 2,4,5-T were withdrawn by EPA in June 1974 pending results of extensive monitoring studies for TCDD. Analytical methods have been developed utilizing gas chromatography/mass spectroscopy which are capable of detecting residues down to 10 parts per *trillion* (0.00001 ppm) or less in milk, fat, and environmental samples (50–53). However, caution is needed in attributing positive responses at this ultralow level as indicating the presence of the highly toxic 2,3,7,8-TCDD isomer. It should also be noted that there are 22 possible tetrachloro isomers and that specific ion masses detected by the method can originate from many unrelated substances in the samples.

The TCDD monitoring program initially called for sampling of many substrates including human milk, fat, and liver. Those most likely to contain residues of TCDD were samples taken from range cattle maintained in areas with a confirmed history of repeated treatment with 2,4,5-T for many years, and slaughtered without the usual fattening period in a feed lot. Samples of fat from these animals were distributed to several cooperating laboratories including EPA and The Dow Chemical Company. Preliminary results from analyses using low resolution mass spectroscopy were erroneously reported in 1975 to indicate "positive" results for about 50% of the initial 34 fat samples analyzed. Furthermore, the findings were not consistent among the laboratories conducting the analyses (Harvard, Wright State University and Dow). As reported by Dr. Ralph T. Ross, Chairman of the EPA Dioxin Project in June 1976 (54), only one of the 85 beef fat samples analyzed showed a positive TCDD level at 60 parts per trillion (ppt), and two samples appeared to have TCDD levels at 20 ppt. He also stated that the analytical method is not valid below 10 ppt. During EPA's 2,4,5-T cancellation hearing in April 1980, it was revealed that the three positive fat samples all came from one farm in Missouri, located near a former trichlorophenol plant from which waste oil had been used erroneously for dust control in several horse arenas and along dirt roads in that vicinity.

The method used in the above analyses involves use of gas chromatographic separation to remove interfering substances, in combination with high-resolution mass spectrometry for quantitation. This method has also been used to analyze other monitoring samples collected by Dow. As reported by Shadoff et al. (55), no residues of TCDD were detected in samples of fish, water and mud from two locations in Arkansas and Texas that have a long history of treatment with 2,4,5-T containing considerably more than the current limit of 0.1 ppm TCDD. Similarly no TCDD was found in surveillance studies on milk (56) or fat (57) of cattle grazing in areas treated with 2,4,5-T for many years. Thus, it seems unlikely that continued use of 2,4,5-T would result in significant residues of TCDD in the environment.

10.5. INTERPRETATION OF RESIDUE DATA

Several basic principles should be applied in the interpretation of positive responses obtained in pesticide residue analyses. For example, a nanogram quantity of a pesticide in a solvent may cause a measurable peak in a gas chromatogram. However, the same quantity may not be distinguishable when superimposed on the "noisy" background obtained from extracts of a substrate. Similarly a small peak from a treated sample should not be compared directly to a large peak from a sample spiked at a considerably higher level. Recovery studies should include complete analyses of several control samples fortified at the claimed sensitivity of the method. Any "positive" responses below that should be reported as "less than" or "traces." A response below the limit of detection should be reported as "none-detected" since it is not possible to demonstrate a zero residue. For example, a residue of only 0.01 ppm DDT still represents 10^{13} molecules per gram of foodstuff (58). Above all, confirmation of residue identity is needed before instrument responses can be translated into positive values.

10.5.1. Variability Among Samples

The inherent biological variation among individuals of the same species can cause differences in response during analysis which might be interpreted as positive residue values unless numerous samples were analyzed. For example, in a recent study reported by Clark et al. (59), residues of several chlorinated hydrocarbon insecticides in fat of groups of feedlot cattle were compared to the residues in the feed given to them for 112 days prior to slaughter. Both feed and fat samples contained detectable levels of lindane, dieldrin, DDT, DDE, and DDD. Except for dieldrin, each insecticide was found at higher levels in fat than in feed. However, the levels were consistently low except for DDT and its degradation products in fat. If only average values are compared, a significant difference appears to exist in the DDT residues found in different groups of animals. However, consideration of the standard deviation within each group shows that they are not really different.

10.5.2. Sensitivity vs. Limit of Detection

As discussed recently by the Federal Working Group on Pest Management, the terms sensitivity and limit of detection do not have the same meaning (43). *Sensitivity* depends on the response of an instrument such as a gas chromatograph under a specific set of conditions. However, as the instrument sensitivity is adjusted to a maximum, the baseline "noise" level generally increases too. The response due to a pesticidal residue can be considered significant only if the peak height at the maximum usable sensitivity is twice the peak-to-peak height of the noise on the chart. Operation of the instrument at less than maximum sensitivity may be preferable, particularly for comparative analyses in different laboratories. For example, FDA analysts set instruments to give 50% of full scale recorder deflection on injection of one nanogram of heptachlor epoxide or two nanograms of parathion to assure that each laboratory operates at the same sensitivity.

On the other hand, the *limit of detection* of an analytical method may be defined as the concentration of pesticide above which a given sample of material can be said, with a high degree of confidence, to actually contain the chemical analyzed. This limit of detection depends upon a number of factors in addition to the instrument sensitivity, including sample size, adequacy of extraction and cleanup, size of aliquot injected, and knowledge and experience of the analyst. Deviations in peaks or baselines caused by inadequate cleanup are more evident when the instrument is operated at maximum sensitivity. These deviations may cast doubt on the validity of the peaks of interest in assigning positive values for specific residues.

Sutherland has reviewed the analytical limit of detectability for residues (60). For all practical purposes, the response of control samples to a specified analytical method is the sole determinant of the limit of detectability. However, these control samples must be taken from plants or animals exactly comparable to the treated samples. They must be sampled at the same time, from identical varieties of the same degree of maturity, and must have been treated in all respects the same as the treated samples except that the chemical to be determined must not have been used on them. Furthermore, the samples must be independent and normally distributed according to statistical definitions.

The responses given by, or apparent residues found in these control samples will usually follow a bell-shaped distribution curve. The average of all samples analyzed will be the true mean. An old rule-of-thumb is that any uncorrected residue value greater than twice the average control represents a real residue (58). However, this is not necessarily so depending on the spread of the frequency distribution around the mean for that substrate. Hahn has recently discussed the consequences of incorrectly assuming that distribution is normal about the mean (61).

Residue data frequently do not conform to normal distribution. In many instances, use of the geometric mean or the log of results provides a better fit to normal distribution. Also, developing statistics around the transformed mean may provide a more accurate estimate of the true residue and/or exposure to a pesticide.

Even greater care must be taken in interpreting data for extremely low residues of highly toxic chemicals in the environment. This was demonstrated following collaborative studies on TCDD conducted as part of the EPA Dioxin Implementation Plan in 1976. Agreement was reached among scientists from EPA, USDA, Harvard, University of Utah, and Dow that it takes a response of 2.5 times background noise to indicate a positive finding. Furthermore, it takes a response of 10 times background noise before quantitative values at low parts per *trillion* can be assigned.

10.5.3. Interpretation of Monitoring Data

Unfortunately, analysts conducting monitoring studies do not have the benefit of comparisons to control samples because samples reliably free from pesticides are usually not available. In such cases, the limit of detection is based on the analytical variability for each pesticide on each substrate, generally by carrying through the entire method at least six replicate samples containing levels near the estimated limit of detection of the pesticide (43). Under these conditions, the limit of detection is considered equal to two standard deviations (approximately 95% confidence level) calculated from the replicated results. The identity of any peak larger than this should be subject to confirmation as discussed in Section 10.5.4, but practical limitations preclude confirmation of most "positive" values.

Although variability among analyses of six replicates may be adequate for samples of certain species of plants or animal tissues, it may not be adequate to establish a reliable baseline for samples of soil or water from widely divergent geographical locations subjected to contamination by a great variety of chemicals. Unfortunately, some chemists do not recognize such limitations and interpret any positive response similar to that produced by a pesticide as proof that a residue of that chemical is present. Furthermore, reports of monitoring studies tend to emphasize the *frequency* of positive responses without regard to the significance of the "residues" found.

One such report issued in 1973 was entitled *Pesticides in Selected Western Streams—1968–1971* (62). It gave results of analyses by the U.S. Geological Survey for nine chlorinated hydrocarbon insecticides and three phenoxy herbicides in water from 20 locations in 14 states. The authors reported "total occurrence (of phenoxy residues) reached a peak of 106 during 1968–69 and declined sharply to 54 during 1970–71. This decrease of about 50% was due to 'reduced occurrences of 2,4-D and 2,4,5-T'." Examination of the lengthy data tables revealed that traces of phenoxy herbicides amounting to less than 0.0005 ppm (0.5 ppb) appeared to be present in about 16% of the samples for 2,4-D and 2,4,5-T, and 7% of the samples for silvex. Furthermore, 80% of the positive findings were at levels below 0.1 ppb. The detection limits were listed at 0.005 μg/liter for silvex and 2,4,5-T and 0.02 μg/liter for 2,4-D (i.e., a sensitivity of 5 to 20 parts per *trillion*). Recoveries of 80–100% were claimed for validation studies but the spiking levels were not specified and may not have approached the claimed sensitivity. (It was established during the 1980 2,4,5-T cancellation hearing that the reliable limit of detection for 2,4,5-T was only 1 ppb prior to 1978.) Strangely, confirmatory analyses for phenoxy compounds were not mentioned although other methods or other columns were discussed for confirmation of the insecticides found.

The report also stated that 64% of the silvex occurrences were at one station in Nevada and 47% of the 2,4,5-T occurrences were in two associated sites in Arkansas and Oklahoma. In view of the localization of positive findings, it is possible that the parts per trillion residues "found" may have been due to the presence of interfering substances in samples from these locations rather than to actual residues of silvex or 2,4,5-T in the water. Unfortunately, the report has been cited as evidence of widespread contamination of water by these controversial herbicides in Western States..

Another example is a paper in the September 1974 issue of *Bulletin of Environmental Contamination and Toxicology* (63). A total of 13 authors summarized studies on the distribution of pesticides in randomly selected soil samples taken from diverse environments in five west Alabama counties. Pesticide residues were found in all samples examined, even soil samples from areas where records indicated that no pesticides had been applied directly. The pesticides included eleven chlorinated hydrocarbon insecticides, six nonchlorinated insecticides, and two herbicides (atrazine and 2,4,5-T). They reported finding 2,4,5-T in 24% of the samples of top soil analyzed. Examination of the data revealed that 37 samples of top soil contained no detectable residues (sensitivity not specified), eight were deemed positive, and only one sample contained more than 0.1 ppm 2,4,5-T. The samples were analyzed twice by GLC using two columns but no analytical data were provided to indicate reliability of responses reported as "positive" at <0.1 ppm. Unfortunately, such data are used as evidence of persistence of pesticides in the environment.

The above examples also demonstrate the pitfalls in drawing conclusions from reports which emphasize the maximum residue found or the percent "positive" samples without adequate allowance for variations in background "noise." This becomes even more important in special monitoring programs for highly toxic chemicals at extremely low levels. An example discussed previously was the premature release of preliminary data from analyses for TCDD at levels around 10 parts per *trillion* in beef fat which could be construed by some as sufficient evidence that 2,4,5-T and related pesticides should be banned (54).

10.5.4. Confirmation of Residue Identity

Residue analyses without confirmatory tests provide only qualitative information. It is not safe to assume that a response similar to that produced by a known pesticide means the pesticide is actually present. Many instances have been noted in which one material masquerades as another in any one test (e.g., PCBs, phthalate esters, etc.). The certainty of identification is increased when the behavior of the unknown and the standard is the same in a number of tests.

Subscribers to *Pesticides Monitoring Journal* received a June 1975 issue entitled *Guidelines on Analytical Methodology for Pesticide Residue Monitoring* (43). The following specific guidelines on confirmation of residue identity were listed:

1. Residues reported should be confirmed by tests such as TLC, element specific GC, *p*-value determinations, derivatization, ultraviolet photolysis, etc., in addition to EC/GC.

2. PCBs or other suspected complex interfering substances must be separated and confirmed.

3. Combinations of confirmatory tests that measure the same physical property parameters or tests that measure the same parameters as the initial analysis, although they may help to support the identification, are not the best means of doing so. If sufficient material is available, excellent identifications can be made by infrared spectrometry, particularly with the use of microcell techniques or Attenuated Total Reflectance.

4. Mass spectrometry (MS) or combined GC/MS is recommended for identifying or confirming identity of important residues not adequately identified by other techniques.

5. The methods used for confirming residue identity should always be described in the report of residue results.

Although some laboratories may not have time to confirm every residue tentatively identified by GC, confirmatory analyses should be conducted at intervals such as one in every five samples when the same residues are apparently present throughout the group. When insufficient sample is available, confirmatory analyses may be conducted on a pool of several cleaned up sample extracts.

10.5.5. Statistical Evaluation of Results

The Federal Working Group on Pest Management also differentiates between accuracy and precision of pesticide residue analyses (43). *Accuracy* involves statistical measurements that relate to the differences between the test results and the true result when the latter is known or assumed, and may be expressed as the mean error or relative error. *Precision* involves statistical measurements that relate to the variation among test results themselves—that is, the scatter or dispersion of a series of test results without assumption of any prior information as to the true result. Precision may be expressed in terms of the variance, standard deviation, relative standard deviation, range, etc. Values from single or duplicate injections of extracts from duplicate samples are more meaningful than values from multiple injections from single samples. These should be differentiated in evaluating differences among groups of data.

As noted in Section 10.3.1.1, pesticide residue methods are often subjected to collaborative studies through the Association of Official Analytical Chemists (AOAC). These studies evaluate specificity, repeatability, and reproducibility of independent analyses for various samples. *Repeatability* is a measure of how well an analyst can expect to agree with himself from day to day, whereas *reproducibility* is how well analyses in one laboratory are likely to agree with results from other laboratories. The *Statistical Manual of the AOAC* (23) presents excellent information on formulating and statistically analyzing collaborative studies (23).

According to a 1978 report by the Committee on Environmental Improvement of the American Chemical Society (64), the acceptability of a method for general use is best evaluated on the basis of the precision and accuracy of the results obtained with it by a number of analysts in different laboratories. When the method is applied in this way to a series of samples containing the same concentrations of the substance to be measured, *precision* describes the variation or scatter among the results, irrespective of the true composition of the sample, whereas *accuracy* describes the difference between the average results and the correct results. Accuracy is especially difficult to achieve with authentic environmental samples, whose complexity can lead to insidious errors that may defy anticipation (64).

10.6. CONCLUSION

The regulation of pesticides involves a complex maze of rules implemented by various government agencies in their everchanging interpretation of laws enacted and amended by the U.S. Congress that are intended to control the use of such chemicals. Two basic elements in these laws are that all pesticide products must be registered and that tolerances must be set for safe residue limits of all pesticides used in food or feed crops. Requirements for residue data in support of these tolerances are discussed in Section 10.3. Monitoring studies are conducted by groups within several government agencies to assure that excessive residues are not present in foods or accumulating in the environment, as discussed in Section 10.4. The importance of careful interpretation of residue data is discussed in Section 10.5.

Concern about the significance of monitoring data was expressed at a 1973 hearing on EPA studies of specific chemicals in estuarine and freshwater environments. At that time the agency's Hazardous Materials Advisory Committee was told that many measurements of DDT in the environment were suspect due to interferences in the analyses (65). Early data did not distinguish between DDT and residues of PCBs used extensively in transformer fluids, and so on (37, 43).

Furthermore, phthalate esters used widely as plasticizers, accompany DDT through many cleanup procedures (39, 40). Positive responses for phthalates can be due to their presence in environmental samples or to laboratory contamination during workup of the samples. In addition, PCB-like responses can be obtained in GC/EC analyses due to contamination of glassware with an oil film from exposure to air. Interferences have also been a major factor in chemical analyses for arsenic residues, mercury, and phenoxy herbicides such as 2,4-D and 2,4,5-T (39).

Responsibilities in the use and misuse of scientific data were discussed during a symposium of the American Association for the Advancement of Science (AAAS) at their national meeting in New York in 1975. It was concluded that responsibility for proper reporting of data lies with individual scientists, scientific societies, and publishers. One of the speakers, Dr. Richard W. Roberts, director of the National Bureau of Standards, estimated that at least half of the data reported in scientific literature is unusable, often because too little information has been provided for independent evaluations (66). Dr. Mary L. Good, Chairperson of the Board of the American Chemical Society, cited four areas in need of responsible attention: (1) determination of permissible levels of potentially dangerous materials in the biosphere, (2) lack of information on error limits and credibility of published data, (3) overmassage or excessive manipulation of raw data by computer techniques, and (4) evaluation and use of computerized data banks. Also, care must be taken to delineate between those parts of reports that are factual and can be repeated by other workers and those reflecting interpretation of the data (67).

Dr. Bernard L. Oser entitled his presentation *Exaggerated Conclusions Based on Inadequate Data* (68). He defined scientific data as observations and findings generally expressed in numercial or descriptive terms and stated that, even when correctly reported, "data" are not necessarily equatable with "facts." Implicit in the term "facts" are the accuracy and reproducibility of findings and the competence and integrity of those responsible for the design, execution, and interpretation of the studies. Data are often statistically manipulated or "staticulated," such as by use of average values without expressing the range or distribution around the mean. Scientists may also draw false or misleading conclusions from their own data or from that of others. This may take the form of overestimating the importance of uncorroborated findings, or as one writer expressed it, extrapolation of unproven speculations. Thus, problems arise from **misuse** of scientific data as well as from **use** of unscientific data.

Emphasis in this chapter has been on the care needed in conducting analyses for pesticide residues at sub-part-per-million levels in a variety of substrates, of the need for adequate reporting of such residue data, and of pitfalls in interpretation of data at levels below the validated sensitivity of the method. More specificity is needed rather than "picogram" measurements as often pushed by the scientific community. Premature or unwise interpretation of data can lead to precipitous action by regulatory agencies who have to respond to public pressures from individuals or groups of persons. Unfortunately, most pressure comes from lawyers and politicians who do not comprehend the limitations of pesticide residue analyses, particularly when the identity of the residues "found" has not been adequately confirmed.

ACKNOWLEDGMENT

The author gratefully acknowledges the many suggestions and contributions made by George E. Lynn, Milton E. Getzendaner, and Wendell F. Phillips in their painstaking review of the original manuscript for this chapter.

REFERENCES

Note: See references 9–12, 14–18, 25–30, 32, 34. The asterisks indicate the following: (No.) CFR is Title (No.) of the Code of Federal Regulations; (No.) FR is Volume (No.) of the Federal Register.

1. F. A. Gunther and R. C. Blinn, *Analysis of Insecticides and Acaricides*, Vol. VI. *Chemical Analysis Series*, Wiley-Interscience, New York-London, 1955.
2. R. W. Fogelman, *J. Agr. Food Chem.*, **4,** 410 (1956).
3. T. H. Harris, *J. Agr. Food Chem.*, **4,** 413 (1956).

4. F. A. Vorhes, Jr., *J. Agr. Food Chem.*, **4,** 415 (1956).

5. H. Frehse, *Residue Rev.*, **5,** 1 (1964).

6. T. H. Harris and J. G. Cummings, *Residue Rev.*, **6,** 104 (1964); also W. I. Patterson and A. J. Lehman, *Assoc. Food Drug Officials U.S.*, **XVII**(1), 3–12, (Jan. 1953).

7. A. Bevenue and Y. Kawano, *Residue Rev.*, **35,** 103 (1971).

8. FDA Guidelines for Chemistry and Residue Data Requirements of Pesticide Petitions, Bureau of Science, Food and Drug Administration, Dept. of Health, Education, and Welfare, Washington, D.C. 20204, March 1968.

9. Environmental Protection Agency: 40 CFR* Part 162: Regulations for the Enforcement of the Federal Insecticide, Fungicide, and Rodenticide Act; Subpart A: Registration, Reregistration and Classification Procedures, Rules and Regulations, 40 FR* 28242–28286, July 3, 1975.

10. Environmental Protection Agency: 40 CFR* Part 162.41 through 162.51: Guidelines for Registering Pesticides: Registration Procedures. Rules and Regulations, 40 FR* 41788–41793, September 9, 1975.

11. Environmental Protection Agency: 40 CFR* Part 162.83 through 162.93: Guidelines for Registering Pesticides: Petitions for Tolerances, unpublished draft, April 1975. (Not issued as of May 1980.)

12. Environmental Protection Agency: 40 CFR* Part 162: Guidelines for Registering Pesticides in the United States. Proposed Rules, 40 FR* 26802–26928, June 25, 1975. (Also see reference 34.)

13. Public Law 94-140, November 28, 1975 (To amend the Federal Insecticide, Fungicide, and Rodenticide Act).

14. U.S. Department of Agriculture, NRC Pesticide Residues Committee—Statement for Implementation of Report on No Residue and Zero Tolerance, 31 FR* 5723, April 13, 1966.

15. Food and Drug Administration: 21 CFR* Part 120: Tolerances and Exemptions from Tolerances for Pesticide Chemicals in or on Raw Agricultural Commodities—Negligible Residues on Commodity Group Basis, U.S. Department of Health, Education and Welfare. Proposed Rule Making, 32 FR* 6059, April 15, 1967. Subsequently codified as 40 CFR* Part 180.34(f). Amended 40 FR* 6972, February 18, 1975.

16. Food and Drug Administration: 21 CFR* Part 120: Milk, Eggs, Meat, and/or Poultry—Pesticide Tolerances; Policy Statement, U.S. Department of Health, Education, and Welfare. Proposed Rule Making, 33 FR* 3438, February 28, 1968.

17. Environmental Protection Agency: 40 CFR* Part 180: Tolerances and Exemptions from Tolerances for Pesticide Chemicals in or on Raw Agricultural Commodities—Part 180.33—Fees. 37 FR* 13978, July 15, 1972.

18. 40 CFR* Part 180.319—Interim tolerances.

19. (*a*) D. P. Schultz, *J. Agr. Food Chem.*, **21,** 186 (1973); (*b*) D. P. Schultz and P. D. Harman, *Pest. Monit. J.*, **8,** 173 (1974).

20. D. L. Stalling and J. N. Huckins, *J. Agr. Food Chem.*, **26**, 447 (1978).

21. W. Horwitz and H. L. Reynolds, *The AOAC Style Manual, Assoc. Offic. Anal. Chem.*, (1972).

22. M. C. Bowman and M. Beroza, *J. Assoc. Offic. Agr. Chem.*, **48**, 943 (1965).

23. W. J. Youden and E. H. Steiner, *Statistical Manual of the AOAC, Assoc. Offic. Anal. Chem.*, (1975).

24. M. E. Getzendaner, *Down to Earth*, **28**(1), 24–29 (Summer 1972).

25. 39 FR* 43289: Tolerances for dinoseb in raw agricultural commodities, 40 CFR* 180.281, December 12, 1974.

26. 40 FR* 51045: Tolerances for 2-methyl-4-chlorophenoxyacetic acid (MCPA) in raw agricultural commodities and for MCPA and its 2-methyl-4-chlorophenol metabolite in foods of animal origin, 40 CFR* 180.339, November 3, 1975.

27. 41 FR* 11514: Tolerances for 2,4-D in raw agricultural commodities and of 2,4-D and its 2,4-dichlorophenol metabolite in food products of animal origin, 40 CFR* 180.142(b) and (h), March 19,1976.

28. Environmental Protection Agency: 40 CFR* Part 172: Experimental Use Permits, Revocation of Part 162.17 of Section 5 of Regulations and addition of new part 172 of Chapter 1 of Title 40 of the Code of Federal Regulations, 40 FR* 18780, April 30, 1975.

29. 36 FR* 24234, December 22, 1971.

30. 38 FR* 24233, September 6, 1973.

31. M. L. Leng, unpublished Dow Analytical Method ACR 58.5 for the Determination of Trace Amounts of Dow ET-57 in Milk and Animal Fats, August 1, 1958.

32. 38 FR* 21686, August 10, 1973.

33. U.S. Department of Agriculture, *PR Notice 70-15* on Guidelines for Studies to Determine the Impact of Pesticides on the Environment, June 23, 1970.

34. Environmental Protection Agency: 40 CFR* Parts 162, 163, 181: Proposed Guidelines for Registering Pesticides in the United States. Proposed Rules, 43 FR* 29696–29741, July 10, 1978.

35. R. R. Rathbone, *Communicating Technical Information—A Guide to Current Uses and Abuses in Scientific and Engineering Writing*, Addison-Wesley, Reading, MA, Palo Alto, CA (1966).

36. *Handbook for Authors of Papers in the Journals of the American Chemical Society*, American Chemical Society Publications, Washington, D.C. (1967).

37. J. H. Ruzicka, "Methods and Problems in Analyzing for Pesticide Residues in the Environment" in *Environmental Pollution by Pesticides* (C. A. Edwards, Ed.), Plenum, New York, 1973.

38. D. C. Staiff, G. E. Quinby, D. L. Spencer, and H. G. Starr, Jr., *Bull. Environ. Contam. Toxicol.*, **12**(4), 455 (1974).

39. D. G. Crosby, Statement before EPA's Hazardous Materials Advisory Committee, *Pest. Chem. News*, **1**(45), 6 (October 17, 1973).

40. J. A. Singmaster III and D. G. Crosby, *Bull. Environ. Contam. Toxicol.*, **16,** 291 (1976).

41. National Pesticides Monitoring Program (Revised), *Pest. Monit. J.*, **5,** 35 (1971).

42. Federal Working Group on Pest Management, *Guidelines on Sampling and Statistical Methodologies for Ambient Pesticide Monitoring*, Washington, D.C., October 1974. (U.S. Government Printing Office 1975-626-592/216 3-1.)

43. Federal Working Group on Pest Management, *Guidelines on Analytical Methodology for Pesticide Monitoring*, Washington, D.C., June 1975.

44. S. D. Fine, Statement before Subcommitee on Agricultural Research and General Legislation, Committee on Agriculture and Forestry, U.S. Senate, January 27, 1976.

45. Bureau of Veterinary Medicine Memo 19: Monitoring for Residues in Food—Animal Tissues, DHEW Pub. No. (FDA) 76-6001, July 1975.

46. R. D. Johnson and D. D. Manske, *Pest. Monit. J.*, **9,** 157 (1976).

47. F. C. Lu, *Toxicological evaluation of food additives and pesticide residues and their "acceptable daily intakes" for man: the role of WHO, in conjunction with FAO. Residue Rev.*, **45,** 81 (1973).

48. National Academy of Sciences, *Principles for Evaluating Chemicals in the Environment*, a report of the Committee for the Working Conferences on Principles of Protocols for Evaluating Chemicals in the Environment, Environmental Studies Board, National Academy of Sciences, National Academy of Engineering and Committee on Toxicology, National Research Council, Washington, D.C., 1975.

49. K. D. Courtney et al., *Science*, **168,** 864 (1970).

50. R. Baughman and M. Meselson, *Environmental Health Perspectives*, Experimental Issue No. 5, September 1973. (DHEW Publication No. (NIH) 74-218.)

51. W. B. Crummett and R. H. Stehl, *Environmental Health Perspectives*, Experimental Issue No. 5, September 1973. (DHEW Publication No. (NIH) 74-218).

52. L. A. Shadoff and R. A. Hummel, *Biomed. Mass Spectr.* **5**(1), 7 (1978).

53. R. A. Hummel, *J. Agr. Food Chem.*, **25**(5), 1049 (1977).

54. R. T. Ross, Statement on U.S. Environmental Protection Agency's Dioxin Implementation Plan (2,4,5-T/2,3,7,8-Tetrachlorodibenzo-*p*-dioxin), June 21, 1976.

55. L. A. Shadoff, R. A. Hummel, L. Lamparski, and J. H. Davidson, *Bull. Environ. Contam. Toxicol.*, **18,** 478 (1977).

56. N. H. Mahle, H. S. Higgins, and M. E. Getzendaner, *Bull. Environ. Contam. Toxicol.*, **18,** 123 (1977).

57. C. W. Kocher, N. H. Mahle, R. A. Hummel, L. A. Shadoff, and M. E. Getzendaner, *Bull. Environ. Contam. Toxicol.*, **19,** 229 (1978).

58. F. A. Gunther, *Adv. Pest. Control. Res.*, **5**, 222 (1962).

59. D. E. Clark, H. E. Smalley, H. R. Crookshank, and F. M. Farr, *Pest. Monit. J.*, **8**, 180 (1974).

60. G. L. Sutherland, *Residue Rev.*, **10**, 85 (1965).

61. G. J. Han, *Chemtech*, Aug. 1976, 530.

62. J. A. Schulze, D. B. Manigold, and F. L. Andrews, *Pest. Monit. J.*, **7**, 73 (1973).

63. R. Albright et al., *Bull. Environ. Contam. Toxicol.*, **12**, 378 (1974).

64. Committee on Environmental Improvement, *Cleaning Our Environment: A Chemical Perspective*, 2nd ed., American Chemical Society, Washington, D.C., October 1978.

65. *Pest. Chem. News*, **1** (45), 6 (Oct. 17, 1973).

66. *Chem. Eng. News*, **Feb. 1975**, 17.

67. *Chem. Eng. News*, **March 1975**, 4.

68. B. L. Oser, *Int. Flavours Food Add.*, **6**, 163 (1975).

69. N. J. Karch, "Explicit Criteria and Principles for Identifying Carcinogens: A Focus of Controversy at the Environmental Protection Agency" in: Vol. IIa of *Decision Making in the Environmental Protection Agency: Case Studies*, the National Research Council, National Academy of Sciences, Washington, D.C., 1977.

INDEX

449